FIRSTSTEPS
A Guide to Social Research

FIFTH EDITION

Michael Del Balso
Dawson College

Alan D. Lewis
Dawson College

NELSON / EDUCATION

First Steps: A Guide to Social Research, Fifth Edition
by Michael Del Balso and Alan D. Lewis

Associate Vice President, Editorial Director:
Anne Williams

Acquisitions Editor:
Maya B. Castle

Marketing Manager:
Terry Fedorkiw

Developmental Editor:
Liisa Kelly

Permissions Coordinator:
Melody Tolson

Senior Content Production Manager:
Imoinda Romain

Production Service:
Integra India

Copy Editor:
Marcia Gallego

Proofreader:
Integra Software Services Pvt.Ltd

Indexer:
Integra Software Services Pvt.Ltd

Production Coordinator:
Ferial Suleman

Design Director:
Ken Phipps

Managing Designer:
Franca Amore

Interior Design:
Courtney Hellam

Interior Design Modifications:
Dianna Little

Cover Design:
Dianna Little

Cover Image:
N-trash/Shutterstock

Compositor:
Integra Software Services Pvt.Ltd

Library and Archives Canada Cataloguing in Publication

Del Balso, Michael, 1953–

First steps: a guide to social research/Michael Del Balso, Alan D. Lewis.—5th ed.

Includes bibliographical references and index.
ISBN 978-0-17-650414-4

1. Social sciences—Research. I. Lewis, Alan D. (Alan David), 1943– II. Title.

H62.D45 2011 300'.72 C2011-904351-3

ISBN-13: 978-0-17-650414-4
ISBN-10: 0-17-650414-1

Contents

Chapter 3   Finding and Refining the Topic   45

Chapter 4   Sampling: Choosing Who or What to Study   72

Chapter 5   Social Survey   97

Chapter 6   Experimental Research   130

Chapter 10   The Research Report: The End and the Beginning   243

Preface

Early in our planning for this fifth edition of *First Steps: A Guide to Social Research,* we had a discussion about what we felt were the key ideas shaping our approach to the writing of this textbook. We came up with four basic components of our approach:

1. To write in as accessible, clear, and nontechnical a manner as possible.
2. To cover the major types of research methods and data-gathering techniques used across the social sciences, without an exclusive or unbalanced focus on any one discipline or discipline-based approach.
3. To be as open as possible to students with many different academic backgrounds and aims. We assume neither social science knowledge nor an intention of becoming professional social scientists on the part of students.
4. To write a basic introduction that teachers from different backgrounds, working in different educational contexts, and with different research-teaching interests can amplify, develop, illustrate, and build on according to their needs.

We were later relieved to find that these aims had been identified by reviewers of previous editions! However, despite the constancy of our commitment we have found much to revise, modify, and add in this new edition.

- The continual improvement in information access resulting from the rise of the Internet has led to major reworking of Chapter 3, "Finding and Refining the Topic," to reflect the importance of academic libraries as "portals" to essential research resources.
- New material guiding students on the development of research references and in preparing annotated bibliographies has been added to Chapter 3, and Chapter 10 includes a new section on oral communication of research reports, including suggestions on how to use audiovisual resources in support of oral reporting.
- New material has been added to Chapter 7 on the case study method and to Chapter 8 on qualitative and historical documentary data analysis.
- The section on natural experiments in Chapter 6 has been rewritten to emphasize and illustrate the way that experimental logic is fundamental to scientific thought and is shared by all the social sciences.
- Citation management has become an essential concern for college and university students and teachers, and we have revised our presentation of APA and MLA format conventions for greater clarity.
- We have both updated the Weblinks at the end of each chapter and organized them so that they more clearly reflect and connect back to chapter topics.
- Discussion of ethical issues with sampling methods has been introduced in Chapter 4.

The pedagogical elements in the text serve to guide students through the text and to provide instructors with material that can be used in the classroom:

What You Will Learn: Each chapter opens with learning outcomes that can be used as a road map for critical reading.

What You Have Learned: Each chapter closes with a summary of the key concepts covered in the chapter.

Focus Questions: Each chapter culminates with questions that can be incorporated into class discussion or assigned as short-answer questions.

Review Exercises: Each chapter closes with exercises that encourage students to apply what they have learned in the chapter. These exercises can be used for essays and projects.

Weblinks: Each chapter includes a list of relevant websites, with descriptions, that can be used to complete the Review Exercises.

ANCILLARIES

About NETA

The **Nelson Education Teaching Advantage (NETA)** program delivers research-based instructor resources that promote student engagement and higher-order thinking to enable the success of Canadian students and educators.

Instructors today face many challenges. Resources are limited, time is scarce, and a new kind of student has emerged: one who is juggling school with work, has gaps in his or her basic knowledge, and is immersed in technology in a way that has led to a completely new style of learning. In response, Nelson Education has gathered a group of dedicated instructors to advise us on the creation of richer and more flexible ancillaries that respond to the needs of today's teaching environments.

The members of our editorial advisory board have experience across a variety of disciplines and are recognized for their commitment to teaching. They include:

- **Norman Althouse,** Haskayne School of Business, University of Calgary
- **Brenda Chant-Smith,** Department of Psychology, Trent University
- **Scott Follows,** Manning School of Business Administration, Acadia University
- **Jon Houseman,** Department of Biology, University of Ottawa
- **Glen Loppnow,** Department of Chemistry, University of Alberta
- **Tanya Noel,** Department of Biology, York University
- **Gary Poole,** Director, Centre for Teaching and Academic Growth and School of Population and Public Health, University of British Columbia
- **Dan Pratt,** Department of Educational Studies, University of British Columbia
- **Mercedes Rowinsky-Geurts,** Department of Languages and Literatures, Wilfrid Laurier University
- **David DiBattista,** Department of Psychology, Brock University
- **Roger Fisher,** PhD

In consultation with the editorial advisory board, Nelson Education has completely rethought the structure, approaches, and formats of our key textbook ancillaries. We've also increased our investment in editorial support for our ancillary authors. The result is the Nelson Education Teaching Advantage and its key components: *NETA Engagement, NETA Assessment,* and *NETA Presentation.* Each component includes one or more ancillaries prepared according to our best practices, and a document explaining the theory behind the practices.

For *First Steps: A Guide to Social Research*, Fifth Edition, we have chosen to focus on *NETA Assessment* to exemplify the Nelson Education Teaching Advantage. *NETA Assessment* relates to testing materials: not just Nelson's Test Banks and Computerized Test Banks, but also in-text self-tests, Study Guides and web quizzes, and homework programs like CNOW. Under *NETA Assessment*, Nelson's authors create multiple-choice questions that reflect research-based best practices for constructing effective questions and testing not just recall but also higher-order thinking. Our guidelines were developed by David DiBattista, a 3M National Teaching Fellow whose recent research as a professor of psychology at Brock University has focused on multiple-choice testing. All Test Bank authors receive training at workshops conducted by Professor DiBattista, as do the copyeditors assigned to each Test Bank. A copy of *Multiple Choice Tests: Getting Beyond Remembering*, Professor DiBattista's guide to writing effective tests, is included with every Nelson Test Bank/Computerized Test Bank package.

Instructor Ancillaries

Key instructor ancillaries are provided on the companion website for *First Steps: A Guide to Social Research*, Fifth Edition, available at **http://www.firststeps5e.nelson.com:**

- **Instructor's Manual:** The Instructor's Manual to accompany *First Steps: A Guide to Social Research* has been prepared by text authors Michael Del Balso and Alan Lewis. This manual contains learning objectives, five to ten suggested classroom activities, and key terms for each chapter to give you the support you need to engage your students within the classroom.
- **NETA Assessment:** The Test Bank was also written by Michael Del Balso and Alan Lewis. It includes over 250 multiple-choice questions written according to NETA guidelines for effective construction and development of higher-order questions. Test Bank files are provided in Word format for easy editing and in PDF format for convenient printing whatever your system.

The Computerized Test Bank by ExamView® includes all the questions from the Test Bank. The easy-to-use ExamView software is compatible with Microsoft Windows and Mac. Create tests by selecting questions from the question bank, modifying these questions as desired, and adding new questions you write yourself. You can administer quizzes online and export tests to WebCT, Blackboard, and other formats.

- **Microsoft® PowerPoint®:** Short presentations on topics relevant to each of the book's ten chapters are presented in PowerPoint format. The PowerPoint presentations were also created by Michael Del Balso and Alan Lewis.
- **Image Library:** This resource consists of digital copies of figures, short tables, and photographs used in the book. Instructors may use these jpegs to create their own PowerPoint presentations.
- **DayOne:** Day One—Prof InClass is a PowerPoint presentation that you can customize to orient your students to the class and their text at the beginning of the course.

Resources for Students

- *First Steps* **on the Web:** Aside from providing access for educators to the book's Instructor's Manual, NETA Test Bank, PowerPoints, and other instructors' resources, the companion website for *First Steps: A Guide to Social Research*, Fifth Edition, also supplies students with study resources such as the following: more than 120 Test Yourself questions, including true/false and multiple choice formats; helpful web links, as highlighted in the text; glossary flashcards and puzzles; and helpful reference materials that span the social sciences. See **http://www.firststeps5e.nelson.com.**

INFOTRAC® 🛰 COLLEGE EDITION

Ignite discussions or augment your lectures with the latest developments in the social sciences. Create your own course reader by selecting articles or by using the search keywords provided at the end of each chapter. InfoTrac® College Edition, gives you and your students four months of free access to an easy-to-use online database of reliable, full-length articles (not abstracts) from hundreds of top academic journals and popular sources in the social sciences. Contact your Nelson representative for more information.

How to Retrieve Articles from InfoTrac® College Edition

Throughout the book we have referred to the availability of certain full-text articles from the InfoTrac® College Edition database, for which a free temporary subscription is included with the *First Steps* textbook. The InfoTrac database includes nearly 20 million articles from about 6000 journals. For a guide on the various ways to search and retrieve articles, go to the InfoTrac home page and click on "User Guide."

To simplify the process of retrieving referenced articles, after the general information about each article, we include its InfoTrac database record number. For example:

> K D. Schwartz & G. T/ Fouts. (2003). Music preferences, personality style, and developmental issues of adolescents. *Journal of Youth and Adolescence, 32*(3), 205–214. InfoTrac record number: **A99985824**.

Illustrated below is a quick and easy way of searching and retrieving an article.

Step 1: After logging on, click on the "Advanced search" option to view the following:

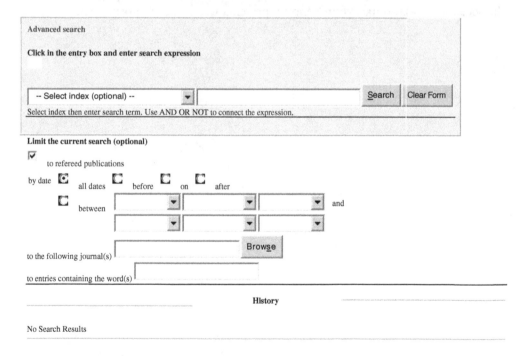

Step 2: In the "Select index" box, click on the arrow and choose "Record Num. (rn)."
Step 3: The letters "rn" will appear in the entry box. Next to "rn," type in the InfoTrac record number of the article. For example, for the article given as an example above, type in A99985824.

Step 4: Click on "Search."

Step 5: After the search is completed you will be returned to the same page. However, in the section entitled "History," the record number of the article will appear, with the option to view it. For example:

History

R1   (RN A99985824) and Re Ref)

<u>View</u> 1 citation; <u>Modify</u> Search

Step 6: Click on "View" and the article will appear. You can read the article online or use the three retrieval options that are indicated at the end of the article—you can print the text, retrieve it as a PDF file with Acrobat Reader, or e-mail the article to yourself or others. **Note:** The InfoTrac website also offers other useful information. For example, on the home page, click on "Student Resources" for access to assistance on research writing, a dictionary, and many helpful hints on adjusting to college life, time management, studying for exams, and more.

Acknowledgements

In preparing this fifth edition of *First Steps* we are deeply indebted to our colleagues for their encouragement and astute criticisms, and to our students who, often unwittingly, create interesting challenges and problems for us to ponder in our research methods courses. We have also always found that reviewers' evaluations of our successive drafts were immensely fruitful and stimulating.

Special appreciation is extended to the many who have provided detailed reviews for the various editions, particularly Paul Angelini, Pearl Crichton, Anna De Aquayo, Nick Gabriolopolous, Enrico Museo, Hans Bakker, Mike Burke, Joan Collins, Barry Green, Vadney Haynes, Susan H. Nelson, Gary Parkinson, Fay Rogers, Sandra Ward, and Jan Warnke. The editorial and production teams at Nelson have been and continue to be a great pleasure to work with over the years. We would like to thank Maya B. Castle, Liisa Kelly, Terry Fedorkiw, Imoinda Romain, Ferial Suleman, Melody Tolson, Franca Amore, Dianna Little, and Marcia Gallego.

Finally, we dedicate this book to our families and thank them for their support and understanding: Roberta and Daniel, and Deborah, Jonah, and Huw.

Michael Del Balso
Alan Lewis
Dawson College

Why Know about Social Science Research?

What You Will Learn

- what is particular about social research
- why social research is important
- how the social science disciplines developed
- what the basic steps are in carrying out social research

INTRODUCTION

Drinking to get high at age 18 was associated with heavy episodic drinking and alcohol use disorders symptoms at age 35. Using alcohol because of boredom at age 18 also was associated with alcohol use disorders symptoms at age 35.

M. E. Patrick, J. E. Schulenberg, P. M. O'Malley, L. D. Johnston, & J. G. Bachman. (2011). Adolescents' reported reasons for alcohol and marijuana use as predictors of substance use and problems in adulthood. *Journal of Studies on Alcohol and Drugs, 72*, 106–127. InfoTrac record number: **A249876964**.*

Canadians are reading more about climate change in their national print news than ever before, but in a manner that is proportionately more skewed toward everyday political and business issues and isolated from discussions of science and direct impacts.

N. Young & E. Dugas. (2011). Representations of climate change in Canadian national print media: The banalization of global warming. *Canadian Review of Sociology, 48*, 1–23. InfoTrac record number: **A252447220**.

[V]olleyball, wrestling, and hockey officials all possess the same degree of personality characteristics as the normal population. However, they are perceived by athletes and fans as being deficient in all of the domains of Extraversion, Openness, Agreeableness and Conscientiousness, while possessing an excess of Neuroticism.

M. J. Balch & D. Scott. (2007). Contrary to popular belief, refs are people too! Personality and perceptions of officials. *Journal of Sport Behavior, 30*, 3–21. InfoTrac record number: **A159644775**.

[C]oncerns about weight and shape emerge as early as the third grade. Hence, it may be argued that prevention programs for eating disorders should begin as early as the third grade rather than in adolescence when the behaviors may be more difficult to modify.... [F]athers

---

*Throughout this book we have included references to articles available in full on the InfoTrac® College Edition database. A free temporary subscription to the database is included with this textbook. Please see the Preface for a quick and easy way to retrieve articles from InfoTrac using the record number.

are an important factor in the development of their daughters' views about their weight and shape. Thus, the role of fathers cannot be neglected in prevention or treatment programs.
W. S Agras, S. Bryson, L. D. Hammer, & H. C. Kraemer. (2007). Childhood risk factors for thin body preoccupation and social pressure to be thin. *Journal of the American Academy of Child and Adolescent Psychiatry, 46,* 171–179. InfoTrac record number: **A159183349**.

The above statements are examples of results obtained from social science research studies using diverse methods. Practically every day we read or hear of results achieved through social science research carried out by researchers in academia, as well as in public and private organizations. The purpose of this book is to show you how social scientists undertake research, and to explain why these procedures are appropriate ways of discovering the facts of social life and explaining them.

We are all interested in questions about our lives and the world around us, which is one of the many important reasons why we should understand research methods. Questions about human beings and their social world have been raised since antiquity. However, for countless generations the answers were based on superstition, intuition, and speculation. This situation has changed over the last two centuries or so, as more systematic modes of inquiry have developed. Intuition and speculation have given way increasingly to a scientific way of collecting verifiable facts. The result has been a remarkable expansion of our knowledge about ourselves and about our social world. With the development of scientific modes of inquiry, social science disciplines gradually emerged that focused on studying human beings and their societies. The broad category of **social sciences** includes anthropology, economics, political science, psychology, sociology, history, geography, and related areas such as business studies and communications.

Social science research—and scientific research in general—has become part of our daily life. The mass media are full of reports and debates about scientific research. Have crime rates increased in the past year? Is there a rise in the rate of high-school dropouts? Has the income gap between males and females narrowed in the past decade? Is the family falling apart? The reported findings influence our lives in far more ways than we may realize.

Consider, for example, how social research helps to shape some of our opinions and behaviour. Suppose this morning's paper reports that the unemployment rate sharply increased last month. Another article cites a study showing that workers are generally pessimistic about their future employment. Worse still, those working in the very occupation you were hoping to enter in the future are the most concerned. These articles, all of which would likely be based on social research, will undoubtedly have an impact on what you think about the state of the economy and about whether you should pursue a particular career.

Peruse any major newspaper or magazine and you are bound to find articles that cite findings generated by social research. Some findings are even headline news; this is particularly true of the latest information on Canada's unemployment rate, the gross domestic product, and the consumer price index. Furthermore, politicians who design social policies that have a profound impact on our lives will often defend these policies by citing findings from social research studies. These studies, it would seem, give scientific justification for the policies. But are you able to critically assess the reported findings? Do you know what the purposes of the cited studies were? Are the conclusions and the weaknesses of these studies assessed in media reports?

Businesses, political parties, and other organizations also use the techniques and findings of social research to assist them in decision making, policy development, and influencing the public. When it suits their purposes, businesses, political parties, and other groups will challenge, distort, or ignore social science research findings. Because of these widespread uses and misuses of social research, it is clearly in our best interest to understand its benefits and to critically evaluate social research. In this book, we will explore how certain types of high-quality information are possible as a result of social science research. By considering the research process used to collect that information, you will be able to critically evaluate the quality of the information. But why do we need this high-quality information? What is wrong with common sense or nonscientific information? To answer these questions, we need to look at information generally and to explore the contrasts between common sense and science.

FACTUAL INFORMATION AND CASUAL RESEARCH

In our everyday lives we accumulate a lot of information. In our social activities we need to make sense of or be informed about our surroundings in order to act effectively and reasonably. We need to know what is going on and why, and what people expect us to do. We often refer to this information or knowledge as our beliefs, opinions, and values. In certain areas we think we know the "facts" because we have read a magazine or newspaper article or have seen a documentary. Or we may have looked up information on the Internet or read one or two books on a subject. As we grow older this knowledge increases, we gain experience, and we may trust our insights and intuitions more. It may come as a great shock to learn that all this "knowledge" we base our lives on is questioned by science! For people engaged in scientific research, all of the socially acceptable ways of attaining knowledge in everyday life may be flawed.

Much of this potentially flawed information comes from parents, teachers, and other **moral authorities**. Such information is *traditional* or *customary*—it indicates the right way to proceed because it has always been right. Questioning or criticizing this type of information is considered blasphemous or deeply disrespectful. In earlier times this kind of information would have been almost all that was available. However, we are now exposed to the Internet, television, radio, newspapers, magazines, and books. When we go to school we find that teachers have different ideas from our parents and that our schoolmates have still other ideas. We soon discover that different people or information sources present different ideas of what is real, appropriate, or good.

How do we choose when there are conflicting or alternative ideas? Most of us become good at interpreting which rules apply to a particular situation, and we adjust our behaviour and attitudes accordingly. When we are with our friends and intimates we are relaxed and follow different codes of behaviour than we would, say, in a classroom or a work setting. In other words, we use social information *practically*, to cope with life. We do not think too deeply about the information itself—about whether it is logically consistent or is really "true."

What are the characteristics of this social information? Essentially, it consists of faith and folkways, or stereotyped rules of thumb. **Faith** means that something is unquestionably right and that it is simply wrong to disbelieve, criticize, or try to explore and evaluate a belief or an action. **Folkways** are working assumptions about ways of doing things and thinking about things, based on social routines and habits. These assumptions stabilize our experience and give it meaning and organization, whether or not reality actually works in accordance with the stereotypes. Prejudices are an extreme form of stereotyped rules; they identify a particular group or institution in a generalized way and define how an individual or a community should relate to that group. If someone does not fit the stereotype, that person becomes the "exception that proves the rule," rather than a reason for reexamining the stereotype. In ordinary, everyday life we use a lot of unquestioned information that *shapes our observations,* rather than being *shaped by observations.*

For example, assume you were brought up believing that members of a certain ethnic group with whom you have little contact are generally lazy and live on welfare. Would you question your view if you met one who was a successful professional? The tendency would be to see that person as an exception, rather than to question your view. Yet to truly know that group we would need to move beyond the stereotype and actually observe and study its members. Likewise, we may dismiss criticisms of our cherished views by pointing to a case without recognizing that special factors may be involved. An example is someone who has always believed that smoking is not a health hazard and dismisses contrary views by pointing out that his healthy 95-year-old grandfather has been a heavy smoker since his teens.

Today, we also live in a world that values *factual information,* and we get much of this from media sources such as television and radio news documentaries, as well as from newspapers and magazines. You are aware, no doubt, that this information has its limitations. Newspapers and newsmagazines tend to have political biases that may colour the way news stories are written. Furthermore, the most successful media do not present much detail because that would reduce their ability to provide a variety of news items and might reduce audience interest. Instead of analyzing and going into the specific reasons for why things happen, television and radio newscasts in particular tend to briefly describe what happens, present the most dramatic aspects of an event, and focus on a particular personality connected with the event. There may be little follow-up on previous news stories, so we do not get the context that might help us to understand more about what is happening and why. Therefore, "news" often tends to become entertainment; only the highlights of the moment are presented, and these quickly disappear when another story emerges. It requires a lot of effort on our part to put together a coherent and meaningful story from this continuous stream of unrelated bits and pieces. Most of us do not have the time or the tools to do this. As well, we lack essential information that would allow us to do so.

The Internet is a recent addition to our sources of information. It is a communications technology that combines elements of the mass media, popular entertainment, the shopping mall, and the library in a form that is easily accessible from your home. Through the Internet you can download music, read newspaper and magazine articles, view films and TV programs, listen to radio broadcasts, shop online, check out what books are in your college library, and even take college courses and download class notes. You can also wander through innumerable chat rooms, newsgroups, and blogs. Perhaps the greatest problem with the Internet is the sheer volume of "stuff" that is accessible, and it suffers from the same problems and biases found in other media and entertainment sources. And of course the personal communication in discussion groups consists generally of opinions. Even much of the material found by means of search engines is often irrelevant, and thus searching for information may be time-consuming. The Internet is unlike the traditional library, which has professional "information specialists" (i.e., librarians) to help guide your search for information. However, when used properly the Internet is a valuable means of obtaining social science information, as we shall see throughout this book.

We have already identified some problems with various forms of everyday knowledge: faith cannot be questioned or analyzed; stereotyped routines shape our observations; and most news is fragmented and superficial. But what about factual knowledge drawn from the "serious" media or from encyclopedias and books? These sources do at least delve more deeply into particular subjects and are less superficial than ordinary news. Most of us treat newsmagazines, in-depth reporting, and nonfiction books with a special kind of respect. We assume that these sources are written by experts who know what they are talking about and who are telling us the truth. We do not normally ask ourselves exactly how the information was obtained, or whether the right people were interviewed. We assume that the reporters and writers have done their research properly and are presenting the facts correctly. But this means we are taking these reports on faith, unquestioningly.

This approach to factual information—taking the work of experts at face value—is the way most of us deal with experts in our lives. We believe our doctors, bank managers, and others without much questioning or cross-examining or looking around for second opinions. When we do research, for example, to buy a new car, latest tablet computer, or portable media player, we tend to do very quick checks using the most accessible sources of information. This way of proceeding might be called **casual** or **lay research**. How is it different from scientific research?

SCIENTIFIC RESEARCH

Science differs from other ways of knowing and doing research because of its aims and the procedures it uses to pursue those aims. Faith, folkways, and everyday factual information give us moral and social guidelines for solving immediate, practical problems in our lives. Science aims at objective knowledge—that is, knowledge that is not tied to immediate personal or practical problems, but is sought for its own sake.

To understand the procedures science uses, it is necessary to understand a little about its history. Scientific research is a recent development in human history. Ideas associated with science, such as experimentation, precise measurement, and the application of mathematics to description and explanation, developed mainly in the period between 1600 and 1800 in Western Europe (see Box 1.1).

BOX 1.1   THINKING SCIENTIFICALLY: THE ORIGINS

The *scientific revolution* refers to the period in European history when the traditional framework of thought of intellectual elites was challenged by new observations (e.g., the discovery of the New World), new techniques of observation (e.g., Galileo's telescope), and new interpretations and philosophical discussions of these novelties. Prior to the rise of the natural sciences, learned and literate elites thought within a Christian religious framework, supplemented by ideas drawn from Greek and Roman thinkers approved by the church. Various social changes contributed to the disruption of the church's authority and permitted a growing number of challenges to the closed religious-based worldview. This revolution in thought is normally identified with the period between the publication of Copernicus' heliocentric theory in 1543 and Newton's major work in 1687.

The *Enlightenment* refers to the ideas and debates that emerged in Europe from the late 1600s to the late 1700s, an era that some historians call the "Age of Reason." Philosophical discussions originating with the scientific revolution concerning the origins and nature of knowledge came to be applied to politics, history, and social issues. Whereas the scientific revolution involved an investigation of natural phenomena, the Enlightenment thinkers debated about the "nature" of human nature, the "nature" of politics, and the "nature" of history. That is, Enlightenment thinkers were philosophical and did not engage in close and systematic observation of human life. However, their ideas were important in stimulating later generations of scholars to develop the social sciences.

During this period when people began to question long-cherished ideas that were at odds with their own observations, they were quickly challenged by the church and other authorities, who argued that the claims were fraudulent, or worse, blasphemous. Pioneer scientists had two strong ways to defend their claims. The first was to persuade critics to look at the evidence and at how it was obtained. If cardinals and bishops looked through the telescope, they too could see mountains on the moon; it was not just a subjective claim. Therefore, the more open scientists were about their findings and about the ways they had achieved these findings, the more likely they were to be believed. The second strong defence against their critics was to present the findings clearly and logically. Using closely reasoned logical arguments to justify their interpretations of their discoveries, scientists made it harder for traditional religious and moral objections to stand up. Logic could be used both to defend explanations of what they had found and to criticize those who argued against their reasoning. These two ways of challenging accepted ideas soon became a basic part of the **scientific method**. A major feature of the scientific method is the importance given to observation. As well, for research to be acceptable, it has to be logically reasoned, follow explicit procedures, and be open to inspection by others, even to the point of being repeatable by other researchers.

Above all, to be scientific means to carry out **systematic empirical research**. And to be systematic means to use specially developed procedures for gathering information and to use these procedures carefully and thoroughly. The aim of systematic empirical research is to observe, describe, explain, and predict in order to improve our understanding of reality. This is in contrast to casual research, which quickly gathers the most convenient and accessible information, often in a haphazard way, in order to solve a pressing problem. To be empirical means that you do not assume that you already know the answers; your research is not shaped by moral convictions or other traditional cultural assumptions that operate elsewhere in your life. On the contrary, you have to set these assumptions aside, be open-minded, and operate according to the special assumptions of scientific research. If we rely only on the "common sense" provided by faith and folkways, we have no way of knowing whether our assumptions are correct. For centuries people believed it was "common sense" that the earth was flat or that the earth was the centre of the universe. Only when individuals raised questions about these cherished beliefs and started looking for verifiable evidence were these views eventually dismissed. A simple "map" of some key differences between science and common sense is provided for your guidance in Table 1.1.

Table 1.1 Science and Common Sense

	Common Sense	**Science**
Basic Foundation	Faith in unquestioned and unchanging truths	Working assumption of order in the universe
Examples	"Violent images in the media are always bad."	"Media images and audience responses are complex; so are the relations between them."
Community Rules	Folkways—commonly accepted ways of believing and behaving that are taken for granted	Objectivity, empirical verification, and open and clear communication, which are subject to critical examination
Examples	"We should ban violence in the media."	"We need to do research on: 1. The variety of media images: cartoons, horror, psychological thrillers, etc. 2. Circumstances in which people view these different images 3. The range of responses of different groups and individuals
Searching for Knowledge	Casual research—what is needed to make a specific decision or to solve a particular problem	Systematic research—conscious attention to rules of description and explanation
Examples	"My children were so upset when they saw that movie. I'm never going to let them see another one like it."	Experiments—exposing groups of children in controlled ways to media images and then observing their behaviour

THE EMERGENCE OF THE SOCIAL SCIENCES

The emergence of the social sciences was stimulated by the dramatic changes in European societies from the Renaissance onward: the discovery of sea routes to Asia and the Americas; the Protestant Reformation; and the scientific, commercial, and industrial revolutions. However, the social sciences first developed as a set of philosophical ideas and only later involved systematic empirical research. (See Table 1.2 for some early examples of pioneering empirical research. Note that most examples are from the 20th century.)

The evolution of the social sciences into research-based, empirical disciplines involves several stages. One stage is the separation of a discipline from predecessors, such as philosophy, or from nonacademic fields, such as art or journalism. A second stage is the development of the discipline in its own right, with a focus of study and an approach to its field that are distinct from those of other disciplines. Associated with these developments is the emergence of institutions that reinforce the practice of research as an open, shared process based on common interests and standards. Professional associations, journals in which research can be published, and universities where researchers can be trained and research can be sustained are the most important of these institutions.

Table 1.2 Examples of Pioneering Empirical Social Research

Principal Researcher	Research	Method
Charles Booth	Conducted several surveys on poverty and society in London. (1886–1903)	Survey (see Chapter 5)
Wilhelm Wundt	Set up the first experimental laboratory. It was used to study human sensory experience such as vision and touch. (1879)	Laboratory experiment (see Chapter 6)
Elton Mayo	With others, used different approaches to study the behaviour of industrial workers in their workplace, including the influence of illumination on productivity. (1927–1932)	Field experiment/ open-ended interviews (see Chapters 6 and 7)
Bronislaw Malinowski	Studied the day-to-day life of the Trobriand Islanders of New Guinea by living with them. (1915–1918)	Fieldwork (see Chapter 7)
Nels Anderson	Lived for some time as a homeless person during his studies of the urban homeless. (1921–1923)	Urban fieldwork (see Chapter 7)
Harold Lasswell	Studied mass communication processes. One study focused on radio and print reports of enemy leaders' speeches. (1942)	Content analysis (see Chapter 8)

Social Science Disciplines

Just as there is no single "science of nature" but rather several different natural sciences specializing in the study of various aspects of nature (physics, astronomy, chemistry, botany, etc.), so too are there different social science disciplines. Each discipline focuses on particular aspects of human life, develops its own theories in a language of its own making, and often uses certain **research methods** more than other social sciences do. Table 1.3 illustrates how different social science disciplines might undertake research on family planning as a result of the differences in their perspectives. Clearly, each discipline and different research approach makes important contributions to our better understanding of the phenomenon.

Anthropology studies the origin and varieties of human beings and their societies. Physical anthropology focuses on human evolution and variation and is closely tied to genetics and other life sciences. Cultural anthropology focuses on ways of life and varieties of beliefs, customs, language, and artifacts. Field research, whereby the social scientist lives with and participates in the life of the people studied, is a major research technique used in cultural anthropology.

Table 1.3 Social Sciences Perspectives: Examples of Research Approaches to the Phenomenon of Family Planning

Discipline	Emphasis	Family Planning Research
Anthropology	Systems of culture, including material artifacts	Differing patterns of population control in tribal and peasant societies
Economics	Rational choice in context of limited resources	Rise in family planning as a response to increasing costs of educating children as economy shifts from agriculture to industry
Geography	Spatial distribution	Patterns of spread of contraception and shifts in family size across rural versus urban areas
History	Time sequences	Rise in middle-class reform movements in industrial cities and key figures spearheading these movements
Political Science	Political, legal, and governmental organization	Variations in strengths of reform movements reflecting class, ideological, and political factors
Psychology	Personal and interpersonal feelings, ideas, and relationships	Patterns of trust, communication, and bonding between couples in relation to family planning
Sociology	Norms and pressures of group life	Class, ethnic, religious, and other community factors in relation to contraception

The predecessors of anthropology in Canada were essentially the missionaries, explorers, and travellers of the past. But in 1910 a federal government department of anthropology was set up, headed by the famous American linguistic anthropologist Edward Sapir. He recruited various Canadians, including one who went on to establish the first centre for ethnography at Université Laval, which has specialized in the study of Quebec rural life. The first university department of anthropology was established in 1925 at the University of Toronto. The main anthropological association in Canada is the Canadian Anthropology Society/La Société canadienne d'anthropologie, whose official publication is the semi-annual journal *Anthropologica*, which publishes articles in French and English.

Economics studies how human beings allocate scarce resources to produce goods and services, and how these goods and services are distributed for consumption. It has two main branches: macroeconomics, which studies entire economies, and microeconomics, which focuses on individual firms, consumers, and other economic actors. Much of its research is based on economic and demographic information available from government sources such as Statistics Canada.

In Canada, economics was at first closely associated with amateur writing on policy issues by government officials and businesspeople and was known as *political economy*. But in 1918 the Dominion Bureau of Statistics (Statistics Canada's predecessor) was founded and provided standardized national and provincial statistics for economic analysis. University departments were created soon after. Economists were formerly part of the Canadian Political Science Association, which published from 1935 to 1967 the *Canadian Journal of Economics and Political Science*. But in 1967 economists set up the Canadian Economics Association/Association canadienne d'économique and a year later began publication of the *Canadian Journal of Economics/Revue canadienne d'économique*. For many years economic research among francophone academics was concentrated at École des hautes études commerciales de Montréal (HEC Montréal), founded in 1907, which began to publish *L'actualité économique* in 1925. The journal is nowadays co-directed by the HEC Montréal and *La Société canadienne de science économique*, created in 1960.

Geography is the study of physical environmental phenomena and their effects on human society and culture. In the late 20th century, geography became a synthesizing social science linking the physical and social sciences through a focus on spatial patterns and environmental factors shaping human history and culture.

Until World War II, to be a geographer in Canada generally meant being a cartographer trained in civil engineering, geology, or drafting. In 1922 the University of British Columbia established a department of geology and geography, and a first full department of geography was established at the University of Toronto in 1935. Over time, geographic expertise was increasingly demanded by governments concerned with strategic resources and economic development. The federal government first hired a full-time professional geographer in 1943, a practice followed by most provinces in the 1950s. While a nonprofessional Quebec Geographic Society had existed since 1877, the Canadian Association of Geographers/L'Association canadienne des géographes was established by university-based and other professional geographers in 1951. Its publication is *The Canadian Geographer/Le Géographe canadien*.

History is the study of the human past. However, modern social scientific history is traced largely to the 18th and 19th centuries. At that time, the practice of using original records, documents, and letters and carefully cross-referencing these sources to establish the truth of facts or events and the accuracy of dates became the normal practice for historical research. Historians stress the importance of documentary evidence and its interpretation as their basic research technique.

In the 19th century much historical writing in Canada was mainly semifictional romantic or moralistic storytelling. The creation of the Public Archives of Canada in 1872, however, contributed to the compiling of objective records and common sources for historical analysis. By the 20th century some chairs of history had been established at various universities. In 1922 the Canadian Historical Association/Société historique du Canada was founded. It publishes the *Journal of the Canadian Historical Association/Revue de la Société historique du Canada* and oversees various other publications. In francophone universities, departments of history were mainly set up after World War II. However, as early as 1915, the influential and controversial Father Lionel Groulx had set up a chair in Canadian history at the Montreal campus (present-day Université de Montréal) of Université Laval. In 1947 he started the journal *La Revue d'histoire de l'Amérique française*, which publishes articles on Quebec history.

Political science studies the processes, institutions, and activities of governments and groups reacting to and involved in these processes. Political science mainly emerged as a discipline in the 18th and 19th centuries, when scholars attempted to classify the varieties of governments and explain why they existed. With the growth of democracy, the analysis of elections and voting patterns emerged. Consequently, much political science research involves the use of polls and other survey research techniques.

The Canadian Political Science Association/L'Association canadienne de science politique (CPSA/ACSP) was founded in 1913. In Canadian anglophone universities, political science was dominated by economics, or "political economy," until after World War II as reflected in the title of the association's journal, the *Canadian Journal of Economics and Political Science*, noted earlier. But when economists formed their own association and started publishing their own journal, the CPSA/ACSP founded the *Canadian Journal of Political Science/Revue canadienne de science politique*. The development of political science in francophone universities largely coincided with the rapid social and political changes in Quebec in the 1950s and 1960s. Political science expanded along with the various francophone universities, and a francophone association of political scientists was set up in the early 1960s, now known as the Société québécoise de science politique, which is affiliated with the CPSA/ACSP and publishes the journal *Politique et sociétés*.

Psychology is the study of behaviour and thought processes. Psychologists do research in such areas as perceptions, memory, problem solving, learning and using language, adjusting to the physical and social environment, and normal and abnormal development of these processes from infancy to old age. The two most widely used techniques for studying behaviour are observing the behaviour of humans and animals and conducting experimental studies on the effects of various stimuli on behaviour.

In Canada before the 20th century, psychology was seen as a subdivision of philosophy, and the first experimental laboratory was set up in 1889 by James Mark Baldwin of the University of Toronto philosophy department. University departments of psychology began to be created in Canada mainly during the 1920s. In 1939 the Canadian Psychology Association/Société canadienne de psychologie was established. In partnership with the American Psychological Association it publishes the *Canadian Journal of Behavioural*

Science/Revue canadienne des sciences du comportement, *Canadian Journal of Experimental Psychology/Revue canadienne de psychologie expérimentale*, and *Canadian Psychology/Psychologie canadienne*. Psychology departments in francophone universities were established in the early 1940s and were at first dominated by a Roman Catholic philosophical orientation until about the late 1950s, when the focus turned to basic research.

Sociology emerged in the 19th century as the study of all aspects of social life in industrial or modern societies. It is now defined as the study of human social relationships, the rules and ideas that guide these relationships, and the development of institutions and movements that conserve and change society. Sociology is a broad discipline with many subdivisions and many overlaps with the other social sciences. Because of its breadth and variety, sociology does not lean toward any one research technique.

Sociology was a late developer in Canada despite the establishment of departments at McGill University and the University of Toronto in the 1920s and 1930s, respectively. In 1941 the prominent economist Harold Innis described it as "the Cinderella of the social sciences." But in Canadian universities, sociology began to emerge as an independent discipline in the 1940s. Until the 1960s sociologists mainly joined the Canadian Political Science Association, but in 1966 sociologists and anthropologists founded the Canadian Sociology and Anthropology Association. In 2006 the name of the association was changed to the Canadian Sociological Association/Société canadienne de sociologie, and its journal, formerly called *Canadian Review of Sociology and Anthropology* was renamed *Canadian Review of Sociology/Revue canadienne de sociologie*. In francophone universities sociology was at first under the influence of the Roman Catholic Church, but over the years the church's influence diminished. Sociology expanded and new journals were created, including *Sociologie et Société*, which began publishing in 1969.

SOCIAL SCIENCE RESEARCH

While the foundation of the social science disciplines rests on research, sharing the basic perspective and approach of the natural sciences, some features are unique to the social sciences. Whereas natural scientists study inanimate objects—chemicals, physical forces, stars, and so on—or animate but nonhuman beings, social scientists mainly study human beings with whom they can communicate. This focus gives social scientists special advantages and challenges in devising methods of research. Natural scientists may carefully and systematically observe and experiment but they cannot communicate with what they study. Social scientists can perform interviews and experiments, conduct surveys, read documents and letters written by long-dead individuals, and learn about cultures by participating or sharing in other people's lives and experiences. Social scientists, then, have a broader variety of research techniques available to them than do natural scientists.

However, since much of social science research involves the direct study of human beings with whom they can communicate, it brings up the possibility that the research may be coloured by emotions, political concerns, and other sources of bias. Social reality appears to be a much more difficult matter to investigate and about which to produce strong, firm conclusions and explanations. Experimentation in social life is much harder, so a researcher often cannot pin down exactly how one thing affects another. It is also much harder to identify clearly measurable qualities and characteristics in social life than it is in the natural world. Social phenomena often tend to be qualitative—describable in words but not as clearly and easily represented by numbers.

It is also difficult to pin down beginnings and endings of social phenomena. When did the Quiet Revolution in Quebec begin? When did it end? These points cannot be identified with any great precision. Social life is also continually and rapidly changing. The speed of social change makes it difficult for social scientists to keep up with and accurately describe the changes, let alone interpret and explain them. These complications mean that the social sciences are less able to produce small numbers of powerful explanations or general theories that apply to a large number of different phenomena. Instead, generalizations and explanations in social science apply to a small range of times and places, seem to have many exceptions, and are often intensely disputed. These characteristics continue to generate debates about the nature of the social sciences, and their progress, usefulness, and comparison with the natural sciences.

THE RESEARCH PROCESS

We have said that the scientific method involves following procedures that are acceptable to other scientific researchers. This book introduces you to these procedures so that you can understand how social science research takes place. With this understanding, you can become an informed and critical reader of social science literature, including what is reported in the media (e.g., see Reading 1.1), and you can begin to learn how to do your own research.

 READING 1.1

Study Links Video Games to Reckless Driving: Canadian Death Cited

Every day the media report on social research of general public interest. Usually the reporting is brief and highlights the main results. Likewise, we tend to retain what was said about the results and pay little or no attention to the aim and purpose of the research and how it was conducted. As a beginning step in understanding social science research and becoming a more critical thinker of what is reported in the media, consider the following questions. In addition to the results, are we told about the objective of the study? How was the study carried out (i.e., the methodology)? Are we told about any weaknesses of the study? What was the researchers' general conclusion?

As you become more familiar with social science research, you will have more questions about the media reports. Read the newspaper article below and critically assess the information about the objective, methodology, results, and conclusion of the reported study. In addition, compare the newspaper article with the cited academic journal article whose reference is indicated below, and for now take note of the steps in the research process.

Inspired by last year's death of Toronto taxi driver Tahir Khan, who was hit on a winding ravine road by a teenage street racer with a copy of the video game Need for Speed in his car, German psychologists have compiled the most extensive case yet that racing games cause reckless driving.

Playing such titles as Burnout, Midnight Racer and Need for Speed "increases risk-taking behaviour in critical road traffic situations," the team led by Peter Fischer reports today in the *Journal of Experimental Psychology: Applied.*

Writing about the Toronto case, in light of earlier examples of people reenacting video game scenarios to lethal effect, such as the Columbine school shootings, Prof. Fischer writes: "What if players of racing games similarly model their actual road traffic behaviour on their behaviour during these games?"

One of the team's experiments largely replicates a survey done last month by BSM, a British driving school.

Both showed that the more a person plays a racing game, the more likely he is to drive in an "obtrusive and competitive" manner. Both also come with the caveat that self-reporting is notoriously unreliable.

But the German team, from the elite Ludwig-Maximilians University in Munich, went further to look at whether playing a racing game can directly "prime" someone for risk-taking.

They had 68 university students play one of the three racing games mentioned above or one of three "neutral" games (Tak, Crash Bandicoot, or FIFA 2005). In keeping with previous research on video game psychology, subjects were allowed at least 20 minutes of playing to ensure they were sufficiently "in" the game.

The subjects then took a "Vienna Test," which measures risk-taking in actual road traffic situations, and is, in effect, another sort of video game. They watched reenactment videos of risky driving situations, such as passing on the highway and crossing train tracks, and were instructed to press a button at the moment they would abandon the manoeuvre.

"The time that had elapsed since participants gained their driver's licence, the number of accidents reported, sensation-seeking and enjoyment of the games had no significant effect on risk-taking behaviour," Prof. Fischer writes.

But simply playing the games seemed somehow to increase a subject's readiness to take "real-life" risks, a result that was "especially pronounced" for men.

"Practitioners in the field of road traffic safety should bear in mind the possibility that racing games indeed make road traffic less safe, not least because game players are mostly young adults, acknowledged as the highest accident rate group," Prof. Fischer writes.

Source: Joseph Brean. "Study links video games to reckless driving:
Canadian death cited." National Post, Pg. A3. (19 March 2007).
Material reprinted with the express permission of National Post Inc.

A more elaborate and detailed analysis than what was reported in the newspaper article is available in the following academic article:

P. Fischer, J. Kubitzki, S. Guter, & D. Frey. (2007). Virtual driving and risk taking:
Do racing games increase risk-taking cognitions, affect, and behaviors?
Journal of Experimental Psychology: Applied, 13, *22–31.*

Figure 1.1 Research Process

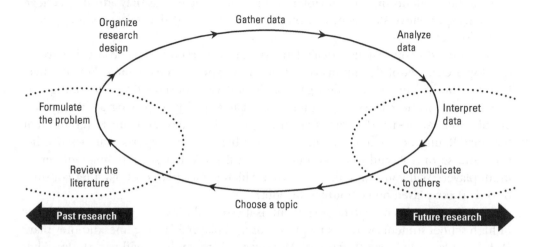

Organize research design

Gather data

Analyze data

Formulate the problem

Interpret data

Review the literature

Communicate to others

Choose a topic

◀ **Past research**

Future research ▶

Social research that is acceptable to other social scientists follows certain steps, which are presented in Figure 1.1 and elaborated on in the following chapters. However, it should be noted that researchers develop their own styles of following this **research process.** Here, we provide only an overview, and in Chapter 2 we elaborate on the process and introduce more terminology.

Social research requires careful preparation. The research process begins when we *choose a topic* from our knowledge base—that is, the general knowledge we have about an area. For example, you might decide to carry out a study to find out about the effects of playing video games among college students who were avid players of violent video games in high school. Your interest in this subject may have been aroused from hearing parents and teachers claim that spending hours playing violent video games results in poor academic achievement and more aggressive behaviour. You wonder who actually purchases and plays violent video games, which are the more popular ones, whether some games have any educational purpose, and so on. You also know that some of your friends have had more exposure to such games than others, and that they do not all play the same games. These general ideas form the beginning of a research topic or issue.

Your next goal is to be more precise about the issue you want to examine. To help you in this effort, you concentrate on relevant background information, including scholarly studies on the topic. In other words, you *review the literature.* In the process, you become aware of what specific issues other researchers have examined, what techniques they used to collect facts, how they defined certain terms to distinguish different types of video games, information on the video games industry, and much more. The review of other studies and discussions on the topic will help you to carefully *formulate the problem* that you intend to research. The stated problem has to be a clear, focused statement that will guide your search for evidence. For example, after reading and reviewing works on violent video games and aggressiveness, you may decide that the literature ignores whether college students who were avid violent video game players in high school lose interest in such games in college. You develop an idea that students' interest in violent video games declines as they adjust to college life. This speculative statement can be put to the test, and that becomes the goal of your research.

But how do you test the idea? This will involve several steps. You will have to develop a clear set of definitions of the different types of video games. For instance, how do you define violent video games? Should you consider different ratings for violent video games? Does playing an online game qualify as playing a video game? Should you categorize the games according to price, sales, and the Entertainment Software Rating Board content rating, or by subjects such as sports, war, science fiction, and so on? Should you take into account if the video game is single-player or multi-player? Will you only consider violent video games played on gaming consoles or only those played on computers, or both?

Other issues also have to be sorted out. Is it essential for you to determine whether in high school students were casual players or avid players? What games did they play? Did they play alone or with friends? Moreover, which students will you study? What year? What program? How will they be selected to be part of your study? How will you carry out the study? It is *your* study and therefore you must decide how to answer these questions. In other words, you need to *develop the research design* and justify it in light of what you want to study.

The next step is to *gather the data,* as spelled out in your research design. After you have collected and recorded the information, you need to *analyze the data.* The data have to be sorted, organized, and analyzed in order for you to interpret them. Do the data show the patterns of change in playing video games? Do the patterns confirm or refute what you expected to find? What can you conclude from the data? How do the data fit with the academic literature you have read on video games?

Finally, you will want to *communicate to others* your data and conclusions. This is often done in a research report that traces each step of the research process. In other words, you should explain how you chose and developed the topic, how you conducted the research, what you found, and your interpretations of the data, and then draw a conclusion. In addition, you need to point out the limitations of your study and suggest areas for future research.

As you can see, social research is a process. It begins with a general question, usually one in which we have a personal interest, that we then make more specific in order to be able to search for an answer. We develop a research design, including special tools and techniques to seek out the desired data. We gather, organize, and evaluate data to answer the specific question and thereby shed light on our general question. Research contributes to the accumulation of knowledge about the issue and suggests further avenues to research. Consequently, research never really ends. As one study comes to a close, another begins and continues the process of expanding our knowledge about ourselves and our social world.

In Table 1.4 we indicate which chapters of this book deal with the various steps of the research process. Throughout, we present the steps in a simplified way, but comprehensively enough for you to carry out your own research or to assess the research of others. The book provides a guided tour of the social research process, but only by doing your own social research can you capture the excitement of this process.

Table 1.4 Steps in the Research Process

1. Overview of the research process.	Chapter 2	What Is Social Science Research?
2. Choose a topic and review the literature.	Chapter 3	Finding and Refining the Topic
3. Choose whom or what to study.	Chapter 4	Sampling: Choosing Who or What to Study
4. Gather the data.	Chapter 5	Social Survey
	Chapter 6	Experimental Research
	Chapter 7	Field Research
	Chapter 8	Indirect or Nonreactive Methods
5. Analyze and interpret the data.	Chapter 9	What Are the Results?
6. Communicate to others.	Chapter 10	The Research Report: The End and the Beginning

What You Have Learned

- There are some important differences between faith, folkways, and casual knowledge, on one hand, and scientific knowledge on the other.
- Science is concerned with systematic empirical observation and logical analysis in order to understand and interpret observations.
- Several social science disciplines have developed, which apply a variety of empirical research techniques to different aspects of human life.
- We should be aware of the possibility that social science research may be affected by emotions, political concerns, and other sources of bias on the part of the researchers.
- Research generally proceeds in a patterned series of steps and stages.

Focus Questions

1. What are the characteristics of faith and folkways as sources of information?
2. What are the problems associated with media-based information?
3. How have the social sciences evolved as they have become research-based disciplines?
4. What distinguishes casual from scientific research?
5. What are the main steps in the scientific research process?

Review Exercises

1. Follow a major news story for two weeks, through the reports on radio and television, and in one newspaper. Identify how you can tell that the story is a major one. How does its presentation shift over the course of the two weeks?

2. Identify a social issue about which you believe you have some knowledge or a strong opinion. For example, "Why do students drop out of high school?" or "What are the characteristics of people addicted to drugs?" With the help of your instructor, a librarian, or a search on the InfoTrac database, try to find some social scientific research on the topic. Compare the ideas and findings of the researchers with your own knowledge or opinion. What have you learned in this process?

3. Select one of the articles cited at the beginning of this chapter and retrieve the full text from the InfoTrac database (see the Preface for instructions on retrieving articles from InfoTrac). In your own words, state the issue that was examined and how the research was designed. Don't worry if you cannot fully understand the content of the article, particularly the jargon and statistical analysis. For now, simply identify the steps in the research process.

Weblinks

The Internet is an invaluable tool for social scientists. But because of the abundance of websites and the vast quantity of information, it is often time-consuming to find relevant material. At the end of each chapter we suggest links to websites that make your search for material easier. (In Chapter 3 we offer advice on how to develop Internet research skills for finding appropriate sources for research and course assignments.)

You can find these and other links at the companion website for First Steps:
A Guide to Social Research, Fifth Edition:
http://www.firststeps5e.nelson.com.

General Resources for the Social Sciences
The sites below include general information of interest to social science students.

Research Resources for the Social Sciences
http://www.socsciresearch.com/index.html

Canadian Social Research Links
http://www.canadiansocialresearch.net/

Dictionary of the Social Sciences
http://bitbucket.athabascau.ca/dict.pl

Scholarly Societies
Academic and professional associations have websites, which generally contain relevant material of interest to social science students. You will find information about the association, as well as about the discipline, the association's publications, including journals, and links to relevant sites. Below we provide links to the Canadian academic associations and to a website with links to thousands of scholarly societies worldwide.

Canadian Anthropology Society
http://www.cas-sca.ca/

Canadian Association of Geographers
>   http://www.cag-acg.ca

Canadian Economics Association
>   http://www.economics.ca

Canadian Historical Association
>   http://www.cha-shc.ca

Canadian Political Science Association
>   http://www.cpsa-acsp.ca

Canadian Psychological Association
>   http://www.cpa.ca

Canadian Sociological Association
>   http://www.csaa.ca

Scholarly Societies Project
>   http://www.scholarly-societies.org

General Interest about Social Sciences

Below are links to reports and videos that illustrate the contribution of the social science disciplines and the impact of social science research on public policies.

World Social Science Report 2010
>   http://www.unesco.org/new/en/social-and-human-sciences/resources/reports/world-social-science-report

Social Science Research Council—YouTube channel
>   http://www.youtube.com/user/SSRCorg

Humanities Matter
>   http://humanitiesmatter.wordpress.com/

InfoTrac College Edition
>   http://infotrac.thomsonlearning.com

>   The InfoTrac College Edition is an "online university library" that offers access to nearly 20 million full-text articles from about 6000 academic journals and popular publications. Four months of unlimited access to the database are included with this textbook. Throughout the book we have referred to articles available on the database. See our Preface for more on how to access the cited articles. For a more detailed guide on the various ways to search and retrieve articles, go to the InfoTrac home page and click on "User Guide."

2 What Is Social Science Research?

What You Will Learn

- how research begins
- the skeptical or critical values of research
- the scientific and the humanistic approaches to research
- basic steps in the research process and their pitfalls

INTRODUCTION

Imagine you met a student from a country with a different language and culture from your own. Before coming to Canada the student learned some English, read up on the history and politics of Canada, and became quite informed about the climate. Yet in his first week in Canada he feels disoriented, lost in the crowd, unsure of what to expect. He is particularly puzzled about a gesture that he has witnessed on a few occasions. At times when two people approach each other, they partly extend their right arms, open their hands, take each other's hand, and shake it. He is baffled by this gesture, especially since someone tried to do it with him and he did not know what to do.

Fortunately, the newcomer has become your friend and feels that he can turn to you for help. He asks, "What is that gesture?" Perhaps you are shocked that someone would have to ask, and so you answer, "It's a handshake." But your friend is still at a loss. He now asks, "How do you do it? When do you do it? Why do you do it?" You try your best to answer his questions, but it seems that every answer elicits more questions: "How far should I extend my arm? How strong should my grip be? Should I do the same thing with both males and females? Should I do it the same with different age groups? Should I…?" You are stunned by the complications your friend sees in what to you, until now at least, was such a simple, common gesture. How would you answer these questions? What other questions do you think he might ask?

Now imagine yourself in a foreign country with different customs and language from your own. Suppose you too read up on the country and learned some of the language. But nothing prepared you for the first encounter with your host in the new country. Before you even have time to greet him with a handshake, he sticks out his tongue at you. Fortunately, he seems to mean well. You quickly offer him your hand, but he turns around. You are baffled by his behaviour, so you turn to him for help. His immediate response is to say, "It's an extended tongue." But you are still at a loss, and you ask: "What is that? How do you do it? Why do you do it?" Consider other questions you would want to ask your host. For example, how far should the tongue be extended?

None of us have taken Handshake 101, yet we know what it is, when to do it, and why we do it. But it can be quite bewildering for someone who is unfamiliar with the custom.

Likewise, if we moved to another country, we would need to know more about the people and their customs. We would need to be able to communicate with them, so we would learn the language. However, we would also note many activities that they generally take for granted that would be strange to us. We would have many questions about these activities, and we would want to learn how, when, why, and so on.

At first, social scientific research may seem as baffling as trying to adapt to the ways of a new country. But with time, the logic and use of social research concepts will become routine, like greeting someone with a handshake.

As with the foreign student who already knew something about Canada but was unfamiliar with the handshake, you already know something about social research but you are not yet acquainted with many aspects of the process. You have heard of polls in the media, you have used the Internet and the library, and you probably have discussed theories in courses. You have even carried out casual research, such as informing yourself about various colleges and universities before deciding where to apply. But, as you may have realized after reading Chapter 1, these activities are not scientific research. As we noted in Chapter 1, science is a way of knowing that is distinguishable from other socially acceptable ways of knowing. Therefore, social scientific research is different from casual research. But how is it different?

To answer this question, let us consider two basic questions about the research process. How do ideas for research begin or emerge? What are the aims and rules of scientific research? After answering these questions, we will elaborate on the steps in the research process.

Because we emphasize those aspects of scientific thinking that are different from common sense, you may find parts of this chapter challenging. Think of the ideas discussed here as a different language or as a new place to visit. Take time to become aware of the differences between what you know and the new language or place you do not know. As you become more familiar with a new language or place, you begin to feel more at home. As you read beyond this chapter, you should find that it makes even more sense.

HOW RESEARCH BEGINS

"How did they think of that?" "Where did they get the idea to do *that* research?" These are common reactions when people read about scientific research in the media. It is often hard to imagine how the researchers thought of doing the research in the first place. The popular image of scientists suggests that their research begins with a brilliant flash of insight—that they suddenly have an idea nobody else has thought of. This image of research is misleading; it suggests that all you need for research to begin is to come up with a brilliant, new idea. If this were the case, there would be far less research than there is, because brilliant ideas are always in extremely short supply.

It is true that research begins with an idea or a question that concerns the people willing to do research. But this idea is likely to have complicated origins in the way the researcher's background, current social circumstances, and ideas in the social sciences themselves come together.

Personal experience can play an important role in shaping people's research interests. Refugees from Nazi Germany in the 1930s played a prominent part in conducting social scientific and historical research into totalitarianism and dictatorship in the 1940s and 1950s. Feminist ideas for social research emerged in the 1970s after an increasing number of women began to graduate from universities.

Current social phenomena and social problems are likely to be of interest to many researchers, even if they themselves are not directly affected. For example, since the arrival of the personal computer, there has been a rapid growth of social science research into the impact of computer technology on work, leisure, education, and an emerging "new economy." The expansion of world trade has led to further interest in international relations, the future of nation-states, the influence of multinational firms, and the impacts of "globalization." The attacks of 9/11 in the United States have led to an enormous expansion of studies on the history of Islam and its contemporary social and political effects.

Perhaps most research is stimulated by the research of others. From the point of view of the individual researcher, this is intellectual stimulation—a game of ideas that may not connect directly to personal experience, a major social problem, or a contemporary event, but are nevertheless interesting and worth exploring further. Much historical research is like this. What was life like for the early settlers of Canada? Who supported Louis Riel and why? These are interesting questions in themselves, even though they may have no direct relation to contemporary issues and problems.

Other people's research can be a stimulus to further research in various ways. The research may have gaps or omissions. For example, there is a lot of research on the nursing profession as a "feminized" occupation—as one that has been primarily identified as a female occupation. Much of the research focuses on the way nurses are subordinated to the traditionally male role of doctors and surgeons. However, the numbers of male nurses and female doctors are growing. This growth suggests that doctor–nurse relations may well be changing as the gender compositions of the two occupations shift.

Research may be marked by controversy and disagreement. For example, there is much research into the relationship between media violence and aggressive behaviour, but a great deal of disagreement exists about what the data show or prove. Researchers often discover new information that is surprising but not easily explainable. The statisticians at Statistics Canada recently discovered that police reports indicate a decline in crime rates over the last several years. Apart from contradicting "common-sense" assumptions and impressions created by the media, these statistics were presented as bare facts without any explanation or interpretation. This lack of interpretation provides an opportunity for other researchers to develop satisfactory analyses and explanations.

What we are suggesting here is that the very beginnings of research are complicated. All the factors we mention—personal experience, current social issues or problems, the stimulus of other people's research—often combine to pull people into undertaking social science research. What events or experiences in your own life could be beginning points of research? Can you think of current events that might even connect with your life or that interest you in some way? If you can say yes to any of these questions, you have taken the first step on the road of social research.

The next step is to narrow down your topic. For the research to get anywhere, it has to become focused, explicit, and developed enough to make it clear, communicable, and acceptable to a community of researchers, as well as to others who may be concerned with the practical results of your research. A clear and focused topic allows you to better communicate your research to others who may provide useful resources, inspiration, and aid.

THE AIMS AND RULES OF SCIENTIFIC RESEARCH

Having become interested in an issue or topic, your next problem is deciding how to follow it up. Becoming interested in an issue could lead you into different ways of expressing and exploring this interest. If your interest is in poverty, you could develop a political program to reform society; you could try your hand at a journalistic exposé of the issue; or you might try writing a novel or directing a documentary video about the experience of poverty. Each of these activities has its own aims and rules or procedures. Here, we shall look at the aims and procedures of a social scientific research approach.

If you have ever tried to read an article in a social science journal, you no doubt found it hard going. Even though the subject matter sounds interesting, the writing style and the terminology seem strange and abstract. This quality of writing results partly because each social science has developed special terms to express its ideas. As well, scientific researchers have developed rules and procedures to avoid the limitations and problems with intuition, experience, common sense, tradition, and authority. The terminology and the abstraction serve as reminders to the researchers themselves to be scientific in their outlook.

Since the articles are written for fellow professionals, the basic assumptions of scientific research are not spelled out. Hence, unless you already know the assumptions, the language of research will seem strange and difficult to comprehend. We are going to identify these assumptions and introduce some of the jargon associated with them. Underlying the rules and procedures of social research are four basic values commonly held by researchers in all fields of science.

1. Researchers are committed to **objectivity** in their research work. That is, whatever their personal biases, researchers are committed to using the appropriate procedures for gathering and interpreting data and to presenting their discoveries honestly, even if these discoveries contradict cherished personal beliefs and values.
2. Scientific researchers aim for **empirical verification** of their ideas. That is, since scientific knowledge is based on observation in the world, scientific ideas must at some point be connected to observations to see if researchers' theories or speculations are in line with the facts. Researchers do not depend on personal experience, intuition, faith in authority, or tradition to provide answers to their questions.
3. Consequently, scientific research is designed to add to those discoveries by contributing to and building on other scientific knowledge. Research is viewed as a cooperative or *collective* endeavour; it builds on past research and lays the foundations for future research.
4. Finally, researchers can make a contribution only if they communicate their research clearly, honestly, and in enough detail that other researchers can fully understand how the research was carried out and how the data were interpreted. The logic and methods of research must be *transparent*, not hidden.

These basic values or aims can be summed up as **skepticism** (see Box 2.1). The value system of research leads researchers to disbelieve any statement that cannot be "proved" or supported by empirical evidence that has been discovered by following the accepted ways of doing research. Researchers, therefore, have developed procedures to ensure that their ideas are clearly and logically developed from previous research; that these ideas can become the basis of research plans that ensure empirical evidence is systematically gathered; and that the evidence is carefully analyzed and interpreted according to accepted rules and not according to the whims and personal beliefs of the researcher. These procedures will be outlined in the section on steps in the research process.

Skepticism could be called the negative or critical side of scientific inquiry. But research is also founded on some positive assumptions. The first assumption is that the universe is orderly or, as Einstein once put it, "God is not a madman." If there are regularities and patterns—events or processes that happen over and over again in nature and society—we should be able to discover them. This idea was advanced by ancient Greek thinkers. It distinguished them from their neighbours, who believed that nature was ruled by the whims and passions of gods, goddesses, spirits, and demons, and that the only way to cope with the resulting chaos was by following religious and magical rituals. These thinkers saw the universe as a disorderly place ruled by mysterious forces; it was impossible to understand how and why those forces worked the way they did, so magic was the key to survival.

BOX 2.1   DOES SANTA CLAUS EXIST? ARE YOU SKEPTICAL?

Do you believe in Santa Claus? Recall the times you sat on Santa's lap and he asked you what presents you wanted, or you sent a letter or an e-mail to the North Pole. There was no doubt that Santa existed and you had proof; he brought you gifts! So when did you stop believing? Why? Was it because you could prove that he did not exist? Or was it because you could not prove that he did exist?

But while you now believe that Santa is just an imaginary construct, what about the claims of those who say they have supernatural powers? There are those who say they can solve crimes more easily than police investigators. One person has even become world-famous for supposedly bending spoons and keys with his mind.

Some psychics maintain that they can communicate with the dead. They have had their own TV shows and best-selling books, and have been featured on major television network programs. Indeed, one psychic had a TV show on which she claimed to have the power to tell to her guests the thoughts of their living or dead pets. And her guests were moved by what she told them.

Many faith healers claim to be able to jolt God into performing miracles. Thousands of people attend their services and many leave believing they *have* been healed, including from serious ailments such as cancer.

Today, as in the past, many continue to believe in astrology. Indeed, almost every major newspaper has a horoscope column. (When was the last time you read your horoscope?) There are also those who claim that by reading palms or tarot cards, looking into a crystal ball, or some other method, they can "read" your personality and foretell your future.

continued

Can people really communicate with our dead relatives and pets? Can faith healers really heal us, and the astrologers and other psychics foretell our future or bend spoons with their minds? Rather than blindly accepting their claims we want proof. More specifically, we want to know if there is any reliable scientific evidence. In short, we are *skeptical.*

The word *skeptic* should not be confused with *cynic.* Whereas the cynic is simply distrustful, the skeptic demands to see the evidence. As noted by the editors of *Skeptic* magazine, "Ideally, skeptics do not go into an investigation closed to the possibility that a phenomenon might be real or that a claim might be true. When we say we are 'skeptical,' we mean that we must see compelling evidence before we believe." (See http://www.skeptic.com/about_us/).

Thus, by being skeptical we are prepared to change our minds, but only if there is reliable evidence that can be acquired using the scientific method. Being skeptical is therefore a common part of scientific research. Whatever the research, being scientific means there has to be an objective, rigorous, and reasoned search for verifiable evidence. We do not simply dismiss the claims, but rather seek appropriate, credible data.

The ancient Greeks also saw that if the universe is orderly, human thinking needs to reflect and express that order. They developed philosophical and mathematical ideas as the basis of rules for describing the orderliness of the universe, and emphasized the importance of **deductive reasoning**—that is, drawing particular conclusions from more general assumptions. For the Greeks, illogical thought and vague, unclear language were the main obstacles to scientific knowledge. Much later, during the scientific revolution in 17th-century Europe, other thinkers stressed the importance of empirical observations and **inductive reasoning**, which begins with particular observations and infers general patterns or principles from them. For these thinkers, purely logical ideas that had no reference to observation or that did not help to steer observation had no part in science. Scientific research is now understood to involve both types of reasoning.

Table 2.1 indicates how both types of logic fit with the research process. Deductive logic—thinking in terms of the relationships between ideas—is used when we move from a broad focus on the ideas of a theory to a narrow focus and then further refine the theory into a hypothesis. Inductive logic, by contrast, refers to thinking that organizes, summarizes, and interprets factual information and tries to draw implications or conclusions from observations. We use this kind of logic when we look for patterns in the data we have gathered. Both kinds of logic, then, are essential to the research process.

Another assumption, related to the idea of orderliness in nature and logic in thought, is the idea that order in nature can be explained. If you can clearly describe *how* things are the way they are, you should be able to develop ideas that explain *why* things are that way. The ultimate aim of research is to provide explanations, and the group of systematic, logically connected ideas that explain a factual discovery is called a **theory**. Research is

Table 2.1 Logic, Theory, and Research

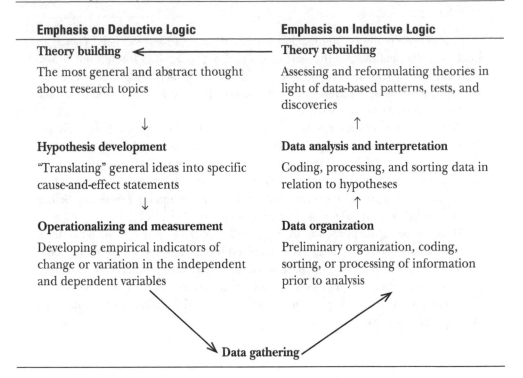

Emphasis on Deductive Logic	Emphasis on Inductive Logic
Theory building ⟵	**Theory rebuilding**
The most general and abstract thought about research topics	Assessing and reformulating theories in light of data-based patterns, tests, and discoveries
↓	↑
Hypothesis development	**Data analysis and interpretation**
"Translating" general ideas into specific cause-and-effect statements	Coding, processing, and sorting data in relation to hypotheses
↓	↑
Operationalizing and measurement	**Data organization**
Developing empirical indicators of change or variation in the independent and dependent variables	Preliminary organization, coding, sorting, or processing of information prior to analysis

Data gathering

never simply an accumulation of facts. Facts are important and meaningful only in relation to questions and ideas. Theories are the general ideas that give meaning to facts. They summarize and interrelate empirical data by explaining how and why things happen and by suggesting logical extensions and implications of these explanations. Stated differently, theories explain why things happen by linking **cause and effect**; they specify the circumstances or conditions under which things happen, and they permit us to make a **prediction** about events because we can look for those circumstances and conditions in advance of the outcome suggested by the theory.

Theories provide us with coherent, systematic syntheses of how we understand particular areas of reality. Yet scientific theories, unlike faith, are flexible. They are organized so that they are open to modification and development based on new empirical information arising from research. Such organization allows for the logical drawing of hypotheses from theories. **Hypotheses** are statements that express cause-and-effect relationships in a special language that indicates that they are not yet as supported by data as are theories. Testing hypotheses through research extends the theory to new circumstances, conditions, and events. Without the process of hypothesis development and hypothesis testing, scientific theories would become closed, dogmatic, and unaffected by new research.

For example, theories of stratification or social class argue that economically based inequalities shape or determine political power and ideas, cultural patterns, family relationships, and so on. Such theories are very broad, general interpretations, which are not easy to test directly in relation to empirical evidence. For theory to be useful in guiding research, two things have to be done. First, we have to break the theory down into a more narrowly focused set of ideas; second, we need to reformulate these ideas into data-oriented, empirically testable hypotheses. For example:

Step 1

General theory: Economically based inequalities determine or shape the major features of social life.

Specific theory: Economically based inequalities shape the lifestyle (tastes, cultural expression, and consumption patterns) of individuals and groups.

Step 2

General hypothesis: Economic inequality determines consumption patterns.

Specific hypothesis: As income increases, the level of spending on entertainment increases.

Hypotheses are ideas put forward to be tested; they are designed to direct our attention to the "facts." Instead of speaking in terms of causes and effects, hypotheses refer to variables. **Variables** are parts or aspects of reality that can be seen to change or vary. Temperature, speed, and volume are variables in the physical world; social class, nationality, standard of living, and sexism are variables in the social world. Hypotheses usually take the form of "if … then" statements, such as "if the temperature of water decreases to the freezing point, its volume will increase," or "if income increases then the amount spent on entertainment will increase." In these hypotheses, temperature and income are called **independent variables**. They are identified as changing, but the causes of those changes are not specified; for the purpose of testing the hypothesis they are viewed as changing by themselves, *independently* of other factors. Volume and spending on entertainment are called **dependent variables** because their changes are linked to, or *depend on,* the changes in the independent variables. In other words, in our earlier hypothesis spending on entertainment "depends" on the level of income. Thus, the hypotheses assume that if there are no changes in the independent variables, there will be no changes in the dependent variables.

How can we tell whether there have been changes in any variable? We detect change by setting up ways of measuring or identifying changes. Setting up systems of measurement or identification of change is called **operationalizing**. Essentially, operationalizing means clearly laying out the rules for establishing changes in variables. In the physical sciences the rules have developed with instruments of measurement: temperature is measured by thermometers, speed by accelerometers, and so on. In the social sciences our instruments tend to be such things as sets of questions specially designed to establish how religious we are, how socially involved, or how satisfied with our work, or psychological tests designed to measure how intelligent, creative, depressed, addicted, and so forth, we are. Other measures are developed from available statistical data to measure unemployment, crime rates, and the cost of living, for example.

The process of operationalizing is one of moving from the *idea* of a phenomenon to a *definition* of the specific ways you intend to measure the existence and variation of a phenomenon in your research. For example, the idea of "intelligence" refers to a combination of skills and abilities, including linguistic, mathematical, and problem-solving or analytical reasoning. An **operational definition** of intelligence is "the skills and abilities measured by this (the researcher's) intelligence test." In other words, a researcher can show us what intelligence is by the operations of developing and applying questions that call for intelligent reasoning.

Take again our earlier hypothesis in step 2 above that states as income increases, the level of spending on entertainment increases. Our hypothesis is in the form of an "if … then" statement connecting variables, that is, if income increases then spending on entertainment increases. But we must now clarify what we mean by "increase in income" and "spending on entertainment." This has to be done by developing clear operational definitions. For example:

Step 3

Operational definition:
a. By "income" we mean … (What do you mean by "income"? The definition must hold for the duration of your study and must resolve issues such as pre-tax or net income figures, earnings only or other income and wealth, individual or family incomes, the time period involved, etc.)
b. By "spending on entertainment" we mean … (What do you mean by "spending on entertainment"? The definition must hold for the duration of your study and must resolve such issues as: Do we include spending on tobacco and alcohol? How do we distinguish between "luxury clothing" worn at parties and regular or necessary clothing? Is eating out a necessity or a luxury/entertainment item?)

Let us take another example, that of "work satisfaction." The idea of work satisfaction might have been taken from general theories of social class and experience of social inequalities. Operationalizing the idea of work satisfaction requires that it be defined in such a way that the researcher can gather empirical evidence of work satisfaction levels through such measures as rates of absenteeism, rates of job turnover, measures of quality of job performance and productivity, and responses to questions about satisfaction with the job. Of course, these operational definitions and measures must be logically consistent with the underlying concepts and the broader general ideas or theories from which they are developed.

Moving from general, theoretical concepts to operationalized concepts of measurable variables is at times difficult and involves complications. Some of these are mentioned in later chapters in relation to specific research methods. A brief and simplified overview of these issues is presented in Box 2.2.

All the measures identified so far have one thing in common: whether you are dealing with temperature, speed, IQ, work satisfaction, or the suicide rate, you can measure these variables in precise quantitative terms. For example, an IQ of 115, a temperature of 21 degrees Celsius, a work satisfaction rating of 7 out of 10, and a suicide rate of 3.5 per 100 000 are **quantitative variables**. But such quantitative precision may not be possible with many social characteristics that are treated as variables—religious differences, nationality, or sexual orientation, for example. These kinds of variations are classifiable as different and you can describe the differences, but you cannot numerically *measure* or rate such differences. Social scientists often deal with such **qualitative variables**, as well as with quantitatively measurable ones. Therefore, hypotheses in the social sciences are

BOX 2.2 OPERATIONALIZATION AND MEASUREMENT ISSUES

ACCURACY

Does your measure produce accurate results or is there some built-in bias that leads to under- or overestimation?

For example, mainly asking questions on what people find unsatisfactory and negative about their jobs and work environment in a questionnaire on work satisfaction might bias the results.

PRECISION

Is it possible to operationalize the variable so that it can be measured with a high level of refinement or precision?

For example, measuring "size of firm" by "number of employees" is more precise than distinguishing between "small," "medium," and "large" firms. How much precision is needed for your research?

RELIABILITY

Will the method used to collect the data lead to consistent results? That is, does the measure produce similar results with repeated use?

For example, ambiguous or unclear questions in a standardized questionnaire may lead to inconsistent responses over several surveys using these same questions.

VALIDITY

Do the operationalized concepts of variables reflect the meaning of the broader theory?

For example, is work absenteeism a useful or realistic way of operationalizing the concept of work "alienation"?

Do the operationalized concepts cover the full range of the phenomena implied by the broader theory or research question?

For example, do measures of "prejudice" apply to a researcher's concern with ethnic, racial, and religious prejudice, or do they really only measure ethnic prejudice?

Are the operationalized concepts plausible or reasonable reflections of the broader ideas in this field of research?

For example, does the operationalized concept of class include measures of economic wealth as well as measures of social status?

often less precise than those in the natural sciences. (This difference in the measurability of variables is discussed in Chapter 9.)

Statements linking independent and dependent variables translate abstract, theoretical ideas into specific, testable ideas. These statements focus on a small number of key factors (variables), and thus the researchers are forced to develop their ideas in terms of observation or data gathering (operationalizing). Thinking in terms of hypotheses requires researchers to identify how variables change, which can be done only by identifying what will be seen as variables change.

Such hypotheses are like pieces of a jigsaw puzzle. As hypotheses are tested, and assuming they survive the tests of observation and data gathering, they begin to accumulate and interconnect into a bigger picture covering more and more factors, events, and processes. This big picture is interpreted and described through theories that represent an overall understanding and explanation of the causes at work.

CAUSAL ANALYSIS

In everyday conversation we are quick to point to a cause. We may have an easy explanation for why there is a high dropout rate among high-school students, why there is air pollution, or why the price of gas may have increased, but to determine cause in research is complicated. The idea of cause is a difficult and controversial one that has long been debated by philosophers. Simply put, there are two basic positions in these debates. One side sees causality as a real, objective characteristic of the universe and considers that the researchers' task is to find evidence of this reality. The other side argues that cause is simply an idea—a logical principle that is a convenient way of organizing or interpreting our information or discoveries. Under this view, we do not find cause in reality; rather, we impose causes on our findings through our analysis and interpretations. Wherever their sympathies may lie in this debate, researchers tend to agree that four conditions need to be met in order to accept a causal interpretation of data:

1. *Temporal Order:* A cause must come before an effect.

   Students who work part-time will have lower grades.

   "A" (part-time work)   ***comes before***   "B" (lower grades)

   Independent Variable   Dependent Variable

   This condition sounds obvious, and in many research situations it is. In the above situation, the student has to first work part-time to see if it affects grade performance. But social science research often involves chicken-and-egg puzzles, in which the sequence is not easily determined. For example, the "differential association" theory of juvenile delinquency argues that delinquency is a result of joining the wrong crowd. Yet it could just as easily be argued that the wrong crowd will attract a certain kind of individual. In which case being recruited into the wrong crowd is not a cause of delinquency, but rather is itself an **intervening** (in-between) **variable**. Furthermore, you could argue that wrong crowds and the problem personalities attracted to them are products of certain kinds of environments—they are both effects of a common cause. Sorting out sequences of causes and effects, then, is not always an easy task.

2. *Association or Correlation:* The researcher must show that the two variables change together in a consistent way.

The more hours a student works, the lower the grades.

"A" changes (increase in hours)　　***occur with***　　"B" changes (drop in grade performance)

Independent Variable　　　　　　　　　　　　Dependent Variable

An increase in the number of hours worked will result in a drop in grade performance. That is, in most cases students who work more hours will have lower grades than students who work less. If there is no consistent pattern of association—that is, too many exceptions—it is unlikely there is a causal connection. How many exceptions still allow for a causal relationship to be inferred requires careful assessment, by statistical analysis if possible. Finding an association by itself does not mean that a causal relationship exists, and the other conditions outlined here must be satisfied.

3. *Elimination of Alternative Causal Influences:*

In our example there might be many other factors influencing grade performance that are unrelated to whether the student works part-time, such as course content. Causal or explanatory research requires controlling alternative causal factors so that clear conclusions can be drawn about the cause actually at work. Different research techniques use different means to establish such controls. The laboratory experiment is the most effective technique for eliminating alternative causal factors, but its use is often limited in social research. Consequently, other means, such as systematic comparison and statistical analysis, have been developed to perform this control function. These techniques, however, are much weaker than experimental controls. Hence, social science research is dogged by many disputes over causal explanations.

4. *Theoretical Consistency:*

Finally, the discovery or interpretation of causal linkages must make sense or be acceptable in terms of the researchers' theoretical framework or assumptions. This is a difficult issue, because the researchers cannot be so attached to their theories that they "explain away" any results that do not fit. Such results or anomalies may be valuable in provoking the development of more adequate theories to replace the existing ones. However, much research consolidates and extends existing theoretical understanding rather than overthrowing it. Assessing the theoretical significance of research data—for example, whether it is an extension or an anomaly—can be very difficult.

 READING 2.1

Read 'Em and Weep: Teenagers' Perusing of Diet Articles Linked to Eating Disorders Later

Social science research can raise important issues about our well-being, and sometimes this research gets noted by the popular media. The newspaper article below reports on a study about teenagers reading magazine articles about dieting. What were the objective, the method, and conclusion of the study? What other questions do you have about the research that are not reported in the newspaper article? What practical advice does the study suggest? Given what you have read in the above section on causal analysis, can we say that reading magazine articles is the cause of unhealthy weight loss behaviour? Why or why not?

After reading the newspaper article you may want to read the original academic journal article on which it is based. The bibliographic information is listed below the newspaper article.

Magazine headlines entice teenage girls with promises: "Get the body you want" and "Hit your dream weight now!" But a new study suggests reading articles about diet and weight loss could have unhealthy consequences later.

Teenage girls who often read magazine articles about dieting were more likely five years later to practise extreme weight-loss measures such as vomiting than girls who never read such articles, the University of Minnesota study found.

It didn't seem to matter whether the girls were overweight when they started reading about weight loss, nor whether they considered their weight important. After taking those factors into account, researchers still found reading articles about dieting predicted later unhealthy weight loss behaviour.

Girls in middle school who read dieting articles were twice as likely five years later to try to lose weight by fasting or smoking cigarettes, compared with girls who never read such articles. They were three times more likely to use measures such as vomiting or taking laxatives, the study found.

"The articles may be offering advice such as cutting out trans fats and soda, and those are good ideas for everybody," said Alison Field of Harvard Medical School, who has done similar research but wasn't involved in the new study. "But the underlying messages these articles send are 'You should be concerned about your weight and you should be doing something.'"

The study appears in January's issue of the journal *Pediatrics*. Its findings were based on surveys and weight-height measurements of 2,516 middle school students in 1999 and again in 2004. About 45 per cent of the students were boys.

Only 14 per cent of boys reported reading diet articles often, compared with 44 per cent of girls. For those boys who did read about weight loss, there was no similar lasting effect.

In the new study, it was unclear whether it was the diet articles themselves or accompanying photographs of thin models that made a difference. The study didn't ask teenagers which magazines they read, only how often they read magazine articles "in which dieting or weight loss are discussed."

The study was based on students' self-reports about their behaviour and, like all surveys, could be skewed by teenagers telling researchers what they think they want to hear, said study co-author Patricia van den Berg.

She said parents should carefully consider whether they want their daughters reading about weight loss.

"It possibly would be helpful to teen girls if their mothers didn't have those types of magazines around," van den Berg said.

Parents also should discuss magazines' messages with their daughters, she said.

"Talk to your kids about where these messages are coming from," she said.

Doctors' waiting rooms are no place for magazines promoting diet and weight loss, she said, "in the same way you don't have materials promoting smoking in waiting rooms."

Source: Carla K. Johnson. "Read 'em and weep: Teenagers' perusing of diet articles linked to eating disorders later."
The Gazette. *Pg. A14. (2 Jan. 2007.)*

Used with permission of The Associated Press. © 2007. All rights reserved.

For the academic journal article on which the newspaper article is based, see:
P. van den Berg, D. Neumark-Sztainer, P. J. Hannan, & J. Haines. (2007). Is dieting advice from magazines helpful or harmful? Five-year associations with weight-control behaviors and psychological outcomes in adolescents.
Pediatrics, 119, *e30–e37.* **InfoTrac record number: A157361009.**

ALTERNATIVE VIEWS OF SOCIAL SCIENCE RESEARCH

In Chapter 1 we pointed out that the social sciences differ from the natural sciences primarily in using research techniques such as fieldwork and interviews that involve communication and social interaction with those whose lives are studied. As well, social scientists often use more qualitative means of organizing and interpreting their data rather than developing precisely focused quantitative measures and concepts. These features are pointed to by some social scientists as reasons for *not* following too closely the assumptions and logic of natural science. They argue that social research is at least as **humanistic** as it is scientific. For these researchers, the task of social research is to bring out the uniquely human qualities of psychological and social reality. This task does not involve abandoning the assumption of an orderly universe or violating rules of logic, nor does it mean that social researchers should not care about empirical evidence. Rather, these researchers argue, social research should be devoted to careful description to bring out the richness and complexity of social and psychological life, and it should pay attention to the unique and the singular, not just to what can be generalized. Furthermore, social and psychological explanations have to be framed in terms of human motives and meanings, rather than focusing on the impersonal, mechanical causality favoured by natural science.

Every social science is divided, to some extent, between the scientific and humanistic approaches, and there appears to be no end in sight to this division. Rather than take sides, we would like to point out that, as far as social research is concerned, both approaches are thoroughly empirical, and their work is generally *complementary* and not mutually exclusive. Valuable information and discoveries have been and continue to be contributed by both approaches.

STEPS IN THE RESEARCH PROCESS

Essentially, doing research involves, first, defining a topic in such a way that it is meaningful, significant, and researchable; second, developing an organized, systematic plan for researching the topic with reference to empirical data; third, following through on this plan; fourth, analyzing the data gathered; and fifth, communicating the results of the research in full—including the definition of the topic, the research plan, how the plan was implemented through gathering data, the results, and your interpretation of the results.

In this section we shall briefly look at this process step by step to give you a sense of how to proceed. Research is in fact more complicated than this, with lots of two-steps-forward–one-step-back movements. Lest we give you the wrong impression, scientific research is never problem-free. Researchers are human, after all; they make mistakes, they misinterpret their results, things fail to work out in the expected way, and all sorts of confusion may occur. Different kinds of things can go wrong at various stages of the research process, so along with identifying the steps themselves, we shall discuss some of the mistakes and problems that may occur at each step.

Step 1: Defining the Topic

As we have already stated, a researchable topic is not just a smart idea no one else has thought of. If you think you have an original idea, you first have to see whether it is really so original. As a basis for research, your idea also has to connect with empirical information. Consequently, the first step is to explore other people's research in order to find out what other people have thought and discovered and to see how your ideas relate or connect to these thoughts and discoveries. In the process, you will be able to clarify your initial idea, to rework it so that it takes advantage of and builds on other people's research. By fitting in with previous research, your ideas do not lose their originality or significance; rather, they become *more* important by becoming part of an ongoing set of discoveries and debates. The process of developing a research topic through examining other people's research is discussed in detail in Chapter 3.

Of course, things can go wrong even at this early stage. The area you are interested in may be controversial, such as the effects of violence in the media on children's behaviour. You may have strong ideas about the subject and want your research to support those ideas. Since the research on this topic is inconclusive, it is open to interpretation and selective reading. In most areas of social science, controversies are plentiful, so you will have to learn to be open-minded and fair in your reading. Any conclusions you draw should be made on the basis of the logical and empirical strengths and weaknesses of the research—they should not arise from your biases. Consequently, this initial step also involves identifying your own biases on the topic.

Misinterpretation of others' research also may occur through illogical reasoning, hasty reading, and omission of important or up-to-date information. These errors will undermine your research by introducing biases, one-sided interpretations, and mistaken interpretations, which will lessen the value of the entire research project. It is essential, then, that you launch the research properly, through a careful, critical examination of your initial ideas and the way they relate to the ideas and discoveries of other researchers in your chosen field.

Let us suppose that you have heard recent media reports about the squeegee kids who clean motorists' windshields when they are waiting at traffic lights. Perhaps you have heard about the incidents in several cities that were apparently provoked by the police crackdown on these activities. Some radio interviews indicated that most squeegee kids are young high-school dropouts. You are aware that, over the past few years, the media have reported periodically on the phenomenon of dropping out of high school. Going beyond the media reports you have, with the help of your college librarian you read a number of government and academic reports on the issue. But now you have questions that these reports do not fully answer. How widespread is the problem? Has it grown worse in the past few years? Is it expected to worsen in the future? Who drops out? Why do they do so?

Step 2: Designing the Research Plan

Research begins with some questions that you want to find answers for, questions that may originate as personal curiosity but are refined and focused by reading previous research. What kind of study you should plan on doing depends on the kinds of questions you wish to pursue. In turn, the nature of your study determines the kinds of research methods you will be using.

First, there is the issue of the general approach to be taken. Is your study to be exploratory, descriptive, or explanatory? You may conclude from your reading of others' research that there is so little information that what is needed is a probe into society to see what the reality is. Alternatively, you may find that there is some information on dropping out but that it is incomplete in certain ways. An **exploratory study** is designed to find out what things are like in situations where there is little firm information. "Is the rising dropout rate real, or is it a media myth?" is the kind of question asked in such a study. An exploratory study is designed to seek out as much information as possible from whatever sources are available about the subject.

Such a study might be qualitative or quantitative, or it might combine both types of information. An example of the former would be a **field study** (see Chapter 7) in which, to make contact with teenagers living on the streets, you do volunteer work at a drop-in centre or a community youth centre. You might also ask friends and teachers if they know individuals who have dropped out who would be willing to talk to you. As you make contacts and build up trust, you can find out about their lives and how they came to be in their current situation, and you might get leads to others in similar circumstances. In addition, you might be able to interview fellow volunteers and community-work professionals and find out what they know about the problem of dropouts. They too could pass you on to other experts or professionals, such as social workers or teachers. From these diverse sources you might put together a picture of dropping out combining both information on the experiences and lives of teenagers themselves, and various perspectives from people close to and informed about the problem.

The distinction between an exploratory and a **descriptive study** is not clear or rigid. However, it is logical to think of detailed description as something that follows from exploration. Once you have discovered some things about dropping out, the next step is to describe it clearly in all its variations. It is possible that the task of detailed and extensive description, then, is a project that follows from the initial exploratory research. Thus, you might go on to do detailed studies of male versus female dropouts, differences between runaways and dropouts living at home, differences in selected communities, and so on. Such studies would probably attempt to use different data-gathering techniques than those used in the initial, exploratory research. The exploratory researcher might use field research, case studies, expert testimony, documentary, and available data. In a descriptive study, the researcher would be more likely to use surveys by distributing questionnaires to community volunteers and professionals, which they would give to their clients, and to survey the volunteers and professionals themselves. Documentary research might be extended to cover a longer time period or to give greater range and detail and so on.

An **explanatory study** attempts to go beyond description by collecting data to show *why* things are the way they are. In our example, such a study would try to explain why dropping out has increased or decreased over a certain time period, or has increased more in some areas than in others; why youths with certain background characteristics seem to be more vulnerable to dropping out than others; or what specific conditions and circumstances seem common among dropouts. Explanatory studies tend to be more quantitative in their approach, to use specific techniques to select their informants, and to use systematic, questionnaire-based interviews or survey techniques and experimental procedures. These procedures allow researchers to draw conclusions more reliably than qualitative approaches do. However, explanatory research often includes exploratory, descriptive, and qualitative material to illustrate and support the arguments made on the basis of quantitative methods.

Consequently, deciding on whether your research approach is going to be exploratory–descriptive or explanatory will lead you to rely on different methods of observation or data gathering and will also tend to determine the relative balance between qualitative and quantitative analysis in your research.

Step 3: Following the Research Plan (Data Gathering)

Once you have developed your research plan and have decided on the aim of the research (exploratory–descriptive or explanatory) and the procedures you will be using to gather your data, you have to follow through on it. The bulk of this textbook (Chapters 5 to 8) discusses these procedures in detail. Here, we should just alert you to two important problems: sampling and errors in data gathering.

Most research involves sampling—that is, you will study only a small part of the reality you are interested in. You cannot track down and interview all dropouts; you will not have time to visit all drop-in centres or to talk to all experts; the government may have useful statistics since 1971 but not before; and so forth.

For various reasons, then, any factual study is restricted to a **sample**, or a part of what the researcher is concerned with. How that sample is made available or chosen for study is an important issue, as we shall see in Chapter 4. What is critical with respect to sampling is how confident you can be that the sample you have studied is representative of the whole. Without reasonable grounds for assuming your sample is representative, your study can be criticized as being potentially biased and as providing a distorted, selective view of reality.

The second problem is that all research procedures have the potential for creating **errors**. Questionnaires asking people to tick off an appropriate box provide opportunities for people to accidentally check off the wrong box, to skip the question, or to deliberately give the wrong reply. Experimenters cannot stop people from trying to guess the purpose of the experiment and altering their behaviour to help or hinder the experimenter from achieving the assumed purpose. Even the best participant observers may slightly alter the behaviour of the people they are observing. All these research procedures are necessary to gather data, but they also have the potential of producing erroneous information. Since these methods are the only ways we have of gathering data, the best we can do is try to be aware of their problems and limitations and attempt to minimize their sources of error.

Step 4: Analyzing and Interpreting the Data

Facts rarely, if ever, speak for themselves. The facts or data that are collected in the course of research always need organizing and analyzing before their patterns and implications can be seen or understood. Notes on fieldwork have to be read and reread before you can see what values, interests, and perceptions people share; the answers to dozens or even hundreds of questionnaires have to be coded and counted before you can identify patterns of opinions and connect these patterns with the characteristics of people with different opinions; statistics from government documents have to be sifted through and reorganized in order to discover trends or other patterns. These processes of analysis, interpretation, reorganization, and sifting and sorting make up the stage of **data analysis**, in which you find out what data you have gathered and what sense you can make of them. As we shall see in Chapter 9, data analysis takes different forms in quantitative and qualitative styles of social research.

Organizing the data badly can give unclear and misleading impressions. Two major problems encountered at this stage of research are drawing improper conclusions and "fudging" unexpected or surprising discoveries. We are often tempted to overgeneralize and to make strong claims on the basis of too few or too weak examples. Our tendency to want to fit observations into clear and distinct patterns may lead us to oversimplify and to force evidence into logical groupings in order to draw clear conclusions. With respect to the second problem, unexpected findings break through the logical groupings expressed in our hypotheses and research questions and leave us with the problem of how to account for these surprising facts. In our concern to explain all of our data, we may improvise new explanations to fit the facts without fully thinking through how these explanations fit with our initial ideas.

Step 5: Writing the Research Report

Research is meaningless if it remains known only by the original researcher. The entire point of research is to add to existing knowledge and, in the case of applied research, to improve our ability to do things effectively—to cure mental illness, reduce poverty, or improve job satisfaction. If no one else knows what you have found out, past mistakes go uncorrected, new conditions are not understood, and attempts to improve or cope with social problems are less effective than they might otherwise be.

Hence, the final step in the research process is writing and circulating a research report. This may take the form of an essay for your teacher, a paper written for publication in a journal, or a report submitted to a client or an organization. Regardless of the report's intended audience, it has to communicate clearly why you selected your problem, what your research design was and how you carried it out, and what the results were and how you interpreted them. Each of these points has to be presented with sufficient detail and clarity so that, in principle, other researchers could **replicate**, or repeat, your research if they so wished. In other words, other researchers may not be entirely convinced of your findings, and they should be able to try their hand at the same kind of research. If others come up with similar results, your research will be accepted as valid and as a foundation for others to build on. Of course, your research may be sufficiently convincing not to need repeating. In either case, at this point it is no longer *your* research. In a sense, it now belongs to all researchers; it has become "what everybody knows."

QUANTITATIVE AND QUALITATIVE RESEARCH

We have made a number of references to differences between quantitative and qualitative research in this chapter, since both approaches are present in each of the social science disciplines. Both are important to social research, and have developed as part of the increasing emphasis on the scientific nature of social science. In fact, many studies combine quantitative and qualitative research methods.

The two approaches are summarized in Table 2.2. Note that although the research strategies differ in some respects, the basic stages of quantitative and qualitative research are similar. Both deliberately and systematically connect their research to previous research, and present the results in such a way that subsequent researchers could further develop the study. Which approach is used depends on the kinds of questions asked, the kind of study that is feasible, and the inclinations and skills of the researcher. Indeed, much research will combine quantitative as well as qualitative approaches to take advantage of their respective strengths.

Table 2.2   Quantitative and Qualitative Research

Step 1: Defining the Topic

Selection of the topic and refining it into a hypothesis or research question

Step 2: Designing the Research Plan

	Quantitative Research	**Qualitative Research**
Aim	Descriptive or explanatory	Exploratory, descriptive, interpretative
Focus	Key variables	Broader interconnected factors
Operationalizing	Identifying or developing numerical empirical indicators for variables focused on	Identifying ranges of relevant qualitative data, open to modification as study continues
Sampling	Random sampling aimed for (see Chapter 4)	Nonrandom samples often used (see Chapter 4)
Instruments	Structured, and can involve experiment, survey, structured observation; the researcher uses inanimate instruments, such as questionnaires and tests (see Chapters 5, 6, 8).	Flexible, and can involve naturalistic observation, fieldwork, intensive interviews, interpretation of documents; the researcher is the main instrument who observes, interviews, and participates in social settings of interest (see Chapter 7).

Step 3: Data Gathering

	Proceeds according to the instrument(s) selected, and is kept separate from analysis and interpretation.	Data gathering is interconnected with data analysis (step 4) and they mutually inform each other.

Step 4: Data Analysis

	Deductive—involves descriptive and inferential statistics.	Inductive—involves largely qualitative interpretation by the researcher.

Step 5: Writing the Research Report

Quantitative research emphasizes precise measurement, the use of numerical data, and, wherever possible, the use of sampling techniques that allow for more valid generalizations (see Chapter 4 for more on sampling). Quantitative research mainly involves using surveys, experimentation, or secondary data analysis in order to gather or produce data (discussed in Chapters 5, 6, and 8). This enables the researcher to describe relationships mathematically in the form of tables and graphs, to use descriptive statistics, and to test hypotheses through statistical analysis.

Qualitative research emphasizes detailed descriptions of people's actions, statements, and ways of life, often including their possessions, cultural objects, and environments (see Chapters 7 and 8). Qualitative research often uses case studies or unusual and "offbeat" situations to throw light on "normal" or general cases and circumstances. Qualitative research methods include naturalistic observation, fieldwork, open-ended interviews, and documentary interpretation. Qualitative research is characterized by a verbal or literary presentation of data, with much less concern for quantification.

In terms of the types of questions asked, quantitative researchers are concerned with establishing and measuring the strengths of the relationships between variables, and representing patterns and social trends in quantitative terms. An example of the first kind of research would be an investigation into the kinds of social and psychological variables that contribute to different patterns of health and illness. An example of the second kind of research would be an analysis of the increasing number of teleworkers. How many such workers are there? How rapid is the rate of growth of this kind of work? What kinds of firms do they work for? What types of work do they do?

Qualitative researchers tend to ask questions about the experiences and subjective meanings of social situations, such as what it is like to be a teleworker rather than a factory or office worker. Qualitative researchers also explore subcultures, such as the informal norms of peer groups in schools or friendships among workmates, and the ways these operate in the broader institutional culture promoted by teachers and managers. On a broader historical level, qualitative researchers are often interested in the social movements and conflicts surrounding changes in the identification and control of various substances such as drugs, or changes to the understanding and treatment of mental illnesses and similar large-scale cultural and institutional changes.

Certain kinds of research are more easily undertaken by one approach than by the other. It is often difficult to sample a large number of cult members or the homeless in order to undertake a quantitative study. Reliable census data are limited to the last century and a half in most industrialized societies, so historical studies have to use fragmentary statistics and combine these with qualitative data, even in studying demographic or economic topics. Studies comparing different time periods or different societies often involve data that have not been collected in similar ways, so that even when the data are quantitative, comparisons require careful qualitative interpretation.

Finally, all research depends on the skills, the creativity, and the interests of the researchers. Some researchers are more interested or excited by one research approach than the other, and develop better skills in their preferred approach.

CONSTRAINTS OF THE RESEARCH PROCESS

So far we have presented social science research as an almost entirely intellectual process. This is only part of the story. Research requires resources of all kinds, and part of the researcher's time, energy, and skills is spent in trying to obtain and manage the needed resources. In addition, research is a human activity—the researcher brings to it a variety of social and personal assumptions and perspectives. Those who are being researched are not always passive subjects opening up their lives for scientific inspection. A variety of social circumstances influence how research issues become important and how they are thought about and researched.

Resources

Doing research requires time, money, and other resources, all of which are in limited supply. Can you finish your report by the deadline? Can you afford to mail out questionnaires to a large enough sample? Will you be allowed to interview residents in an institution? People who undertake research for the first time are shocked to discover that so much of their time and energy is expended organizing these resources so that the research can actually be accomplished. You need to assess the amounts of time, money, and other resources you are going to need prior to starting the research. These considerations must be part of the development of the research plan, the second step in the research process. A *researchable* topic is one that is not only feasible in terms of its meaning and significance in relation to other research but also practical in terms of the resources to which the researcher has access.

Ethics

Social researchers also face a variety of constraints arising from their subject matter—other human beings. The most important of these constraints involves **ethics**: What are researchers' responsibilities to the people with whom they interact in their research? Experiments involving human subjects are obviously limited by basic obligations to treat others with decency and dignity. But almost all social research methods raise ethical questions concerning how much stress the research should be permitted to impose, how much information should be extracted, and how much people's right to privacy should be protected. All professional social science associations have developed ethical guidelines for their researchers (see Box 2.3). Institutions that support research, such as universities and governments, routinely monitor studies by applying these guidelines. Ethical issues will be discussed as part of the presentation of each major research method in subsequent chapters.

BOX 2.3 ETHICAL CONSIDERATIONS

Is it right or wrong to observe people for a research study without them knowing? Is it right or wrong to ask for the racial origin of the respondent in a questionnaire? Is it right or wrong to force students in a university program to participate in a professor's research project? Is it right or wrong for a member of a research team to reveal the partial results of a study without the consent of the other members? For now, consider the following:

- A participant in a research project on attitudes is asked to be part of a half-hour group discussion on immigration. During the discussion, all the other

continued

participants state their strong opposition to immigration and make racist remarks about various ethnic groups. At the end of the discussion, all are asked to complete a questionnaire on their attitudes toward the ethnic groups mentioned in the discussion. When the participant hands in the questionnaire, the researcher points out that the others in the group were actors. They had been told to state they were opposed to immigration and to make racist remarks about different ethnic groups. Are there ethical issues with this study? What should the researcher do or not do with the participant before, during, and after the group discussion? What else should the researcher do or not do?

- A researcher is carrying out a study of youth gangs. Part of the study involves the researcher spending time with the gangs, and the researcher becomes aware of some illegal activities. The police find out about the research and request the researcher to provide them with the names of the gang members. What should the researcher do or not do?

- Researchers carry out a study on the sexual behaviour of college students. In their introduction to their questionnaire they clearly state that the respondent's answers will be kept confidential. They do not ask for a name or any other information that they believe could possibly identify the student. However, in compiling the information from the questionnaires, they realize that because of a question related to age, they are able to identify some respondents. What should the researchers do or not do?

- A researcher wants to study the behaviour of children when they participate in certain arts and crafts activities. A friend who owns a summer camp agrees to allow the researcher to carry out the study at the camp. After preparing for the study, the researcher realizes that the parents have not been asked to give their consent, but the friend says it is too much trouble to ask the parents. It will take too much time, and the arts and crafts activities must be completed before the closure of the summer camp. The owner insists that the researcher simply carry out the study. What should the researcher do or not do?

Clearly, ethical concerns are primordial throughout the research process. Indeed, it is for this reason that we include a section on ethical considerations in each subsequent chapter on major research methods. Ethics should be taken into account from the very beginning of the research process and be part of your research design. In other words, before doing any other preparation, consider the possible ethical issues and how they will be handled. A suggestion for recognizing possible issues is to put yourself in the place of the participants in your study.

At times, no set rule exists to determine whether a situation calls for ethical considerations. A helpful source can be the ethical guidelines of a professional association, such as those listed in the Weblinks section at the end of Chapter 1, and the site indicated in the Weblinks section of this chapter. The various social science disciplines share ethical issues that are common to all of them, but each discipline may also encounter ethical issues unique to it. See the information on the home page of the professional associations for documentation on their code of ethics.

Problems with Human Data

People can be troublesome to researchers for a variety of other reasons: they may be unwilling to be interviewed or to allow others to observe their activities; only a minority of people will return mail questionnaires; people in power—and deviants—tend to be secretive about their actions and resist studies of their lives; some people are inarticulate and unable to express their feelings and perceptions clearly; and historians often find it difficult to examine the lives of the humble majority who do not leave written documents and monuments behind as evidence of their actions. Social researchers, then, have to be extremely aware of the many ways human beings can hide and be hidden, deceive, and limit access to their lives and thoughts.

The Social Context and the Personal Equation

Another set of constraints or pressures arises from the social and personal circumstances surrounding research. Social research looks at quickly changing and controversial parts of reality: society, human relations, and human psychological characteristics. Not long ago spousal abuse was not a prominent topic for social research. In the 1970s Canadian social scientists debated and did research on the U.S. domination of the Canadian economy; in the 21st century many of these same social scientists are now doing research on the impact on Canada of the "global economy." Shifts like these arise from changing social conditions, which contribute to altering definitions of what are important issues, ideas, and phenomena in society. These shifts, in turn, encourage researchers to direct their attention to new areas and to lessen their interest in others.

In addition, social research can use a diverse range of designs and methods: laboratory experiments, surveys, fieldwork, analysis of documents, statistical analysis, and so on. There are many different *styles* of social research, ranging from the scientific style of experimentation and statistical analysis to the humanistic style of oral history and of many field studies. Each researcher develops an interest in specific issues and a taste for a certain style of research, because each of us brings our own concerns, outlook, and skills to research. Some may enjoy working with historical documents or personal letters and trying to recreate the lives and times of individuals and groups in a humanistic style. Others will enjoy the challenge of designing experiments or developing statistical interpretations of economic and demographic data.

As you learn the different approaches to research, you too will develop a leaning toward particular approaches, and you will become better at using these approaches than others in which you are less interested. This is not a problem as long as you are aware of the limitations of *all* research methods and realize that your way is not the only way of doing research. Each design and procedure has advantages and limitations; often the strengths of one method compensate for the weaknesses of others. The different research methods are each equally valuable and often can be used in a complementary way. Combining different methods in the same research design is often essential, as is remaining open-minded about all the available methods for obtaining empirical evidence.

What You Have Learned

- Research may begin with questions or topics arising from personal issues or social issues or from the puzzles and questions posed by previous research.

- The initial issue or question needs to be developed and focused by relating it to what is already known and thought in a broader area of social science.
- All research follows basic guidelines requiring objectivity, empirical verification, and openness to inspection and analysis by others.
- These guidelines rest on certain assumptions that emphasize the orderliness of reality and the need for logical thinking.
- Much social science research is developed through testing hypotheses, which extend theories by focusing on a small number of variables that are assumed to be causally connected.
- Variables are defined in terms of the evidence required to show how they change—an approach called operationalization.
- Some social scientists deemphasize hypothesis testing and approach social research in a humanistic manner, stressing description over explanation.
- Research involves several steps: defining the topic appropriately; developing a research design and following it through; analyzing the data; and communicating the results.
- Research is affected by nonscientific factors, such as a scarcity of resources, ethical considerations, the social context, and the personal character of the researchers.

Focus Questions

1. Why might a researcher become interested in a specific research topic? Give several possible reasons.
2. What is scientific skepticism?
3. What are the basic assumptions of scientific inquiry?
4. What is explanatory research and when is it used?
5. What are the differences between the scientific and the humanistic approaches to social research?
6. What are variables? What kinds of variables are there?
7. What is the relationship between theory and hypothesis?
8. What problems may occur at different stages in the research process?
9. What factors hinder social research or make it difficult?
10. What are some of the ethical issues that deserve attention in research? When should researchers focus on the ethical issues in the research process?

Review Exercises

1. Do a subject or keyword search for a research-based article in a database on a topic in which you have a particular interest. Use a database to which your library subscribes or InfoTrac, for which a free limited subscription accompanies this textbook. Then answer the following questions.

    a. What is the issue or research question in the study discussed in the article?

    b. What is the source of the information in the article?

c. Is a hypothesis being investigated or tested? If there is a hypothesis, what is the causal idea behind it? If there is no hypothesis, is there any causal idea? What are the variables in the hypothesis? Identify the dependent and the independent variables. How are these variables measured?

2. Discuss the various questions and issues raised on ethics in Box 2.3. Review and discuss the code of ethics of one of the professional associations listed in the Weblinks section of Chapter 1. Are some ethical issues unique to members of that profession?

Weblinks

You can find these and other links at the companion website for *First Steps: A Guide to Social Research*, Fifth Edition: **http://www.firststeps5e.nelson.com**.

Social Science Research
Each of the following sites offers an advanced overview of social research. While they are intended mainly for upper undergraduate and graduate students, they could be helpful in further understanding issues examined in this chapter.

Web Center for Social Research Methods
> **http://www.socialresearchmethods.net**

Exploring Online Research Methods
> **http://www.geog.le.ac.uk/orm/site/home.htm**

Online Guide to Qualitative vs. Quantitative Social Research
> **http://www.onlineschools.org/resources/qualitative-quantitative-social-research**

On-Line UCI Research Tutorials
> **http://apps.research.uci.edu/tutorial**

Ethics
The sites below cover various issues on research ethics. More information is also available from the scholarly association sites listed in the Weblinks section of Chapter 1.

Social Science Research Ethics
> **http://www.lancaster.ac.uk/researchethics/index.html**

EthicsWeb.ca
> **http://www.ethicsweb.ca**

What Is Ethics in Research and Why Is It Important?
> **http://www.niehs.nih.gov/research/resources/bioethics/whatis.cfm**

Research Ethics in the Social Sciences and Humanities
> **http://www.researchethics.ca/social-science-humanities.htm**

Finding and Refining the Topic

3

What You Will Learn

- how to select a topic for research
- how to use a library and the Internet
- how to expand the number of sources of information
- how to read monographs and academic journal articles
- how to plan and organize a literature review

INTRODUCTION

In the preceding chapters, we have argued that there are important differences between scientific thought and everyday, common-sense thinking. A scientific approach is based on two elements: attention to the structure of the reasoning and ideas used in thinking about an issue, and the use of systematic procedures designed to get at the facts of the matter at issue. In everyday life, even in situations where we need expert advice and have to assess this advice carefully, as when buying a computer or a car, for example, few of us are systematic or methodical. We turn to friends, use the most easily available information sources, and are often at least partly persuaded by advertising or the personal qualities of sales personnel.

How far such casual research is from scientific inquiry is reflected in the demands placed on us at different levels of our education. When you come to college or university after high school, the difference between common sense and science becomes evident in your classroom experience. In high school, teachers generally encourage students to discuss topics, whether they know much about them or not. The emphasis in classroom discussions tends to be on participation. Teachers encourage you to give your opinions, to state these opinions as clearly and reasonably as you can, and perhaps to debate and modify your opinions in response to those of others. At college and university, however, your opinions are no longer so highly valued. Instructors still challenge you to be logical and reasonable, but also to take account of ideas that are accepted in a particular discipline, and to provide evidence to support your arguments. This process forces you to become familiar with the concepts, theories, arguments, methods, and data of entire groups of social science researchers.

At first you may be shocked that your ideas alone are not good enough to be accepted as the basis for class discussions and assignments. In high school you could write essays combining clever ideas from off the top of your head with "data" or "information" from an encyclopedia, a single library book, or some websites. At college or university you are expected to go beyond this and organize your thoughts in terms of the accepted ways of thinking in a discipline. Your instructors may assign essays in which you will be asked primarily to summarize the ideas of specific areas of social science research. An important part of the research process is to discover, summarize, and critically assess previous research on the issue or topic of your concern.

By directing our attention to what has been done, we eliminate wasteful efforts and needless rediscoveries of what is already known. Previous research suggests ideas that may be developed, methods that may be applied, and puzzles that may be investigated, and provides data that may be reanalyzed or added to. Even if you have a completely original idea, looking at past research is likely to show that it connects with other ideas and information, and these connections indicate just how relevant or important an idea or a discovery may be. Scientific research is a cooperative process (although it also involves much competition between individuals or teams of researchers), and it is also a cumulative process, in which each piece of research builds on previous knowledge and leads to new research.

At the moment you may be a long way from thinking of yourself as a social scientist or researcher. However, you are likely at some future time to need to find out more information about a social issue or phenomenon. Perhaps you want to know more about income inequality in Canada. You may simply want to know the extent of income inequalities. But such descriptive information is insufficient to explain income inequality. You will also need to be informed about others' discussions of the phenomenon. What works exist on the subject? How do they define and measure income inequality? What are the main causes of income inequality? What are its social effects?

But how do you obtain information on these issues so that you can answer these various questions? This chapter takes you through the process of becoming knowledgeable about a topic or field and summarizing your knowledge by preparing a literature review.

SELECTING A RESEARCH TOPIC

The starting point in research is the selection of a topic. Whether the topic is assigned by your instructor, proposed by you, or selected from an instructor's list, it is essential that the research be achievable. You would certainly want to take into account the time you have available to do the project, as well as other constraints, including costs. Therefore, no matter how interesting the topic, you need to be sure that you can carry out the research involved, given the time and other resources you have available.

There is no one best strategy for selecting and narrowing the topic for research. No matter where the topic idea came from (it may have been sparked by curiosity, a magazine article, a class discussion, and so on), one way to begin is to write down the research issue as a complete sentence. Say you have read in a newspaper article that income inequality is increasing partly because the number of college dropouts is increasing, and that most dropouts are males. You would not simply write down "dropouts," because you need to narrow the topic to formulate a clear statement of the issue you will be researching. You might, therefore, write the following:

More male students are dropping out of college than female students.

You will now want to narrow down this issue. One way to clarify and limit the issue is to write down questions about the statement. You might begin with the concepts.

What is meant by *dropout?*

What is meant by *college students?*

Next, you will ask yourself what information you want to find out about the issue. For example:

Are they completely breaking their ties with the college?

Did they hold part-time employment before dropping out of college?

You should continue to ask questions that will help make the issue clearer. A key one to ask is:

Why are more male students dropping out of college than female students?

Now, you can write down the possible explanations, for example:

Females have higher academic aspirations than males.

Females invest more time and effort in their studies.

Females are less likely to find full-time employment than are males.

Each of these explanations suggests another research topic. For example, do female college students have higher academic aspirations than male college students do? What are the academic aspirations of college students? Why do college students have certain aspirations? And so on. As you raise questions about the issue, you are also coming up with factors that contribute to the explanation. Take the earlier explanations as an example. Each points to the following factors:

- Academic aspiration
- Amount of time spent on studies
- Employment opportunities

The more questions you ask, the more clearly focused and interesting the topic is likely to be. Your next step is to concentrate on one or two specific questions you want to research. To begin, you would want to find out what others have written about the issue. This information can help you narrow your topic further.

BECOMING MORE INFORMED ABOUT THE TOPIC

In finding out what others have written about the topic, you can begin with what you already have close at hand. For example, you may get some ideas and even data from media reports, since they often identify sources of information, such as Statistics Canada or the name of the researcher and the place of research. But you should not treat these items as scholarly sources. Your research project has to be grounded in firsthand knowledge of social science research, not secondhand media accounts of this research. However, this can be a way to begin collecting ideas and possible sources of information.

As well, class readings or textbooks will have references, footnotes, bibliographies, or "further readings" and "web links" mentioning relevant books, articles, and Internet resources. It is easy to make notes on your class textbooks and handouts and begin to develop a list of sources of information for your topic. If your textbook deals with the topic in some depth, you should be sure to read the appropriate section carefully and check the index for other references to the topic. Textbooks are useful starting points, because they present material in a straightforward, easily understood way. They are also continually revised and updated, so the sources of information they contain are quite current.

Another easily accessible and familiar starting point is a multivolume encyclopedia, such as *The Canadian Encyclopedia* and *Encyclopedia Britannica,* found online. Such works have the advantage of being written for the general reader, so you can obtain an overview of a topic, perhaps accompanied by a brief bibliography or list of related topic entries elsewhere in the encyclopedia. One popular online encyclopedia is *Wikipedia,* whose English edition contains over 3.5 million articles. It is continually evolving, with articles added or updated within minutes. And, given that it is a "wiki" and "open-content" site, the articles are written by several writers or editors and anyone can add and edit information to most articles. But these contributors may be neither scholars nor experts on the topic. As well, *Wikipedia* and general encyclopedias lack depth and detail, and so should be considered as, at best, an initial source only. You might consider consulting specialized encyclopedias that offer articles written by experts on particular topics related to your research interest. There are an increasing number of such encyclopedias, many of which are available online, such as the *Encyclopedia of Economic and Business History,* the *Encyclopedia of Canada's Peoples,* and the *Encyclopedia of Religion and Society.*

The information from textbooks and encyclopedias should overlap to some degree in their presentation of key ideas, most important findings and researchers, and so forth. Both types of sources may have footnotes, references, and a bibliography identifying the most important or up-to-date books and articles on the general topic.

It is useful, even when you are new to social science research, to develop a critical approach to what you read and to continually review your critical assessments. Even though these assessments are likely to change over time, you will be working with a purpose, rather than aimlessly collecting mountains of references.

From this initial search you may have obtained some basic but more focused information on, say, college dropouts. Of course, your search for these initial sources will have been greatly influenced by how much you already know about the topic. Nonetheless, the aim here is to note the recurring major definitions, concepts, and ideas. What are the names of major researchers or research traditions associated with those ideas? What types of empirical data are often noted? In other words, you will have possibly gathered some information on different definitions of "college dropouts," the demographics of college dropouts, reasons given for the phenomenon, and so on.

At this early stage in the literature search you can begin to assess the basic information in terms of your own assumptions and interests. Do the ideas and data confirm or challenge your assumptions, arouse your interest, or puzzle you? Are there gaps or other aspects of the information with which you are dissatisfied in some way? You have now gained some familiarity with the topic and most likely narrowed it down sufficiently to carry out a more systematic search on a specific aspect of the topic. To return to our earlier example, whereas initially you were interested in the broader topic of college dropouts, you now might have narrowed the topic to focus on the differences in academic aspirations of female and male college students. So the search for sources continues, but now with a more specific focus. How do we proceed from here?

THE LIBRARY AND THE INTERNET

Our first thought may be to look for information on a specific issue by doing a general search on the Internet. In seconds we have thousands, if not millions, of links. But links to what? Websites are easily created and accessible; thus most web pages are self-published and intended to promote a product or certain point of view. Aside from, say, the *Wikipedia* link, many others may be to websites that have commercial content, are noncredible, or are outdated. Another challenge we face is that we most likely use the Internet for leisure in communicating with friends, viewing videos, listening to music, playing games, or simply browsing for interesting material. The result is that we have not trained ourselves to critically assess whether a website is trustworthy. As you increasingly use the Internet to gather information for course work and research, you need to cultivate the habit of critically evaluating the information found on the Internet. (See pp. 52–54 for ways to improve your Internet research skills.) But first let us be aware of the essential role of the college or university library in searching for materials (see Box 3.1).

BOX 3.1 WHERE TO BEGIN?

Before you browse or indiscriminately search the Internet for sources we suggest you do the following:

1. Ask your instructor what sources you can use or should use. Are you expected to use only peer-reviewed articles? Can you use material from the popular media? Are there any restrictions of what sources can be used from the Internet? And so on.
2. Take into account your course material and reading lists, and especially library resources.
3. Search for information in databases and other sources to which the library subscribes, as well as links to resources on the Internet selected by librarians.
4. Use academic **Internet directories** that have been specifically created to serve students, faculty, and researchers. These provide a list of links to websites that have been reviewed by librarians or specialists in the subject. These directories also include links to sites of research groups that offer free online research publications or working papers. (See the Weblinks section at the end of this chapter for examples.)

The fastest and easiest way of learning more about your college or university library's holdings is to visit the library's web page. The library's website provides access to many traditional resources, such as the **library catalogue** listing its holdings of books, reference materials, newspapers and periodicals, audiovisual resources, and so on. As well, the website is a gateway to social science research worldwide, much of which is accessible only to registered library users. Most academic libraries are also accessible from home to registered users so that you may use them on and off campus.

Libraries have a large variety of resources, such as books, journals, newspapers, government documents, video recordings, specialized encyclopedias, monographs, and periodicals. Moreover, they often provide research or subject guides prepared by a librarian that offer information on the library's resources and services. Generally, the guides focus on the relevant information on a specific subject or area of interest. They include such information as how to use the library catalogue to search for material, a list of relevant databases (more on what these are later), and selected web links, as well as the name of a reference librarian.

As you become more familiar with your library's resources, you will quickly discover that it has more to offer than what is on the shelves. Libraries allow access to resources available over the Internet, such as books, government documents, academic journals (more on these below), and especially databases. Indeed, much of what a college or university library offers is available only over the Internet through a link or password provided by the library. Consequently, students who neglect using the library or use it only sparingly may be unaware of the myriad of information resources accessible through the library. If you take full advantage of the library's resources you will build a list of accessible, reliable, high-quality references in less time and with less wasted effort than by doing general searches on the Internet.

SEARCHING FOR INFORMATION

The core of an academic library's resources will be listed in the library's catalogue and, whether on the shelves or online, are likely to be arranged according to the **Library of Congress (LC) Classification system**. The LC system groups together books and other material by subject matter, indicated by an alphabetical labelling of 20 broad classes of information, such as H for social science, J for political science, and so on. Each book has a call number combining this main letter with other letters and numbers indicating a subdivision in the knowledge group, the first letter of an author's last name, and the date of publication. Each book, then, has a unique number by which it can be traced in the catalogue. In the United States, books have a heading on the copyright page entitled "Library of Congress Cataloguing-in-Publication Data." Books published in Canada, including this one, have "Library and Archives Canada Cataloguing in Publication." The information helps librarians index the book in their libraries. But this information can also be useful to you. It includes the LC number and appropriate subject headings in order of importance. The following is an example of the information included on a copyright page.

Library and Archives Canada Cataloguing in Publication

[Author]	Quarter, Jack, 1941
[Book title and authors]	Understanding the social economy: a Canadian perspective / Jack Quarter, Laurie Mook, Ann Armstrong
	Includes bibliographical references and index.
[ISBN]	ISBN 978-0-8020-9695-1 (bound) ISBN 978-0-8020-9645-6 (pbk.)
[Possible subject headings]	1. Community development – Canada 2. Cooperative societies – Canada 3. Nonprofit organizations – Canada 4. Canada – Social policy 5. Economics – Canada – Sociological aspects I. Mook, Laurie II. Armstrong, Ann 1951 - III. Title
[LC number]	HD2769.2C3Q37 2009

The above information can help you find other works on the same subject by providing subject headings that you can use as the basis for searching other relevant material for your research or course assignments. For example, you could find out what works exist under the subject headings "community development," "cooperative societies" and "non-profit organizations." The LC number (also known as the call number or stack number) is useful even if you already have the work but the library does not have a copy. (A book can also easily be searched using the ISBN—the International Standard Book Number—which is a number that uniquely identifies a book.) The LC system places works in the stacks according to subject; therefore, you could go to where the work would be located if the library had a copy on hand and browse to see whether there are other relevant works for your research. Some libraries are even set up so that you can browse the stacks online through a "shelf search" option. Moreover, the subject headings may be used to search materials in databases and other electronic data resources, or even to do a general search on the Internet. (See the end of the chapter for a website that lists the various LC subject headings and subheadings.)

So first become familiar with the library's resources both on and off the shelves. And if you need help never hesitate to ask one of the reference librarians; it is their job to help people access information, and they are experts at it!

DATABASES

Nowadays the most widely used library resources for researchers are the scholarly online information resources commonly referred to as **databases**, including the multidisciplinary databases EBSCOhost, PROQUEST, and InfoTrac. These databases provide access to newspapers, magazines, and, most importantly, hundreds of academic journals. Many other databases offer particular services, such as JSTOR, which provides an archive of full-text scholarly journals for past years, or PsycARTICLES, which offers full-text articles from journals published by the American Psychological Association and Canadian Psychological Association. Some databases are highly specialized, such as the database House of Commons Parliamentary Papers—Great Britain, which allows you to access documents from the 18th century onward, or the database Globe and Mail Canada's Heritage from 1844, which provides access to the newspaper's articles. Many databases are accessible over the Internet only if your college or university is a subscriber. An example is the Early Canadian Online database, which provides over three million pages of print from different publications from the time of the first European settlers to Canada until the early 20th century.

Today there are hundreds of databases, from the general to the highly specialized, and their numbers are likely to grow. They make our search for material easier and more convenient, but we need to know they exist. So first begin by checking with your library for information on the databases you can access and which are the most appropriate to search for sources on your research topic, as well as how you can connect to them off campus.

When you first begin your search on a topic you may want to use the multidisciplinary databases. An example is the college edition of InfoTrac, which is bundled with this text-book as a free four-month subscription. Databases usually offer various search options, such as by author, subject, title, journal, or keyword, or by whether you want to search for only peer review articles, articles during a certain time period, and more. Much will depend on the types of work contained in the database. One quick way to learn about searching for material in a database is to use the help key to explore the different options. As you become more familiar with your topic and more focused on a particular area, you may then want to turn to specialized databases. However, you should be cautious. Many topics for research are interdisciplinary, as, for example, topics on the family, education, leisure, and income inequality. Therefore, relevant material may be found in other publications and databases than those in your particular discipline. Clearly, the more familiar you are with the issue you want to research and with different databases, the more efficient you will be at searching for appropriate material.

INTERNET RESOURCES

The wide accessibility of the Internet has resulted in many other possible sources of information than those available through the college or university library. At the end of the chapter we provide websites that can be useful in finding sources online, including those of virtual libraries, reputable organizations, and scholarly societies. Indeed, much of the information is available only online. For example, say you want to research issues related to multiculturalism in Canada. In this case, government sites such as that of Citizenship and Immigration Canada would be useful to explore possible publications and data. There also may be other sites of interest that provide information and publications available over the Internet which are not available on your library's shelves or mentioned in the library's catalogue. Again on the topic of multiculturalism, the websites of the research network Metropolis and its regional centres, for example, have much to offer, since they focus on issues of migration and diversity in Canada and elsewhere and across different disciplines. Yet, you may never come across their publications if you rely only on a library search. Further, many agencies, such as Statistics Canada, increasingly provide much of their data and publications only over the Internet.

So in addition to library resources, we have come to rely on the Internet in finding other sources. And if used properly the Internet can offer access to a variety of appropriate sources and save us time.

INTERNET RESEARCH SKILLS

Although the Internet offers an easy means to obtain and retrieve information, it has to be used cautiously. So let us explore ways to improve our Internet research skills.

What Are the Pitfalls?

Let us first recognize that there are many pitfalls to avoid in using the Internet for gathering information for research and course assignments. A key challenge is to critically evaluate the quality and credibility of the information that is easily available. Nowadays almost anyone can set up a website complete with sophisticated graphics and extensive links. This gives the Internet an entertaining and exciting quality—it is wide open, constantly changing, and always expanding. Consequently, these may create many traps, leaving the unwary with the impression that the information is reliable.

The ease of posting and updating items on the Internet leads to constant change and expansion as material is reorganized and amended to make way for more up-to-date input. On the other hand, some sites have been around a long time but the information has not been updated. Further, a website can be from anywhere and contain questionable and even false information, with the purpose of misleading you. Some sites are actually fakes intended to be parodies, some are designed to look like another but give different information, and some are maliciously intended to provide false material.

A popular search engine will find links to thousands and perhaps millions of websites. But much time and effort can be expended in determining if a site has what we are looking for. Moreover, even the most powerful search engines, such as Google, Yahoo!, or Bing, do not search all existing sites. For example, much of the information available on the databases to which an academic library subscribes does not appear in the results of a search engine. In addition, the Internet is driven by market forces and this sometimes affects the operation of search engines.

In searching the Internet it is easy to get sidetracked and distracted. A suggestion is to take note of the time taken for various searches and to limit the length of time set aside for Internet searching. When you find websites that look promising for future work, use your browser's "Bookmark" or "Favourites" feature to mark the site for easy return later. It is better to work carefully and thoroughly with reliable sources you can obtain in a reasonable length of time than to try to be exhaustive, leaving little time to think and write.

How to Evaluate the Quality of Websites

In using the Internet—as in many other aspects of social science research—it is prudent to have a skeptical attitude. A general rule is that if you are unsure of the credibility of the site or the information provided, do not use it. There are, however, various strategies we can use to assess reliability. We list a few below, and also encourage you to see the online tutorials, especially the "Internet Detective," whose URLs are listed at the end of this chapter, for many other helpful suggestions.

- Is the material provided sponsored by a group or organization? What is their objective in doing so? To know more about the organization, see if there is an "About" or "FAQ" section with information on the purpose and aim of the organization and the site, and who posted the information.
- When the site was last updated? Check for a "last updated" indicated at the bottom of the material presented to assess whether the information is outdated. If no date is indicated, you may not be able to determine if certain material is still relevant.
- Is the site linked to by others that you know to be reputable, such as academic web directories? What is the site's domain? Is it commercial, educational, or some other type?
- Is the author of the material indicated? What are the author's credentials? Is the author affiliated with a university or organization? If you do not know the author or organization, a suggestion is to do an Internet search on both. You also may want to do a search of the author on a periodical database to which your library subscribes. If the initial site is the only one to refer to the author or organization, you may want to discard it. If the author is often cited and has acceptable credentials, you are likely to have come across other useful material.

- Is the text of academic quality? How does it compare with books and articles you are familiar with? Is the writing carefully edited and written in a formal style? Is there documentation and support for statements in terms of references to other works or bodies of research data? If you have doubts about the credibility of the material provided on a website, compare it with material in other sources, including on other sites.
- If "facts," mainly data, are provided and the site is not one you know, or you are unable to assess its credibility, see if other trustworthy sites include the same "facts." If they do not include those "facts," it is probably best to not use them.

DEEPENING YOUR KNOWLEDGE

The breadth and depth of the literature search will depend on the purpose of your review and the time you can spend on it. Among various sources, those you are most likely to obtain for the purposes of your research are scholarly books and journals.

Reading Monographs (Books)

Book-length reports on research can look very intimidating. They are usually long, include many endnotes, and have extensive references. As if that were not enough, they may also contain a preface, acknowledgements, an author's introduction, and several appendixes. Many students assume that such monographs have to be read from cover to cover like a novel and that if you omit any part you will not understand the book. Fortunately, it is possible to benefit from selective reading if you know how to do it, and especially if you have a preliminary sense of the area of research involved and what you want to focus on.

If you want a basic overview of the book, read the introduction or the first chapter for a review of the background of the work, the basic focus, and the intention of the research. Similarly, the final chapter will review the research findings and their implications. Often each chapter will have its own introduction and concluding sections that may be used to get a sense of specific sections of the book. If you already know something about the topic, you can focus further on specific aspects of the book. Are key concepts and theories to be found in the index (or subject index) at the back of the book? If so, use the index to find the pages where the author discusses these terms and ideas. Are key pieces of research identified by their authors in the references or list of sources used? Again, you can use the index (or name index) to locate discussion of specific pieces of research in the main body of the book. Are you concerned with questions about the author's research design and research methods? Check to see if there is a chapter on research design, or an appendix on any specific research methods used. Again, you can look for methods terms in the index.

Are you interested in the author's biases or personal perspectives? Take a look at the foreword or author's introduction or acknowledgements to see whether the author identifies the reasons for the research, or discusses important influences, teachers, or colleagues that have shaped the author's ideas. You can also scan the list of references if you know who's who in the field and look for any leanings toward or omissions of specific research approaches. If you follow all the preceding steps, you will have a good idea of the book's contents and will still have spent less time than it would take to read the entire book!

Reading Academic Journal Articles

Academic journals are publications for professionals or specialists in which fellow experts present and discuss their latest ideas and discoveries. Articles are usually reviewed by a panel of experts in the field before they are accepted for publication. Academic journal articles are important sources of information because they may be more up to date, more focused, and more specific than books. The longest article is also generally much shorter than a book. The full texts of articles in recent and past academic journals are increasingly available on online databases to which the library subscribes. However, the database may not include the full text of the most recent journal issue. The latest academic journal issues to which the library subscribes are usually located in a "current periodicals" area of the library.

A typical social science article that reports on empirical research is organized in four or five standard sections, usually indicated by headings such as Abstract or Summary, Introduction, Method, Results, and Discussion. Steps 2, 3, 4, and 5 below may also apply to reading and summarizing books presenting original research.

1. ABSTRACT OR SUMMARY   Most academic journals require the authors to provide either an **abstract** (a one-paragraph outline at the beginning of the article) or a summary at the end of the article. You should read these sections first. They give you an overview of the article, which allows you to decide quickly whether it is relevant to your research project. In these sections the authors usually briefly state the main purpose of the research, the methods, the main results, and the general conclusion. Therefore, the abstract or summary quickly answers the following questions:

What is the purpose of the research?

What is the main method used to collect the data?

What is the main finding of the research?

What is the author's main conclusion?

2. INTRODUCTION   Once you have concluded from the abstract that the article is worth reading, you should begin with the introduction. (It will not necessarily be labelled as such, but it is generally obvious.) The introduction will provide some information about the general question examined in the article, report on the findings of previous research, state what the particular study attempts to find out, and in the case of many articles, end with the hypothesis to be examined.

The introduction was probably written for readers who have some familiarity with the research topic; thus, the authors are likely to cover certain issues only briefly. For example, they may not fully explain a theory, or they may cite sources without elaborating on their content. Nevertheless, this section contains essential information that can help you increase your understanding of the topic. You will become aware of the major issues raised and of the concepts that have been recognized as important. Moreover, the cited studies can serve two purposes: you can look them up to see if they are relevant sources for your research project, and you can read the articles to gain more background information on the topic.

The introduction is likely to answer the following questions:

What is the research about?

Why is the research relevant?

What is the main hypothesis (if any) or research question?

3. RESEARCH METHOD Next, you should read the methods section to find out how the study was carried out and the data were collected. This section is generally quite detailed, usually providing enough information for the reader to be able to replicate the study. It provides information on the research method used to collect the data (What type of social survey, experiment, or field research was used?); on how the data were gathered (Were interviewers used? How were the respondents contacted?); the population considered for the study and the sampling procedure used (Whom or what were the researchers mainly interested in studying? What procedure was used to select the sample—the participants or material for the study—from the population?); and the dependent and independent variables and operational definitions (What do the authors expect to be the cause and effect? How are the concepts defined and measured?).

Although at this time you may be unfamiliar with some of the terminology used in these articles, you should still gain a general understanding of the following issues:

What method was used to collect the data?

What are the independent and dependent variables?

What are their operational definitions?

4. RESULTS The main findings of the research are contained in the results section. In many articles, this section can be especially intimidating for a beginning student in research methods and for the nonmathematical reader. It probably will contain statistical notations and technical language that summarize the data. Don't despair—the authors will more than likely explain their main findings. However, the reader with little background in statistics will simply have to accept that the appropriate statistical analysis was carried out.

The main findings, as we noted, were briefly stated in the abstract or summary, but the results section is more detailed. Therefore, you may have to disregard some of the explanations of the analyses carried out and limit yourself to the main findings.

Whatever your knowledge of statistics, you should be able to determine the following:

What are the major findings?

5. DISCUSSION The discussion section (at times combined with the results section) is where the authors state their conclusions and the implications of their study for future research. The authors also will offer possible explanations of why they had certain results and will point to similarities or differences between their results and those of other studies. This section will help you answer these questions:

What is the major finding?

What have the authors concluded about their finding?

What should be the direction of future research?

The above is a basic introduction to reading academic journal articles that are based on empirical research. While you may never carry out a research project, after graduating you are likely to read such articles often as you become more focused on a specific discipline. But you are also likely to concentrate on articles covering research issues with which you are familiar. The knowledge you have acquired over time in a discipline, together with your increasing understanding of research methods and possibly statistics, will lead you eventually to make more critical evaluations of such articles.

BOX 3.2 EVALUATING RESEARCH RESOURCES

A. RELEVANCE

- Does the source focus on your topic? Is this focus identical, or broadly similar, or only distantly related?
- Is the source a recent study and, if so, does it include up-to-date references?
- If the source is dated, could it still be useful for comparative purposes to see what has changed over time?
- Is the source understandable or is it too technical for your current knowledge and research skills level?
- Even if the study is technical, does it provide references that might be useful?

B. RELIABILITY

- Where did you find the source? Is it an academic source published in a social science journal, or a book published by a university press?
- Is it an academic source but from an encyclopedia, textbook, or popularized work? These last two sources might be useful but only as an initial overview or introduction.
- Is it a newspaper or popular magazine article? Some of these media may employ high-quality journalists who specialize in particular areas. Nonacademic sources vary considerably in quality and should be viewed as less reliable, although they may be better than no sources at all.
- What are the author's credentials? If these are not apparent, this should raise suspicions about the quality of the source. It may be necessary to do a biographical search in the library or on the Internet if the author's background is not provided by the reference source.
- Is the source organized and presented in a scholarly way? Scholarly sources present the methods of research, identify the author's background and biases, and attempt to justify arguments and procedures by reference to other scholarly work. Even where researchers disagree with others they try to be fair and reasonable in their treatment of alternative viewpoints.

BUILDING A BIBLIOGRAPHY OR REFERENCE LIST

As a student you will be asked to read published social science research in order to write in-depth term papers on specific topics. Such a paper might relate to a particular historical event, a contemporary social phenomenon, or a review of ideas or debates in some area of research. Whatever the particular task, you will have used other people's research and ideas, and these sources need to be appropriately identified at the end of your paper. Such a reference list or **bibliography** should clearly and fairly identify all the sources you have used. The better your sources, the better your paper is likely to be. The creation of a useful, relevant, and credible set of resources, then, is an essential step in the research process as well as in the writing of any essay paper.

References are useful only if you understand them! Fortunately, as noted earlier, most academic journal articles have an abstract, a brief summary preceding the article itself summarizing the main ideas, approach, and findings. Always read this to ensure that you understand it in at least a basic way before including the article in your reference list. In the case of books there is no alternative but to scan the table of contents and the introduction and conclusion sections of key chapters before including these sources in your reference list.

As you search for reference material you need to be constantly checking to ensure its relevance for your research. Does each item focus on or relate to the issue, topic, question, or area that you are concerned with? Do the articles focus on the same populations, time periods, and places, or concepts, theories, and variables that are of concern in your work?

Since your teachers will insist on reputable and accredited social science research sources, you need be aware of the status of the monograph, journal, conference proceedings, and so forth, as well as the author's scholarly credentials. You also need to know that the research methods are indicated and that the accepted conventions and procedures of social science research are being used. These two sets of issues—who are the authors, publishers, and sponsors of the research, and the way the research was conducted—are two key indicators of credibility.

STYLES OF DOCUMENTATION

As we have noted, the research process follows the same basic steps across the various disciplines; however, disciplines use different documentation styles, that is, guidelines on how to document sources. There are many different styles of documentation. Two widely used styles are the APA style, developed by the American Psychological Association, and the MLA style, developed by the Modern Language Association. (The APA and MLA documentation styles are discussed in more detail in Chapter 10.) If the research report is for a course, you should ask whether you need to use a particular style. You can also find out which style of documentation is most widely used in a discipline by noting the style used by the leading academic journals in that discipline.

Nowadays citation management software is available, two popular ones being EndNote and Reference Manager. Such software helps you manage and insert citations and organize bibliographies and reference sources automatically in the particular style of documentation you choose, and some do so as you write your paper. Some citation management software is also available on the Internet, but online software generally has limitations. (See the end of the chapter for websites.)

BOX 3.3   WHAT IS PLAGIARISM?

Plagiarism is the use of other people's ideas without acknowledgement. It is a serious violation of the basic communal and cooperative norms of scientific research. For students it is an act that could result in a grade of zero for an essay or paper, or worse.

continued

Plagiarism includes:

- Copying, downloading, or otherwise incorporating a phrase, sentence, or longer passage without indicating its source
- Summarizing or paraphrasing someone else's ideas without citing the source
- Presenting as your work material that was written by someone else
- Forgetting quotation marks, carelessly failing to identify every source, or simply being unaware of the need to acknowledge sources

The way to avoid plagiarism is to acknowledge all your sources by documenting them fully, using one of the accepted styles of documentation such as the APA or MLA style. See the Weblinks section at the end of the chapter for links to sites on what is plagiarism and how to avoid it.

THE LITERATURE REVIEW

Planning the Literature Review

Now that you have a sense of the range of information available, we turn to the actual process of writing the literature review. This process may vary according to the purpose of the review—and this may alter as you proceed with your research. The literature review is not an annotated bibliography or simply a list and summary of sources. The aim of the literature review is to focus on works relevant to your research, critically analyze them, compare them to each other, and point to how they relate to what you intend to research. (See Research Info 3.1 on a study that is based only on a literature review.)

 Research Info 3.1 *Literature Review: Health Care in Isolated Rural Regions*

As we have noted, the literature review places the research in context in that it critically evaluates previous research and justifies the research that will be carried out. The literature review is usually written to be part of the research report, as we will further illustrate in Chapter 10. However, some works are only literature reviews. These works are often intended to assess the kind of evidence that exists regarding a particular issue or theory, point to gaps in the existing research, and suggest the direction of future research or policies.

For example, it is well documented that people living in isolated, rural areas have less access to public health services. To facilitate the development of effective policies to overcome this problem, a team of researchers decided to do an extensive and thorough search for articles that systematically evaluated government policies for attracting and keeping health care practitioners in rural areas. Dolea, Stormont, and Braichet, of the Department of Human Resources for Health of the World Health Organization, undertook electronic searches of major medical and health policy databases; read through the reference lists of the retrieved articles; requested references from appropriate experts; and searched further through "Google, the Human Resources for Health Global Resource Centre and various websites of government ministries." A total of 14, 746 studies were retrieved from the electronic searches. Two reviewers scanned titles and abstracts, and independently judged the relevance of the scanned material. This was followed by a full review of the texts and further elimination, until both reviewers agreed upon a final list of 27 articles.

The articles were then analyzed in terms of two key issues. The first was to develop an idea of the range of policies that had been developed to attract and retain health care professionals in rural and isolated areas. The second was to evaluate the methodological quality of the studies, and to isolate the strengths and weaknesses of the ways governments measure and evaluate the effectiveness of their policies.

For the complete study see:

Created by authors with information from the following: Carmen Dolea, Laura Stormont, & Jean-Marc Braichet. "Evaluated strategies to increase attraction and retention of health workers in remote and rural areas."

Source: Bulletin World Health Organization, Issue 88, Pg. 379–385. (2010).

At the outset you may not have a very clear idea of what you are looking for, beyond a general area of interest. But as you work your way from initial, general sources to more specialized publications, the sheer amount of information on the general area will seem overwhelming. You will begin to see that there are particular lines of research that have been advanced, some areas of controversy, some puzzling findings, and perhaps even some (you may think) obvious areas of research that need to be explored.

At this point you need to sit down and use pen and paper (or computer) to review where you have been and where, precisely, you need to go. First, you have to summarize what you have learned about the field. What seem to be the agreed-upon perspectives, theories, or concepts? What ideas are in dispute? Who are the major researchers in the field? What are their positions, and what do they believe the evidence shows? What kinds of data, research designs, and procedures are most used in studies in this field? Where does the research seem to be headed? Do there seem to be gaps, puzzles, omissions, or other significant limitations?

At this stage your review of these questions does not have to be detailed or formal: it is an outline to help you move along in your own research. Looking over this review, do you find aspects or issues that intrigue you or that connect with your interests or life experiences? Remember that research is demanding, and unless you have some interest in the topic it will tend to become pure drudgery. Consequently, any area you find interesting is one you would do well to consider seriously as the area to follow.

Having identified aspects of the field that interest you, you must next consider what kinds of research seem to be acceptable and needed, given the field. Are there obvious gaps where no one seems to have done any research? For example, perhaps no one has looked in detail at the growing number of teenagers suffering from stress. A descriptive study filling in this gap is required. Or perhaps some studies have been done but circumstances have changed so that the same kinds of studies need to be repeated (a procedure called *replication*) to produce more up-to-date information. Possibly some studies have been done, but they do not satisfy you in other ways: experiments have been done where you think a survey is more appropriate, or a survey has been used where you believe a qualitative field study would be better.

As you sharpen your focus onto a more specific topic within a field, your interest will move from a vague, general concern to a series of questions about what is known, how it is known, and what needs to be discovered. That is, you will begin to develop a series of questions that relate to previous research that has been done (its achievements, its limitations, its gaps). It is quite likely that you will return to the library, databases, and online sources several times in the course of this process. You will be reviewing some of the literature you have already discovered and looking at specific sections of articles or books more closely—for example, their methods sections, or perhaps their samples or the time period of the study. You are likely to move deeper into the more specialized academic journals and trace interconnecting pieces of research, follow-up studies, critiques, and so on.

Throughout the process of reviewing the literature you need to be systematic. You must be sure to copy down all details of the references as you locate them. It is best to record more than the minimum bibliographic elements needed to form a citation. See Box 3.4 for details on bibliographic references.

BOX 3.4 ESSENTIAL BIBLIOGRAPHIC INFORMATION

A. BOOKS

1. Author(s): Last name, first name, and initials
2. Title: Complete title, including subtitle
3. Edition/volume: Edition (if not the first edition) or volume number
4. Place of publication: City, province or state, and country
5. Publisher
6. Year of publication

B. CHAPTERS IN BOOKS OR READERS

1. Author(s): Last name, first name, and initials
2. Title of chapter
3. Page numbers
4. All of the information on the book, as in A1 to A6 above

C. JOURNAL OR PERIODICAL ARTICLES

1. Author(s): Last name, first name, and initials
2. Title of article
3. Title of journal or periodical
4. Volume and issue number
5. Year of publication
6. Page number

continued

Box 3.4, continued

D. PUBLICATIONS ON THE INTERNET

1. Author(s) or organization
2. Title of article
3. Title of journal, periodical, or document
4. Volume and issue number, if available or relevant
5. Year of publication, if available
6. Page numbers, if available
7. Date retrieved
8. URL (the Internet address from which the document was retrieved)

Organizing the Sources of the Literature Review

Before the arrival of personal computers, researchers kept their references on file cards. The researcher would write bibliographic information on the front of the card, and an abstract, brief notes, and cross-references on the back of it. Now bibliographic files can easily be created on the computer using word-processing programs. It is best to create these files as you gather the references, rather than leaving all of the data inputting to the end. Inputting your references daily minimizes the tedium and the chance of errors creeping into your file. You can copy this file and add brief notes and cross-references to create an annotated bibliography (see Box 3.5). You can also modify your files, dropping works that turn out to be irrelevant, organizing the files into subsections according to priority or to your own classification of subject matter, and so forth.

It is important when photocopying sections of books and articles to remember to write down the full bibliographic information on the photocopy to identify its source; likewise for when you print material from a website. Although photocopying or printing may seem to be a timesaver, it has its drawbacks. It is expensive, it may take a lot of time, it creates a lot of material to keep track of in your files, and it may result in a great collection of material that you never get around to reading. Furthermore, unless you highlight carefully or take good notes, you will often end up rereading the material. For these reasons, it is best to develop your abilities to be a careful and efficient reader and note taker and to be selective in what you photocopy and print.

If you find that many studies have been done on your topic, you will need to be more selective than in an area with fewer studies. There are several ways of being selective. You might choose only those studies that are most directly related to your own research project. That is, you look at those studies that deal with the same kinds of individuals or groups you plan to examine or at those that share your particular theoretical approach, research design, or methods. You could limit the time span of your general literature search and select only the most recent studies. Since scholarly books and articles provide references, you will still be able to target earlier material that appears relevant later on. (At this stage you may also want to find out if your instructor has imposed certain restrictions on the type and number of sources you are to examine.) Finally, some areas of research are dominated by a few authorities, and this permits you to select only material that relates to the work of these individuals.

Organizing the Writing of the Literature Review

You have collected the works and documents you need, or you know where to find them. Your next task is to summarize and evaluate the material, and to do a comparative analysis of the material. Then you need to organize and write the literature review.

After you have completed your search of works related to your study, you need to critically assess these works. You need to move from doing a literature search to writing a literature review. The literature review is likely to make up most of the introduction to your research project. (We will have more to say on this in Chapter 10.)

It is best to note early in your research the main points of each work you examine. Below are various suggestions to help you summarize and assess the main points of a published work. For each work examined:

1. *Write down the name of the author and title of the work in a standard bibliographic form.* If you write down the bibliographic information for the sources as you find them, you will already have the necessary information for your research report and will save time when preparing the report.
2. *Decide what the authors are specifically trying to find out.* If they provide a hypothesis or research question, write it down word for word, place it in quotation marks, and note the page number. If you prefer to paraphrase, you can do so later; at least for now you have the exact wording and know where the hypothesis is located in the work. The focus of the work is often stated as a possible fact, with phrases or terms that are probably clarified in the text.

Suppose you noted that the main focus of the work was the following:

Most Canadians believe that their quality of life has improved.

This statement in itself will raise certain questions in your mind: What do the authors mean by "quality of life has improved"? How will they measure quality of life? And so on.

If the work examined is a book, you may want to concentrate on the sections or chapters that deal with the topic that interests you. For example, if the book is on various issues related to voting behaviour, but you are mainly concerned with the voting behaviour of retirees, you may want to concentrate on the sections or chapters that deal with age and voting behaviour. Be very careful not to take the authors' arguments out of context.

You should also pay attention to the publication date and location. Is the work still useful? Are the arguments and information outdated? Should you look for more recent publications?

The place of the study is also important. Say the work concerns marriage and divorce in the United States. You must be careful to restrict the claims made by the authors to the United States, rather than simply assuming that the same is true in Canada. (However, you may want to consider whether the same arguments hold for Canada.)

3. *Examine how the study was carried out.* Is the work based on empirical research? If so, ask:

What was the research method (survey, experiment, field research, and so on)?

What was the population studied (people, items, or events of interest to the author)?

What was the sample (the portion of the population actually studied)?

Which were the important variables (e.g., quality of life)?

Which were the operational definitions of the variables (e.g., what do the authors mean by "quality of life")?

If the work was not directly based on empirical research, you will still want to note the authors' main arguments and what they mean by the variables. Furthermore, you should note the kind of evidence provided in support of their arguments (i.e., other people's studies and data).

4. *Note the findings.* In the case of a work based on empirical research, you will want to note the specific results. You need to be careful with what you write. For example, assume the main finding was that women between the ages of 35 and 45 expressed more satisfaction about their employment situations than women between the ages of 55 and 65. You would not simply write, "Younger women are happier about their jobs than older women." Instead, you would write down the complete information, including the level of satisfaction, and clearly note the age groups. Such information will help you when you want to compare this study to another study.

If statistical analyses were carried out, you should write down such statistical information as the level of significance. Of course, how much statistical information you will want to note will also depend largely on your level of understanding of statistics. And while your knowledge and interest in statistics may be limited, you should be able to consider the findings in relation to how the study was carried out. Remember, statistics are only a tool that the researchers used to analyze the data, not a measure of the quality of the data collected or an indicator that a study was carried out properly.

5. *Note the conclusion.* What did the authors conclude? For example, did the authors conclude that most Canadians do indeed believe that their quality of life has improved?

6. *Make critical comments.* At the end, leave space to note whatever weaknesses or strengths you believe the study has. You might also want to note any similarities the work has to other studies you may have read.

After you have completed the search and have examined each work to your satisfaction, you will have to organize and write the literature review, at least in draft form. You may want to organize the review around particular subsets of the issue. For example, in studying the concentration of economic power, you may decide first to cover what the publications have to say about the concept of power, and then follow with your critical assessment of the research methods they used to study the phenomenon. Or you may want to discuss the various works chronologically (especially if you want to show that interest in a particular phenomenon varied over time) and note what contributions the works made to the understanding of the issue. For example, say your research was on spousal abuse. A literature review organized chronologically can illustrate what importance social scientists have given to this issue and how they have examined it over time.

Writing the Literature Review

As with all research writing, the literature review needs careful planning and clear writing. All of the rules of good writing (coherent organization, appropriate paragraph structure, logical division of the paper into sections and subsections, an introduction and conclusion, etc.) are as important in the literature review as they are in any other section of a research paper. You will probably go through several drafts before you get the literature review to fit with the other sections of your final research report. It is even possible that you may discover gaps in your review during the evolution of your research. If so, you will have to extend your original literature review. Similarly, some original material may become less relevant and be dropped from consideration.

To write a good literature review you have to bear in mind its purpose. You will find that there are several kinds of literature reviews, apart from those undertaken in the course of empirical research. Journal articles and even entire books may be lengthy review essays that look at the historical development of an issue or a research area over time, that review in depth the different theoretical approaches to a particular issue, or that focus on the methods of research used to investigate a specific topic or the degree of agreement and accumulation of data.

If a literature review is part of an empirical research report, it is supposed to show how your research fits with the research of others. This does not mean that you simply list and summarize a series of earlier research efforts. You have to show how the earlier research develops or ties together as a series of interconnecting and mutually supportive ideas and discoveries, and how your own research will flow out of these interconnections. Usually, doing so involves both a critical assessment and a synthesis of previous research. As we have already suggested, all research has limitations, and a critical purpose of the literature review is to highlight these limitations and problems and suggest ways of overcoming them.

The synthesizing process involves clearly drawing out only the main points of previous research that are relevant to your own study. This means you extract common or complementary ideas, themes, and data; state them clearly; and draw out their implications as the context of and justification for your research. You may conclude with a clear statement of the research question or hypothesis that you intend to examine empirically.

What the Literature Review Contributes to the Research Process

The preceding sections on undertaking, organizing, and writing a literature review indicate how the literature review contributes to the research process. By learning what others have done and what is known, you can become an informed researcher—one who is able to identify the purpose of your research more clearly and to make better research decisions. After surveying the literature, you can see where further research is most needed, what useful methods or procedures have been used, and which ideas represent current thinking about the issues.

As a researcher making informed decisions about your research, you are likely to see the quality and the importance of your research increase. Having understood an issue in depth and being able to link your research to a broader context also make you credible to other researchers. They are likely to take notice of your research if it is published or circulated. Consequently, your research will make its own contribution to the growth of knowledge in a particular subject area.

What You Have Learned

- Social scientific research is cooperative and cumulative: every new piece of research builds on previous research.
- To begin to do research you need to know what other researchers have discovered and what they think, so you can use their ideas and data to focus and develop your own research topic. In other words, you need to do a literature review.
- A literature review is best undertaken with a clear purpose in mind, which may change as you proceed, so you should review and reassess it along the way. It involves moving from general reference sources, such as encyclopedias and handbooks, to recent books and academic articles.
- Finding out about other people's research involves systematic use of different sources of information available from or through academic libraries, including databases. We should use these before turning to a general search on the Internet.
- In doing general searches on the Internet, we should follow various strategies to assess the credibility of the website and information provided.
- The literature review should be written in such a way that it places your research project in a clear context by showing its connections with the research of others. You should use the appropriate style of documentation and must avoid plagiarism.

Focus Questions

1. Why is it essential to find out what previous researchers have discovered and discussed?
2. What can you achieve in a search for information from basic sources such as class notes, textbooks, and encyclopedias?
3. Why is it useful to begin the literature review process by taking into account your course material and reading lists, and library resources?

4. Why should you be concerned with the credibility and reliability of the information available over the Internet? What are some strategies you can use to assess the credibility of the website and the information provided?

5. What are some steps to use in guiding you to reading and summarizing books presenting original research?

6. How can you be selective in an area where there are many studies on your chosen topic?

7. What does a literature review contribute to the research process?

Review Exercises

1. Familiarize yourself with what is available from and through your school library, including the services offered, the periodical databases to which it subscribes that are of specific interest in your area of research, and the information and documents available on the library's website for its students.

2. Select a research topic that interests you. Search for information on the topic on a periodical database to which your school library subscribes, or the online InfoTrac database to which you have a free limited subscription with this textbook. Do a general search on the Internet on the same topic. Contrast the type and quality of material available on the database(s) with that on the Internet.

3. Select an academic article from a journal in a discipline that interests you. Apply the steps and answer the questions stated in the sections above entitled "Reading Academic Journal Articles" and "Organizing the Writing of the Literature Review."

4. Do a general search on the Internet on a research topic that interests you. Examine a website that seems to be useful and evaluate the credibility and reliability of the site and information provided. Apply the suggestions and answer the questions stated in the section above entitled "Searching the Internet."

5. Using the article selected in exercise number 3 here and the website material in number 4, document each source using a style of documentation, such as APA or MLA.

Weblinks

You can find these and other links at the companion website for *First Steps: A Guide to Social Research*, Fifth Edition: **http://www.firststeps5e.nelson.com**.

Virtual "Libraries"

Virtual libraries generally include links that have been selected, and in some cases whose material has been evaluated, by subject experts. They are usually organized according to different main topics, which in turn contain links under various subheadings. An example is Intute (see below), which includes hundreds of annotated links evaluated and categorized by specialists at U.K. universities that can be searched by subject heading and keyword. A keyword search for "happiness" results in a number of annotated links, which can be filtered by resource types such as databases, journals, and primary sources. One link, for instance, is the World Database of Happiness site, which holds an extensive bibliography of scientific publications and a directory of researchers. Another example of a virtual library is Infomine, created by librarians at various American universities, which contains over 100 000 links. For example, a keyword search on "happiness" provides 28 expert-selected resources. While many virtual libraries cover broad categories, some focus on a specific discipline, such as the Virtual Library on History Central Catalogue, or an area of study, such as the Virtual Library on Sustainable Development. One suggestion in finding these sites is to do a search for "virtual library" followed or preceded by your subject or area of interest. Below are some sites of virtual libraries:

Social Sciences Virtual Library
      http://www.vl-site.org/sciences/index.html

Social Sciences (Intute)
      http://www.intute.ac.uk/socialsciences

Scholarly Internet Resource Collections (Infomine)
      http://infomine.ucr.edu

Internet Public Library and the Librarians' Internet Index (ipl2)
      http://www.ipl.org/

Reputable Organizations

Government and nongovernmental agencies, as well as think tanks, are increasingly relying on the Internet to distribute information and publications. While some material may be available only through a service to which a library subscribes, much of the information is available for free online. Such is the case with, for example, government statistical agencies such as Statistics Canada or international organizations such as the United Nations. So if your topic was on human rights, you would certainly find material on the UN website. And if you were more specifically interested in the rights of children, you would benefit from also searching for sources on the UNICEF website. If you are unaware of any organizations related to your topic, you may want to do a general search and use your Internet research skills to assess the quality of the site. And of course you can always ask the reference librarian for assistance in searching for relevant organizations. Below are sites of interest:

Government—International Organizations
      http://dir.yahoo.com/government/international_organizations/

Directory of Development Organizations (lists 65 000 development organizations)
      http://www.devdir.org/

Non Governmental Organizations
      http://www.unodc.org/ngo/list.jsp

Canadian Think Tanks
      http://www.lib.uwo.ca/programs/generalbusiness/webttanks.html

Canadian Policy Organizations
      http://www.policy.ca/policy-directory/Policy-Organizations/index.html

Scholarly Societies

There are thousands of scholarly societies worldwide and most, if not all, have a web page with publications and links. Many also publish journals for which they provide the abstracts, if not the entire texts. For example, say your topic has to do with "bullying." Among the different scholarly societies you may want to consider is that of the American Psychological Association, which has over 150 000 members; publishes numerous journals, books, and reports; and is a provider of various databases. On its website you can obtain free material on bullying and are informed about other material that can be obtained through subscriptions to journals or databases, to which your library possibly subscribes. In addition to the Canadian scholarly websites listed at the end of Chapter 1, see the following site with thousands of links:

Scholarly Societies Project
>   http://www.lib.uwaterloo.ca/society/

College and University Libraries

The websites of college and university libraries generally offer an array of practical information that is accessible to everyone, such as research or subject guides and links to Internet sites, including those of governments, agencies, and institutions. Library sites are also valuable in offering information and guidance on steps to writing a paper, carrying out a literature review, citing sources, avoiding plagiarism, and much more. The first site below consists of links to college and university libraries in Canada. The second offers links to libraries across the world, of which the links to academic libraries in the United States, the United Kingdom, and Australia are especially useful. The third site offers links to thousands of subject guides available from college and university libraries worldwide.

Canadian University and College Libraries
>   http://www.collectionscanada.gc.ca/gateway/s22-210-e.html

Libweb—Library Servers via WWW
>   http://lists.webjunction.org/libweb/

LibGuides Community
>   http://libguides.com/community.php

Academic Journals

An increasing number of academic journals are available for free over the Internet, and many are now available only as e-journals.

The World's Electronic Journals—E.journals.org
>   http://www.e-journals.org

Free Electronic Journals
>   http://knowledgecenter.unr.edu/materials/articles/journals/free.aspx

Directory of Open Access Journals
>   http://www.doaj.org/

Encyclopedias

The links below are to general or specialized encyclopedias that offer free access to their content.

The Canadian Encyclopedia/The Encyclopedia of Music in Canada
> http://www.thecanadianencyclopedia.com/

Citizendium
> http://en.citizendium.org/

Wikipedia (and sister projects)
> http://en.wikipedia.org/wiki/Main_Page

Evaluating Websites

The links below offer useful advice on evaluating the quality of websites, particularly for academic research.

Internet Detective
> http://www.vtstutorials.ac.uk/detective/

Evaluating Web Resources
> http://libraries.dal.ca/using_the_library/tutorials/evaluating_web_resources.html

Working with Websites: Evaluating Search Results (video)
> http://www.youtube.com/user/TheHartnessLibrary#p/c/748297966B083400/16/cVIG-Nsxi_A

Library of Congress Classification

The following site lists the main classes of the Library of Congress Classification. By clicking on a class list we can view its different subclasses and can click on the subclasses for more details.

Library of Congress Classification Outline
> http://www.loc.gov/catdir/cpso/lcco/

Literature Review

The following links provide information on the characteristics of a literature review and tips on conducting and writing a literature review.

Writing a Literature Review (article)
> http://www.lib.uoguelph.ca/assistance/writing_services/components/documents/lit_review.pdf

Writing the Literature Review: Step-by-Step (videos)
> http://www.youtube.com/user/peakdavid#p/search/0/2IUZWZX4OGI

Plagiarism

The following links offer information on determining what is plagiarism and how to avoid it.

How Not to Plagiarize
> http://www.writing.utoronto.ca/advice/using-sources/how-not-to-plagiarize

Avoiding Plagiarism

        http://owl.english.purdue.edu/owl/resource/589/01

Plagiarism

        http://www.mcgill.ca/library/library-assistance/plagiarism/

What Is Plagiarism? (video tutorial)

        http://library.camden.rutgers.edu/EducationalModule/Plagiarism/

Research Papers

The Internet has made research papers and working papers widely available. Many are provided by university department websites or agencies and institutions. There also exist electronic databases maintained by scholars that have allowed a greater distribution of academic papers. These sites are generally a collaborative effort of many scholars worldwide and contain numerous papers either in an array of subject areas or mainly in one subject.

Social Sciences: Academic Departments

        http://www.dmoz.org/Science/Social_Sciences

Social Science Research Network eLibrary Database Search

        http://papers.ssrn.com/sol3/DisplayAbstractSearch.cfm

Research Papers in Economics

        http://repec.org/

Styles of Documentation

The sites below offer information on how to cite sources using different styles of documentation. On the Citation Machine site, you provide the necessary information and the website gives it back according to the selected style of documentation. See also Chapter 10.

Citation Machine

        http://www.citationmachine.net

How to Cite

        http://www.mcgill.ca/library/library-assistance/how-to-cite/

MLA Citation style

        http://library.concordia.ca/help/howto/mla.php

APA Citation Style

        http://library.concordia.ca/help/howto/apa.php

4 Sampling: Choosing Who or What to Study

What You Will Learn

- the concepts of "population" and "sample"
- why sampling is essential
- the difference between random and nonrandom sampling
- how to decide on the size of a sample

INTRODUCTION

Sampling is the most widely used technique for gathering information of any kind. In one sense, sampling is a shortcut to information that we are already familiar with in our everyday life. If we want to know whether an instructor of a particular course is an "easy marker," we may ask a friend who took the course last semester and then judge the instructor from what our friend tells us. Our friend is the sample selected from the population of all the students who took the course. And on the basis of what our friend tells us, we generalize about whether the teacher is an easy marker or not. Sampling and generalization in research follow similar reasoning, except that much care goes into the selection of the sample.

As a fast solution to a particular everyday problem, this convenience sampling is very useful, saving us much time, energy, and anxiety. But things can go wrong even with this simple sampling activity. Your friend may feel personal antagonism toward the instructor, or she may underestimate how much her grades actually reflect her own skills rather than easy marking. These factors will bias or distort the report that you get. Maybe you discover that the way your friend evaluates courses is not the way you do, and that the information you get is not so useful after all.

The point we are making is that sampling, even in its everyday, casual form, is a tricky business. A sample may give you some information, but this information may be distorted or biased, or the information may not be what you are looking for. However, we can rarely avoid sampling—we have to ask some people, not all; we have to look at a few things, not everything. Because of its efficiency, it is used both in everyday life and in scientific research, although in the latter case, sampling has to be done much more carefully and systematically. Scientific researchers have to overcome problems of bias and try to achieve a sample that is *representative* of the population—the shortcut to information is no longer so short!

POPULATIONS AND SAMPLES

In defining the topic of your research, you need to be clear about what it is you are interested in studying. Broadly speaking, this means defining two things. First, you need to be clear about the questions and ideas framing the research and about the range of empirical observations you are going to be making when exploring or researching these questions and ideas. Second, you need to choose who or what to study, which is where sampling is of great importance.

BOX 4.1 QUEBEC SOVEREIGNTY POLLS

The result of the Quebec independence referendum on October 30, 1995, was 50.6 percent for the No side and 49.4 percent for the Yes side. Five polls were carried out in the 10 days preceding the referendum. Below are the results of the five polls. Although each poll placed the Yes side ahead among the decided voters, because of a margin of error of at least 3 percentage points, both sides had about an equal possibility of winning.

There was also the question of what to do with the "Undecided or refused to reply" responses. Various approaches were used to allocate the response of voters who did not state a preference. One polling agency, for example, considered the age and language of these voters, as well as responses to other questions. Nonetheless, for various reasons it was widely recognized that at least 75 percent of respondents who did not state a preference would actually vote No. The table also gives the expected results, including those who did not state a preference, distributed as 75 percent supporting the No side and 25 percent the Yes side. Compare these results with the actual referendum results.

Polling Company	Sample Size	Date Completed	No	Yes	Undecided or Refused to Reply	Breakdown of undecided or refused to reply — No 75% — Expected Results No	Breakdown of undecided or refused to reply — Yes 25% — Expected Results Yes
Leger & Leger	1005	Oct. 20	42.2	45.8	12.0	51.2	48.8
CROP	1072	Oct. 23	42.2	44.5	13.3	52.2	47.8
Angus Reid	1029	Oct. 25	40.0	44.0	16.0	52.0	48.0
SOM	1115	Oct. 25	40.0	46.0	14.0	50.5	49.5
Leger & Leger	1003	Oct. 27	41.4	46.8	11.8	50.2	49.8

			No	Yes			
Referendum result:		Oct. 30	50.6	49.4			

Take note of the number of interviews carried out: a little over 1000 interviews! And yet pollsters were able to come within a fraction of a percentage point in estimating the voting intentions of nearly 4.7 million voters. How was that possible?

Sources: Leger & Leger, Oct 20 poll; CROP, Oct 23 poll; Angus Reid, Oct 25 poll; SOM, Oct 25 poll; Leger & Leger, Oct 27 poll.

In most areas of research we are faced with the kind of choice we discussed in the introduction to this chapter: we need to get information about a group of events, people, actions, and so on, by using a small proportion of this group as our source of information on the whole group. Consequently, we are faced with the problem of selecting who or what will be in the group we actually study. This selection problem is the key issue in sampling.

For example, whom do you want to study? All the students at your college or university? All the clients of a community clinic? All shoppers at a mall? All women over 60 years of age who live on their own in your community? The group you are interested in studying is known in technical terms as the **population**. However, the meaning of the term *population* is different from its everyday usage. In research, you determine which people or objects make up the population about whom or what you want to draw conclusions.

It is usually difficult or impossible to study all the members of the group you are interested in, or, stated differently, to study the whole population. Moreover, doing so may be unnecessary. It is often more desirable to study some individuals from within the population—that is, to select a sample. For example, instead of studying all the students at your college or university, you could examine only a certain number of students, say 300 out of 6000 students. The manner in which you select the sample (the 300 students) will determine the extent to which you can generalize from your results to the population (the 6000 students). Which results would you be more willing to accept as reflecting the 6000 students—the study that gave all 6000 an equal chance of being selected for the sample of 300 students, or the study that had a sample of 300 students, made up of students who lingered in the cafeteria?

Can we generalize from the opinion of only one student, as we did in checking out how easy a marker an instructor was? Can we generalize about 6000 students on the basis of the opinion of 300 students who linger in the cafeteria? In both cases students who were immediately available were used, and we can only hope that their views reflect those of the population. Researchers, however, aim to select a sample that reflects the population as accurately as possible.

Would you be more willing to accept the conclusions if, for example, we asked every fifth student on the class list from last semester for their opinions about a course? Would you be more willing to accept conclusions about students at your college or university if every 20th student who registered this semester were interviewed? Needless to say, these techniques in drawing the sample from the population are more complicated than simply asking a friend or students in the cafeteria. But although these techniques are laborious and time-consuming, they obtain information that more accurately reflects the views of the population. Thus, how we select the sample from the population determines the degree to which we can generalize from the results.

So far we have spoken of a sample and a population as if these terms applied only to survey researchers asking people questions. But almost anything can be sampled: opinions, TV programs, advertisements, historical documents, phone conversations, music on the radio, air, water, stars, rocks, volcanic eruptions, and so on. Populations, then, can be any range of objects, people, or phenomena that are of interest to the researchers.

There are situations where sampling is not an issue. Certain studies may have a population of one and so do not face a problem in choosing a representative group to study. Archaeologists and physical anthropologists, who study ancient civilizations or the remains of early human life, often have so few cases to deal with that they can study all the sites or physical evidence available. Historians often argue that certain events or historical eras are unique and can be studied only as singular events, a population of one. Sociologists may look at completely new phenomena, such as an emerging political movement or a religious cult that is so small it is possible to study everyone involved. Sampling is also not an issue when data are collected on the entire population. Governments throughout the world, for example, undertake various censuses, including censuses of housing and businesses. A key one is the census of population. Usually referred to simply as the census, it collects social data, such as age and gender, from all persons in the country (see Box 4.2 on the census in Canada).

BOX 4.2 THE CENSUS

What is the population of Canada? What are the ethnic–cultural backgrounds of people in Canada? Answers to these and many more questions are obtained by means of a unique survey, the census. The census is unique in that it attempts to collect information on the entire population of the country at a particular time. Hence, sampling is not an issue.

The first official census in Canada, and probably North America, was carried out by Jean Talon in the second half of the 17th century in what was then New France. As Intendant of Justice, Police, and Finance in the colony, Talon was responsible for encouraging the colony's economic expansion and making it self-sufficient. To help him in his effort, Talon conducted a survey, or rather a census.

The aim of the census was to collect information on the settlers and the colony. Talon was able to find out about such factors as the age, occupation, and marital status of the settlers, as well as such information as the price of lumber and the number of government buildings. Talon then used the data to help him better administer and develop the colony.

Thanks to the census, Talon learned that the colony contained 2034 men and 1181 women of European descent. Therefore, he arranged for young single women (*filles du roi*) to come from France in an effort to encourage the growth of the colony. From 1665 to 1673 nearly 900 women came. He also rewarded early marriages and large families, and fined bachelors. The result was the growth of the colony's population of French descent. Meanwhile, he helped set up various businesses to stimulate the economy's growth and make the colony more self-sufficient. A few years later Talon proudly noted that the number of births was on the rise and that the clothes he wore were all made in New France.

Talon and others who followed continued to turn to the census as a means of collecting information on the colony. By the time New France became a British colony nearly a century later, dozens of censuses had been carried out.

Needless to say, the information in Talon's census has become a valuable source of information for historians. Today, the census continues to be the main

continued

means of collecting information on Canada's population. It is carried out by Statistics Canada (an agency of the federal government).

The census collects some information on everyone in Canada and, until recently (see Box 4.4), more detailed information on only 20 percent of the population. The purpose is to determine the number of people in Canada, their age, gender, education, income, ethnic–cultural origins, and many other details.

By comparing the census data of various years, we can discern basic trends in our population, including aging, education, and occupational shifts. The information gathered is useful for governments in determining policies, for businesses in determining the consumer markets, and for researchers in determining a representative sample of the population of a country or region.

More on the history of the census is available from the Statistics Canada web page on Jean Talon at **http://www.statcan.ca/english/freepub/98-187-XIE/ jt.htm**. Information from different censuses is available at **http://www12.statcan. ca/census-recensement/index-eng.cfm**.

If you plan on interviewing only some members of a larger group, however, and want a valid picture of all the members, you need to use a sampling procedure. The essential problem of sampling, then, is to select a sample that is representative of the population from which it is drawn.

How do you select a sample? In formulating a research question or framing a hypothesis, you have already in part identified a population. Suppose you want to find the most popular physical recreational activities among college students. The population would consist of college students, but you would have to be more specific about which population of college students. Do you mean college students in the whole world? In Canada? In your province? City? Community? You will have to limit your population to one from which you can sample. You may decide, because of cost and time, to restrict your study to students at one college. The population would then be all the students at that particular college, and the sample would be selected from that population. Once you clearly determine your population, you can decide how to select the sample. Thus, the population about which you want to generalize influences how you select your sample.

SELECTING THE SAMPLE

There are two broad types of sampling procedures: **random** (or probability) and **nonrandom** (or nonprobability). A sample will be random or nonrandom depending on how it was selected from the population. A random sampling procedure ensures that each member of the population has an equal chance of being selected. The procedure provides reasonable assurance that the information collected reflects the population. (For simplicity's sake, we will avoid discussing the statistical details of sampling.) In contrast, a nonrandom sampling procedure is used when it is impossible to give each member of the population an equal chance of being selected. The information collected cannot be generalized to the population.

Note that the term *random* in research does not mean you can select a sample in any arbitrary way. For instance, suppose you were carrying out a study on the type of summer jobs held by college students. Assume that your sample consisted of 50 students you met or were referred to by friends. Does this compose a random sample? No. People whom you happened to meet, who were recruited through some advertisements, or who were suggested by others make up a nonrandom sample.

BOX 4.3   HOW WAS THE SAMPLE SELECTED?

Consider if two studies have been carried out on the spending habits of university students in Canada. One is based on 3000 interviews and the other on 500 interviews. Could you tell on the basis of the size of the sample which is the more reliable study? Do not be deceived into believing that just because a sample is larger it is more valid than a smaller one. It is worth asking: How was the sample selected?

To illustrate, consider the widely cited case of the now defunct U.S. magazine *The Literary Digest,* which conducted a postcard poll during the 1936 U.S. presidential campaign. The *Digest* mailed out 10 million ballots to telephone subscribers and automobile owners, of whom two million sent in their ballots. From these, the *Digest* predicted that the Republican candidate, Landon, would get 57 percent of the votes, and the Democratic candidate, Roosevelt, 43 percent. The reverse occurred: Roosevelt won by a landslide, with 61 percent of the vote. Clearly, one of the many errors of the *Digest* poll was that its sample consisted mainly of one socioeconomic group: telephone subscribers and automobile owners in 1936, a time when the impact of the Depression was still being felt.

Meanwhile, others, such as George Gallup, were developing new sampling methods aimed at making samples more representative of the electorate. The samples were also far smaller than that of the *Digest.* Mainly because of their sampling technique, Gallup and others correctly predicted that Roosevelt would win.

Thus, it is essential to avoid assessing the reliability of a sample only on the basis of its size, and instead to consider other factors, such as how the sample was selected. The random sample gives everybody the same statistical chance of being selected; its aim is to be representative of the target population. The nonrandom sample is less statistically accurate, and we do not know how representative it is of the target population. But a random sample is nevertheless also prone to providing poor results, so we should not judge the quality of a study solely on the sampling technique.

Further, if the results of the study are intended to reflect the opinion of a larger group than the sample, we need to restrict our findings to the target population. For example, a random sample selected from adults in Toronto can reflect the views of adult Torontonians, but not of all adults in Ontario or Canada.

RANDOM (PROBABILITY) SAMPLING PROCEDURES

Random sampling procedures are designed to give everyone in the population an equal chance of being included in the sample. The aim is to avoid biasing the sample toward any group within the population. Therefore, to achieve a random sample we need a complete list of the population; in technical terms, we need a **sampling frame**. A properly selected sample relies on the accuracy of the sampling frame for its generalizability.

Suppose you want to carry out a study on the religious behaviours of all the students at your college or university. To select a random sample, you will need a list of the students. The only available list is the student directory, and you decide to select the sample from this list. The sampling frame is the student directory.

The sampling frame depends on the study. For example, it may be a list of the community clinics in a city, hockey players in a minor league, veterans who are active in the Canadian Legion, or whatever. In each case the researcher must take into account any problems with the sampling frame. Is the list up to date? Is it complete? How was it compiled, or how will it be compiled? The aim is to have a sampling frame that includes as many members of the population of interest as possible.

Depending on the study, it may be difficult to have a sampling frame that includes the entire population. For example, not all the students are listed in the student directory, and some may have dropped out. Despite the weaknesses of the student directory, it may be the best sampling frame (list) available for the population. Thus, we have to take into account the limitations of the sampling frame in generalizing results to the population of all students. Once you have a sampling frame, you can choose a random sampling technique, of which there are four main types. (For simplicity's sake, in the following discussion we will assume that the sampling frame includes the entire population.)

SIMPLE RANDOM SAMPLE   The **simple random sample** technique ensures that everyone in the population has an equal chance of being selected. Suppose you want to study first-year students in your college or university (population), and you compile a list of such students from the student directory. Let us assume there are 500 names. What approach would you use to select 100 students from the list that ensures that all 500 students have an equal chance of being selected?

One easy but rigorous approach is the simple random sample. For example, if you want to select a random sample of 100 out of a population of 500, you might do the following:

1. Write each of the 500 names on separate pieces of paper of equal size and thickness that can easily be mixed. (Or you assign each student a number and draw the numbers, as in the case of provincial lotteries.)
2. Place the names in a container, mix them, and draw out a name. That student is selected for the sample.
3. To ensure that each student has an equal chance of being selected, place the name back in the container, mix the names, and draw out another name.
4. Continue the process until you have selected the required number for your sample. If a name is drawn twice, it is returned to the container. The first 100 students chosen will make up the sample.

If you did it all over again, the names would not correspond to those of your initial sample, but the second sample would be as reliable as the first.

Another approach in selecting a simple random sample is to use a table of **random numbers** (or you can use computer-generated random numbers). The table of random numbers is made up of numbers that are in no specific order, as if someone else mixed the numbers for you. It is relatively easy to use a random number table. Briefly, you start anywhere on the table of random numbers, then move in a particular direction and select your sample. For example, suppose the following numbers were taken from a table of random numbers (see Table 4.1, row 12, third column):

81300 12228 92726 99026 90308 27638 24570

Table 4.1   Random Numbers

Nowadays, computer spreadsheet programs and statistical packages all have a function that generates random numbers for any range. In case you do not have such a program, below is a list of computer-generated random numbers of five digits. The first column (01 to 40) is included to make it easier to read the numbers in the various rows.

01	09430	73223	70535	18533	79431	66128	23456
02	01563	48261	40090	42529	39321	88719	79737
03	91619	06271	86315	74543	54276	80612	11677
04	99980	97778	72852	85762	64551	10988	06827
05	93140	33141	48741	15264	13407	10486	41691
06	25598	14607	35369	88154	08681	24369	10585
07	68317	20603	02956	37784	61458	55566	78160
08	98846	67347	53576	33548	55446	45871	16090
09	84210	87145	74381	93241	07356	77914	44914
10	14451	23345	48777	76197	18693	73240	76948
11	95528	65911	46410	31626	78562	37872	85112
12	35754	89950	81300	92289	57256	68646	83260
13	80792	75646	12228	30562	72917	65055	97489
14	43644	97983	92726	55145	07336	22470	81040
15	60863	72144	99026	82130	47914	39931	50068
16	13443	60905	90308	37870	46354	71315	87354
17	98789	36197	27638	57789	74508	70089	62507
18	87884	94249	24570	10831	22028	20417	17769
19	65777	71115	18967	51052	93366	87433	21554
20	20891	92655	30080	26634	12020	58419	86744
21	21278	27361	84248	04471	99684	61911	44798
22	03846	55135	48911	91011	41471	70775	16108
23	39437	87310	82972	05570	91964	67980	79659
24	22306	91645	05060	73037	46983	25471	60492
25	13308	23580	93920	79172	44234	45317	11187
26	45281	54523	64743	31021	54823	40253	50301
27	25920	33294	64967	55091	43970	09307	97462

(*continued*)

Table 4.1 (*Continued*)

28	90149	85488	70964	90097	07011	12687	31212
29	87364	41973	79875	43923	72970	83110	48937
30	52631	71564	99594	22685	40374	22024	63070
31	43192	08144	48950	53592	39644	85520	61724
32	11889	30112	17316	22318	69227	63698	99177
33	55463	37010	88934	19056	92129	11090	98311
34	34349	13016	47883	43687	60928	72117	18027
35	87300	72345	49585	40915	96257	24810	26302
36	75181	01577	40904	75792	93851	93925	92334
37	78810	73992	63836	29333	44341	56843	70191
38	32099	75710	05649	39568	68069	10955	73474
39	30219	47586	88962	71482	13743	67061	46713
40	11142	52243	89842	25147	69038	48310	11928

To select our sample of 100 students from the population of 500, we would do the following:

1. List the names of the first-year students and number each name from 1 to 500. For example:

    1. Mary Antoine

    2. Sam Austin

    . . .

    500. Rick Zagreb

2. Since the population is made up of 500 students and you used up to three digits to number each one, you need to select three-digit numbers from the random numbers table to ensure that all have a chance of being selected.

3. Decide which approach you will use to create the three-digit numbers from the five-digit numbers. Will you use the first three, the last three, or another approach? Assume that from the five-digit number you will take only the first three numbers. For the five-digit number 18659, you would read only the first three numbers, 186.

4. Decide on a pattern for selecting the numbers. Will you be moving up or down the column, across from left to right or right to left, or diagonally? Once you decide on the pattern, you need to stay with it. Assume you decided to move down the columns.

5. Pick a starting point anywhere in the table of random numbers. Let us say you closed your eyes and with your pen pointed to the five-digit number 12228 in the table of random numbers. The first three digits are 122, so the student whose number is 122 is selected for your sample.

6. The next five-digit number reads 92726, and since the first three digits are 927 and are greater than 500, you would skip this number. You would continue to skip five-digit numbers until you reach one whose first three digits are in the range from 001 to 500. In the table, that would be 27638, since the first three digits are 276. Thus, the student numbered 276 would be selected for your sample. You would continue this procedure until you had a random sample of 100 students.

If you picked another starting number or selected the same number but moved in a different direction, the 100 students selected would include different students from your earlier selected sample, but the results would be just as reliable. In other words, the aim is to avoid bias in the selection of the sample.

Systematic Random Sample   A **systematic random sample** involves selecting individuals at every fixed interval on your list of the population—for example, selecting every 15th member of an association. Thus, you need to know the total number of the population and the number you want to sample. You simply divide the total number of the population by the number you want in your sample and determine the interval. However, you do not necessarily begin the systematic sampling procedure at the start of the list of names. Instead, you would begin the procedure by randomly selecting a name from those in the first interval of names. The randomly selected name is the starting name, and from then on, every name that falls at the determined interval is selected.

Suppose you have a complete list of all 250 faculty members of a community college and want to sample 75 members. The following steps indicate how to select the sample:

1. Have the complete list of 250 faculty members.

   Sanders

   Trevisano

   Goldberg

   Trudeau

   Brandon

   Handy

   Kim

   Anderson

   Lee … and so on

2. Calculate the interval.

   Population/Sample size = 250/75 = 3 (the interval)

3. Randomly select any name from the first three names on the list. You can close your eyes and point to one of the names, which then becomes your starting point. Or you can place the three names in a box, and the first one selected becomes your starting point.

   Assume the second name of the first three names was randomly selected. Your starting point and the first name selected for the sample would be Trevisano.

   Begin the selection process with Trevisano. Count three more and select Brandon, and continue the technique until you have the 75 names for your sample. (Which would be the third name selected for the sample?)

If one of the faculty members whose name appears on the list is no longer at the community college, select the name just before or after it. One suggestion is to flip a coin, and if it comes up tails, select the name after, and if heads, choose the name before.

Stratified Random Sample   Suppose you want to compare the future aspirations of male and female students in a health science program. Assume you have access to a student directory that also states the gender of the students in the program. How can you ensure that the sample reflects the population in terms of gender? In other words, if females make up 60 percent of the student population, how can you ensure that 60 percent of the sample consists of females?

You can do this by means of the **stratified random sampling** procedure, which is a modification of the simple and systematic random sample procedures. However, you will need a list of the population that can be rearranged according to the necessary groups or strata. Social scientists often choose to stratify, or sort out, the population using a number of variables, such as gender, age, ethnicity, occupation, or income. Which groups or strata are chosen will depend on the research question or hypothesis.

For example, since you want to compare the future aspirations of male and female students, your population will be stratified according to gender. Assume the desired sample size is 100 of 500 first-year students in a program. There are different approaches in selecting a stratified random sample. A common approach involves the following steps:

1. Divide your list of 500 first-year students into two groups: females and males. Say there are 300 females and 200 males.
2. From each list, select systematic random samples. Since the desired sample size is 100, select 60 from the list of females and 40 from the list of males.

To carry out a stratified random sample, it is necessary to determine the proportions of each group in the population. From each list of the group, you then select the number of names for the sample that constitutes the proportion of that group in the population. For example, the proportions of females and males in the student population are 60 percent (300 divided by 500) and 40 percent (200 divided by 500), respectively. Since your desired sample is 100, you would want 60 females and 40 males—the same proportion as in the population. You would then select a random sample of 60 females from the list of females, and a random sample of 40 males from the list of males (see Figure 4.1). The result is a proper representation of each group in the sample. If necessary, you can complicate the stratification method by adding other characteristics. Suppose you know the

Figure 4.1 Stratified Random Sample

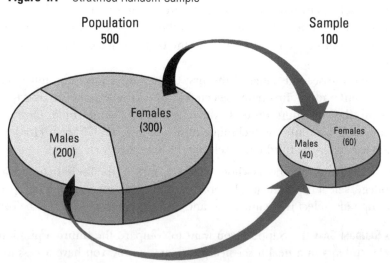

academic discipline of each student; in this case you could organize each group of females and males by academic disciplines.

MULTISTAGE CLUSTER SAMPLING The previous sampling techniques have dealt with selections of samples from lists of populations. But what if you want to carry out a study of a population that cannot easily be listed? Or what if it is difficult and costly to compile a list of the population?

For example, what could you do to get a sample of all the graduating high-school students in the province? Of course, you could get the lists of the graduating students in every high school and then select a random sample. But that would be time-consuming, costly, and difficult. To save time and money you could use **cluster sampling**, whereby you randomly select a sample of high schools (clusters) and then survey all of the graduating students at those schools. Another possibility is to use **multistage cluster sampling**, whereby you randomly select a sample of high schools and then randomly sample students from the lists of graduating students at each of the selected schools.

Cluster sampling involves a single stage, where everyone in the selected clusters is in the sample, such as all the graduating high-school students in the selected sample of high schools. Multistage cluster sampling, as the name implies, involves various stages and selecting subsamples from the clusters, such as a sample of graduating students from selected high schools. An advantage of these procedures is that they save time and money. But a disadvantage is that they can result in a biased sample. The selected sample of high schools may have certain characteristics that are unknown to the researcher and that can have an impact on the results. For example, they might all be located in low-income districts.

Thus, in cluster sampling you select samples in stages. Cluster sampling is especially useful in selecting a sample that covers a population that is geographically dispersed. National samples are usually selected by means of multistage cluster sampling.

The essentials of cluster sampling can be demonstrated even in selecting a sample at a college or university. Suppose you want to carry out a study of first-year students in social science disciplines. Assume the student directory does not exist or is not useful. Now suppose that all first-year social science students are required to take the course Social Research, of which there are 40 sections. You could compile a list of all the students in the 40 sections and then select a simple random sample. But you do not have the time and money. Another approach is to do cluster sampling.

1. Compile the list of the 40 sections of the course. Assume you want a sample of 10 sections. Randomly select 10 sections.
2. Survey the students in each of the 10 sections.

Or you might add another stage in the sampling process:

2. (*Replaces 2 above.*) Compile a list of students for each of the 10 selected sections.
3. Do a random sample of the students in each of the 10 selected sections, whereby one-fourth of the students in each section are selected.

The sample evolved in stages from 40 sections to 10 sections to randomly selected students in each of the 10 sections. Cluster sampling will provide a reasonably representative sample of first-year social science students without requiring you to identify the entire population at the outset.

SAMPLING: CHOOSING WHO OR WHAT TO STUDY

BOX 4.4 SAMPLING ISSUES IN THE 2011 LONG-FORM CENSUS CONTROVERSY

In July 2010 the Canadian government decided to replace the compulsory, 53-question long-form census with a voluntary survey for the decennial census of May 2011. The decision was justified on the grounds that some Canadians found the questions intrusive and the compulsory status offensive. The decision aroused much criticism from business and community groups, academics, and researchers. All argued that the reliability and validity of the data would be seriously weakened by the change to a voluntary response survey.

The sampling method of the long-form census dates from 1941. In that year the Canadian government directed the then Dominion Bureau of Statistics to survey 10 percent of Canadian households through a supplementary household data survey to provide information to guide postwar policies dealing with anticipated housing problems. It was found that a random sample of 1 in 10 households provided high-quality data, adequately representing all households for government purposes. As well, a 10 percent survey involved significantly fewer costs for the government and burdens on the population. In 1951 the supplementary section of the census was expanded to 1 in 5 households in order to achieve greater regional detail. This permitted interprovincial, urban–rural, and other social comparisons while maintaining sufficient size of subgroups for valid generalization and inference. All subsequent long-form, supplementary censuses have been based on a 1 in every 5 households survey.

The key element here is that the selection of these 1 in 5 households is through a *random selection* process, using previous census data as the basis for a sampling frame. Making the response compulsory produces a 94 percent response rate, thus ensuring a nearly complete sample. With the change to voluntary selection, it is projected that the response rate will be, at best, around 50 percent. Even the government's proposed increase in sampling by sending the long form out to 30 percent of households will not fully compensate, so subdividing responses for various comparison analyses will produce weak data for generalizing from. Worse still, volunteering is a nonrandom sampling process, so one can no longer assess the probable margin of error.

In addition, by shifting to a voluntary selection process the sample is less representative but its biases are inestimable. Without compulsion, different social groups have different response rates to social surveys. The very poor, the less educated, residents in economically depressed areas, and the very rich are less likely to respond than others. Groups with specific grievances against government, such as Aboriginals, or fears of government, such as certain immigrants, are less likely to respond. We do not know what the rates of nonresponse of specific groups are, however. Consequently, we have no way of estimating the extent of biases in the survey.

Critics of the decision have pointed out that the overall result of these problems is that the Canadian census no longer provides basic reference points from which major social changes can be measured, and social data from which valuable sampling frames and other points of comparison essential to business decisions, policy guidance, and meaningful knowledge of Canadian society can be obtained.

For more on the recent censuses of population, see the Statistics Canada web page at **http://www.census2011.gc.ca/ccr_r000-eng.htm**.

Random Sampling and Random Assignment

Until now our focus has been on how we randomly *select* a sample of the population we want to study. However, in experiments we may randomly *assign* the sample to different types of groups. Say you have selected a random sample of 100 students from a population of 1000 for a study on stress. Your study involves some students carrying out a stressful activity and the others not carrying out the activity. Your next step is to randomly assign the 100 students, with 50 in one group and 50 in the other. In this case, a random sample was first selected and then the subjects were randomly assigned to different groups. Hence, be careful with the terminology; random sampling involves selecting a sample from the population, while random assignment involves assigning subjects into different groups for research purposes. Also note that for many studies using the experimental design you do not need to select a random sample in order to do a random assignment. We will elaborate on random assignment in Chapter 6, "Experimental Research."

Nonrandom (Nonprobability) Sampling Procedures

Nonrandom sampling procedures provide a weaker basis for generalizing from the sample to the population than random sampling. However, the majority of samples in social science research are nonrandom samples. Historical, anthropological, and archaeological research can study only the traces from the past that have survived many centuries of physical destruction and erosion. Studies of contemporary life often focus on rapidly unfolding events and processes such as political and economic crises, fads, social movements, and technological change. Sampling frames are impossible to generate in these highly fluid situations, since informants and evidence have to be obtained as the opportunities arise. In other instances the use of nonrandom sampling is due to the objectives of the research and the lack of resources. Students are especially likely to lack the time and resources to carry out a study based on a random sample and may therefore have to resort to a nonrandom sampling procedure. In carrying out your own research, you need to first consider what resources (time, money, sampling frame, and so on) you would need to draw a random sample. If you are unable to draw a random sample, for whatever reasons, you should consider a nonrandom sample that best serves the purpose of your study and explain why you are using it.

Five common procedures of nonrandom sampling are described below, along with examples of situations in which you may need to use them.

ACCIDENTAL SAMPLING In one social science discipline, history, probability sampling tends to be the exception rather than the rule. This is because history (and other historical sciences like archaeology and paleontology) works with ready-made samples: the historical remains that have been accidentally left behind. The earlier the time period you study, the fewer the number of items that have escaped wear and tear, destruction, and burial. All you can study is what remains, what has been saved, and what has been discovered from an original population of documents, buildings, monuments, and other artifacts. This is called an **accidental sample**.

An enormous amount of effort goes into describing these remains and establishing as clearly as possible how they were located and discovered. Each piece of information is used to cross-check every other piece and to build up a picture of life in the past. The logic of sampling still holds. That is, historians are trying to draw conclusions about a population from a subgroup of that population. However, for historical work there is no choice but to use whatever evidence has survived over time.

CONVENIENCE SAMPLING   **Convenience sampling** involves selecting for the sample whomever or whatever is convenient for the researcher. This method is roughly equivalent to "person-on-the-street" interviews by reporters and journalists who want to give a sense of what some people think about an issue. Journalists operate under severe time pressure to get a story out while it is still "hot," and consequently they are forced to use shortcuts in gathering information. We cannot and should not assume, however, that the opinions of the persons interviewed reflect those of everyone else. The same can be said about any sample that was conveniently selected.

Suppose you want to do a study on nutrition among college or university students, and you survey only students in one of your classes. Are the students surveyed representative of all college or university students? You have no way of knowing. You selected the students because they were available and because it was convenient for you to carry out the survey on them. The sample results could be highly misleading.

Likewise, assume you want to carry out a study on the opinions of students at your college or university, and you interview every 10th student you meet in the cafeteria on a Wednesday afternoon. But are the opinions of the students you interviewed representative of all the students in the college or university? You have no way of knowing; you selected the sample from whoever happened to be in the cafeteria on a particular day and at a particular time. The information tells you only about the opinions of students in your sample, and you cannot generalize to all the students in the college or university.

Therefore, the results obtained from a convenience sample pertain only to the sample itself. There are many reasons for researchers to resort to convenience sampling. One common purpose is to **pretest** questionnaires. In this case, the researchers are concerned with checking how well the questionnaire works in terms of the clarity of questions and any problems arising from the format. (We cover questionnaires in Chapter 5.) There is no attempt to make generalizations from the results. As long as the questionnaire is tried out with a test group similar to the population being studied, no special sampling techniques beyond convenience need be used.

Convenience sampling is also used in advancing our understanding of particular issues. For example, suppose you want to carry out a study on people who at one time in their lives were homeless. It would be virtually impossible to draw up a list in a large city of all such people from whom you could draw a random sample. Instead, you may resort to a convenience sampling procedure whereby your sample consists of people whom you know have at one time been homeless. Although you cannot generalize to all formerly homeless people, the information gathered is nevertheless useful.

Another area in which convenience samples are often used is psychology. Psychologists often use student volunteers as subjects in their experiments and tests. They argue that the kinds of phenomena they study—cognition, memory, language learning and use, and so on—are found in all healthy, normal human beings. Differences between students and nonstudents, or between students who volunteer for the experiments and those who do not, are viewed as irrelevant to the issues studied in their experiments, so the use of convenient groups of highly accessible students is acceptable. As well, when researchers perform experiments, other procedures improve the validity of their research, including random assignment, so that generalizations from the results are possible.

PURPOSIVE (JUDGMENTAL) SAMPLING **Purposive sampling** involves selecting for the sample whoever or whatever the researcher judges has characteristics that meet the purpose of the study. Suppose you want to study female students who are single parents. It may be difficult and time-consuming to compile a list of all such female students and draw a random sample. Therefore, if you have limited time and resources, you may decide to select all female single parents who are members of, say, the Mature Students Association at your college or university or who have children in the campus daycare, or both. You purposely select these female students because you assume that they are typical of female students who are single parents. You will never know for sure, but the study may still be of practical interest.

Similarly, suppose your study on homeless people concerns single women over 60. For your sample, you purposely select women over 60 who are aided by a service agency in your community, because in your judgment they are typical of homeless women over 60. You may have various reasons for assuming this, but you can never be certain that these women are typical. But representativeness may not be your concern. The aim of your research may be to test a theory or to further our understanding of homeless women and gain insights that challenge widely held assumptions.

QUOTA SAMPLING **Quota sampling** involves selecting quotas (or proportions) of each defined subgroup from the population. To do this we need to know the proportion of each subgroup in the population, such as the proportion of males and females, or of other features, such as age, academic discipline, or place of residence. However, despite this effort, quota sampling is a nonrandom sampling procedure; it should not be confused with stratified random sampling. A list of the population from which to draw the sample is missing in the quota sampling procedure.

Suppose you want to find out which political party is favoured by college or university students, and you have reason to believe that support may vary according to gender. Let's say a directory of students does not exist, or you do not have the time and resources to draw a random sample. But you find out from the registrar that the student population consists of 55 percent females and 45 percent males. Assuming you want a sample of 100, your sample should have 55 females and 45 males. You then seek out in your college or university, such as in the cafeteria on a Wednesday afternoon, 55 females and 45 males for your sample.

The study would be further complicated if some other feature, say an academic year, were also relevant for your study. You would have to determine the relevant proportions of the various subgroups with the desired features—for example, the proportions of first-year males, first-year females, and so forth. Then you would determine the appropriate number of each subgroup needed for your sample that would correspond to the proportion of each subgroup in the population. For instance, if first-year males make up 8 percent of the student population at your college or university and your sample is to consist of 100 students, only eight students in the sample should be first-year male students.

Quota samples are common in market research and opinion polls, mainly because they are quicker, easier, and less expensive to obtain than random samples. But they have various inherent problems. The information from which the proportions are determined may be inaccurate or outdated. And even if the appropriate proportions are selected, biases may exist among the people chosen for the sample. For instance, after having examined the results, you might learn that all eight first-year male students in your sample were members of a specific political party. By contrast, in a stratified random sample this bias would be less likely to occur, since the samples for each stratum are selected randomly.

SNOWBALL SAMPLING **Snowball sampling** involves first selecting people who have characteristics that meet the requirements of the study, and then asking them for further contacts. Like a snowball that grows larger when you roll it in wet snow, a snowball sample grows ever larger from the initial contacts onward.

Suppose you want to do a study on racism and wish to interview college or university students who attended a march in support of white supremacism. You were able to find two students at your school who attended the march and were willing to be interviewed. You might ask each for other names of march participants, and you would do the same with these other subjects. You would continue the process until no more new names were suggested or until you felt the sample was of a sufficient size. You cannot be sure, however, that the sample is reflective of all the students who participated in the march. Moreover, unbeknown to you, a bias may have entered into the sample from the outset. For instance, the first two students could have been members of a clandestine white supremacist organization and suggested names of other students who belonged to the same organization. But despite this drawback, you would have gained information about the racist beliefs of such students that might not have been available to you without snowball sampling.

Snowball sampling is, in fact, essential for studying minority, unpopular, or unofficial beliefs such as racism. When beliefs or actions are controversial, unpopular, marginal, or hidden for other reasons, snowball sampling is often the only way to get to talk with a group of individuals associated with such beliefs and actions. Social scientists interested in studying criminals, cult groups, or political extremists often use snowball sampling techniques. At the other end of the social spectrum, social elites are often wary of publicity; thus, studies of the rich and powerful often need to overcome initial suspicion of the research, and the researcher is able to build up a body of informants only by personal introduction from informant to informant.

Another example would be a study on leadership in ethnic communities. Suppose you need to interview leaders of an ethnic community with which you have little contact. Assume you are able to obtain the names of executives of some associations in that community. You could first interview them and then ask them for the names of other leaders in the community, and so forth. The underlying assumption is that leaders of an ethnic community would know one another. As you can imagine, the snowball sampling procedure might lead you astray and to a biased group of leaders, but you would still have learned about some aspect of leadership in that community. Being referred to the next leader not only helps the researcher to gain access, but also gives the researcher credibility and helps to promote trust between the researcher and the people being interviewed. When trust is an important consideration in research, the snowball technique may be the only practical method of sampling.

BOX 4.5 TYPES OF SAMPLES

RANDOM SAMPLING

Designed to give everyone in the target population an equal chance of being selected. Aims to be representative of the target population. Needs a sampling frame from which to select the sample. Permits the use of various statistical techniques for estimating error and drawing inferences.

Simple Random

- Everyone in the target population has an equal chance of being selected, such as by selecting names out of a hat.
- Generally less used than systematic sampling.

Systematic Random

- A sample is selected from a list at certain intervals, say every 10th person on a list, using a random start.
- An efficient technique, provided that you have an appropriate sampling frame (i.e., a list of the individual units in the population).

Stratified Random

- The target population is divided into groups (strata), such as men and women, and samples are selected from each.
- A useful method when you need to ensure the representation of subgroups in the sample.

Multistage Cluster

- Samples are selected at different stages: for example, the first stage is a sample of city blocks; the second stage, a sample of specific dwellings in those blocks; the third stage, a sample among the people residing in those dwellings.
- A necessary technique when your sample frame does not list individuals but some other kind of unit, such as households or schools.

NONRANDOM SAMPLING

Any sample for which it is impossible to give each person in the target population an equal chance of being selected. The sample is usually biased, since it is not representative of the population. Nevertheless, it may be effective, and it is often the only way to obtain information in certain research situations.

Accidental

- The sample emerges from uncontrolled events, such as historical wear and tear and accidents.

(continued)

Convenient

- Whoever is most convenient to be interviewed is chosen, similar to the person-on-the-street interviews often used by television reporters.
- Used for pretesting survey questionnaires.

Purposive

- Where it is difficult to set up a sampling frame, the researcher judges who best represents the population and selects the sample accordingly.

Quota

- A certain number of respondents within various groups are chosen, such as a certain number of females in different age groups and a certain number of males in different age groups.

Snowball

- People with characteristics that meet the purpose of the survey are chosen, and they then provide names of others like them to contact, and so on.
- Often the only possible way of studying certain social groupings, such as deviants or elites.

DETERMINING THE SIZE OF THE SAMPLE

How large should the sample be? This depends on many factors, including the sampling method, the degree of accuracy you want, and the variation in the population. Thus, the answer will partly depend on the researcher's purpose and on the characteristics of the population. For both nonrandom and random samples, another major factor affecting sample size will be the time and resources available for your research.

For nonrandom samples, the size of the sample is not a major issue, since the results cannot be generalized to a population. Nonetheless, as a student you may find yourself using such a sample for a class project, such as a social survey. Experience tells us that in such a situation it is best to have at least 30 respondents in order to be able to make some simple analyses.

Sample size is of particular concern in random sampling procedures, since you want to generalize to the population. Depending on the research objective, the more characteristics being taken into account about the population, the larger the sample. For example, if your study requires a sample of students from all the different programs at your college or university, then you are likely to require a larger sample than if you were selecting students irrespective of their program or other characteristics. The greater the variation in the population that has to be present in the sample, the larger the sample—or else the lower the accuracy of the results. Various statistical formulas can be used to compute the appropriate size of the sample in random sampling procedures, but these are beyond the scope of this book. Instead, let us explore certain factors to consider in examining sample size.

In a study at a college of 1000 students, which findings would you be more confident in accepting: those based on a random sample of 100 students, or those based on a random sample of 500 students? The random sample of 500 is more accurate. But why? The results from a random sample of 500 are likely to be closer to the "true value" of the characteristics of the population. Thus, we recognize that random samples, despite researchers' attempts to select samples that are representative of the population, are subject to a **margin of error** (or sampling error): a difference between the "true" value of a characteristic of the population and the value estimated from a random sample of the population. There are many kinds of sampling errors. For example, suppose the average age of your sample of 500 students is 19.2 years, but the average age of all 1000 students in the college is 19.8. The sample average age is close to that of the population value, but it is off by 0.6 years from the population value. (See also Box 4.6.)

But, you might rightly ask, how can you know the margin of error if you do not know the true value of the population? With regard to the above example, how could you assess the margin of error if you did not know that the average age of the population is 19.8 years? Researchers deal with this problem by establishing a **confidence interval**, a range in which they expect the true value to fall. Again, take the example above. The sample average age of the students was 19.2, and if the confidence interval is +1.2 to −1.2 from the average age, we expect the average age of the population to fall somewhere between 18 years and 20.4 years.

It is also possible to measure how confident we are that our results will fall in this range; this is known as the **confidence level**. For example, rather than simply saying that we expect the average age of the population to fall between 18 years and 20.4 years, we can be even more precise by noting that we are 95 percent confident. In other words, if various random samples were drawn from the same population of students, we are confident that 95 percent of the samples, or 19 out of 20 samples, would have an average age that would range from 18 years to 20.4 years, while another 5 percent of the sample would have average ages outside the range. Note that we do not know the actual average age of the population, since we collected information only on a sample of the population. Yet, by means of certain computations beyond the scope of this book, we are able to state that we are 95 percent confident that the average age of the population falls in the confidence interval of the sample. (Or we could use another confidence level, such as 99 percent, and so on.)

The rules of probability also tell us that for large populations, say 10 000 or more, we can achieve about the same degree of accuracy with a sample size of about 1000 to 1100, irrespective of whether the population consists of 10 000, 10 million, or 100 million. For example, to achieve a sampling error of plus or minus 3 percent at a confidence level of 95 percent, we would need a sample of about 250 for a population size of 500, a sample of about 1000 for a population size of 10 000, and a sample of only about 1100 for a population size of one million. National polls in Canada usually accept a margin of error (sampling error) of plus or minus 3, at a 95 percent confidence level, and therefore consist of a sample of about 1100 people. Other factors may have to be taken into account, such as provincial or regional social variations, which can affect sample size and confidence level. (For a sample size calculator see the Weblinks at the end of the chapter.)

Thus, especially in the case of student projects, in general the larger the random sample, the more confident you can be with the results. But your confidence in your survey does not depend only on the size of your sample and the procedure used in selecting the sample. It will also depend on the design, construction, and administration of your questionnaire, among other factors in the research design. If, for example, you are studying a sample over a long period of time (such as studying changes in the career plans of students as they go through school), you need to allow for a certain amount of shrinkage in your sample. Students will move away and transfer to other schools, drop out and get jobs, and so on. The longer the time period of your study, the greater the problem of shrinkage, known as **mortality**, will be. Therefore, your initial sample will have to be larger than otherwise necessary.

A similar problem occurs when you are investigating sensitive, intimate, or controversial subjects such as sexual behaviour. You may have a large number of outright refusals to answer, as well as incomplete and even spoiled questionnaires, which will require compensation through a larger initial sample. As you can see, the problem of sample size is a complex one that depends on the aims and methods of the research and on the resources available to the researchers.

BOX 4. 6 WHAT IS THE MARGIN OF ERROR?

The television announcer has just informed us of the latest poll results. Candidate Humble is preferred by 52 percent of voters, while Candidate Vanity is preferred by 48 percent. Who would likely have won if an election had been held when the poll was conducted?

The results in themselves provide insufficient information. After telling us the overall results the television announcer is likely to state the margin of error, and then point out that according to the polling agency the results are too close to call a winner. Why? There are 4 percentage points separating the two candidates! Could it be because of the margin of error?

The best way to understand margin of error is by using examples. Assume we are told that the poll had a margin of error of plus or minus 3 percent, 19 times out of 20. The information provided tells us that the researchers are confident 19 times out of 20 (or have a 95 percent confidence level) that the sample is from plus to minus 3 percentage points of the results they would have obtained if they had interviewed the entire target population. But we need to remember that this does not mean a mistake of 3 percentage points occurred 19 times out of 20.

Again, let us use the example of Candidates Humble and Vanity. With plus or minus 3 percentage points, the expected support for Candidate Humble is between 49 percent (52 – 3) and 55 percent (52 + 3), and for Candidate Vanity between 45 percent (48 – 3) and 51 percent (48 + 3). Hence, we cannot infer from these figures which candidate is more popular. Why? The answer lies in the overlap of possible support.

continued

Another example is the last poll taken before the 1995 Quebec independence referendum with a sample of 1003 (see Box 4.1). A sample of this size has a margin of error of plus or minus 3.1 percentage points, 19 times out of 20. Among those who stated a preference, 41.4 percent said they would vote Yes, and 46.8 percent said they would vote No. But the margin of error tells us what the range of the final results might be. With a margin of error of 3.1 percentage points, the results among those who stated a preference were expected to fall between 38.3 percent and 44.5 percent for the Yes and between 43.7 percent and 49.9 percent for the No. Why was it too close to call? (Researchers may also take into account other factors in determining how close the expected results are. See Box 4.1 for an example.)

When a sample is broken down into subgroups, such as by gender, age, or location, the margin of error rises. Suppose the above results for our two candidates covered the whole country. Now assume that we also had the results by region. In the Atlantic region, support for Candidate Humble is 46 percent and Candidate Vanity 54 percent, with a margin of error of plus or minus 6 percentage points. Therefore, in the Atlantic region, support is expected to be between 40 and 52 percent for Candidate Humble and from 48 percent to 60 percent for Candidate Vanity. Can you tell who is likely to win in the Atlantic region?

Note that the margin of error is determined on the basis of what to expect in a perfect survey. However, perfect surveys do not exist. And if subsamples are considered, their margins of error are greater than that of the overall survey. The margin of error alone is insufficient in judging the results. We need to always interpret the results with caution. There can be many serious flaws in how the study was conducted that make the results difficult to interpret.

(The aim here is to help you grasp what is meant by the concept of margin of error. To produce a confidence statement, it is necessary to have an elementary understanding of statistics.)

READING 4.1

Many Canadians Skeptical of Healthy Food Claims

The media often sponsor research on an issue of general interest whose results can become the topic of public discussion. One such example is the article below. Given what you have learned in this and the previous chapters, consider the purpose of the study. What is the population of the study? How was the sample selected? What other questions would you raise about the sample and the results? What further research do you suggest needs to be pursued to further understand the issue(s)? What hypotheses could be formulated?

Nearly half of Canadians don't believe health claims on food products, according to a new poll released Monday.

The nationwide Ipsos Reid survey conducted last week for Global National and Postmedia News found only a slim majority—53 per cent—believe health claims made on food labels. And most of these people only "somewhat agree" (47 per cent) with the statement that they believe the health claims on food labels, compared to only five per cent of Canadians who "strongly believe" that statement.

Overall, 47 per cent say they don't believe health claims made on food labels, including nine per cent who feel strongly about it. Younger Canadians aged 18–34 are the least likely to be skeptics, while those 55 years old are the most likely to distrust label claims.

Among the regions, residents of Quebec are the most likely to believe health claims (59 per cent), followed by Atlantic Canadians (54 per cent) and Ontarians (53 per cent). Western Canadians are the most skeptical, with fewer than half saying they believe health claims (47 per cent in each of the provinces).

It's not all bad news for food industry, however.

Seven in 10 (72 per cent) responded that they believe probiotics—live micro-organisms added to food products—improve their health, while eight in 10 (79 per cent) believe that the addition of Omega-3 fatty acids to food products makes them healthier.

But only 44 per cent of Canadians are willing to pay more for products that make health claims. The mean increase in food prices acceptable to those willing to pay extra is 12.8 per cent, according to the poll.

The poll results come as Health Canada struggles to respond to what it calls the "proliferation" of health messages on the front of food packages, written by food marketing experts and approved by legal advisers familiar with the federal government's rules for nutrition labelling and health claims. That's why government officials are now reviewing options such as bringing in nutritional eligibility criteria for simplified health claims, symbols and logos highlighted on food packages.

Health Canada requires a minimum or maximum amount of an ingredient or nutrient to make specific composition claims on the front of food packages—such as "low in sugar" or "high in fibre." And unlike some other countries, in Canada a food company has to get approval from Health Canada to use disease risk-reduction claims if it wants to market a product as a way to combat heart disease, for example.

However, there are no specific regulations to deal with simplified nutritional claims—including symbols and logos for front-of-the-package labelling—which means consumers can find a "healthy" product on a store shelf that is high in salt, a "smart" snack high in fat and a "vitamin" drink that's full of sugar.

For its survey, Ipsos Reid conducted a poll Jan. 12–14 of 1,012 adults from an online panel.

The survey has a margin of error of three percentage points, 19 times out of 20.

Source: Sarah Schmidt. "Many Canadians Skeptical of Healthy Food Claims."

Postmedia News. *(January 18, 2011).*

Material reprinted with the express permission of Postmedia News, a division of Postmedia Network Inc.

ETHICAL CONSIDERATIONS

The Canadian government justified the decision to abolish the May 2011 long-form census (see Box 4.4) on political-ideological grounds: that it was wrong to force citizens to answer intrusive questions. Critics of the decision countered that the greater good of effective national policies outweighed citizens' discomfort in this instance. Regardless of the political issues involved, the incident highlights the potential existence of ethical considerations in relation to sampling in social research. The selection of people as participants or informants in social research in effect establishes a relationship of mutual obligation between the researcher and the researched. The researcher wants to obtain data through a survey, set of tests, in-depth interviews, observations, or access to some aspects of life. This requires at least some commitment of time, focus of energy, memory, or other psychological and social resources. The subject or participant does not want to be overly stressed, or to have her reputation damaged or information given in confidence disclosed. The initial act of sampling, then, may be viewed as a contract between the researcher and the researched. Hence, in the process of sampling the researcher should be prepared to make clear to would-be participants the purpose of the research, the identity of sponsors and funders, the time and other commitments probably required of the participants, and what safeguards are in place protecting anonymity or confidentiality. If there are particularly high or unusual commitments or risks involved, these should be made clear and there probably should be some compensation available. Although these issues arise at the point of sampling for research, the way to deal with these issues largely depends on the specific method for gathering or generating data and is therefore explored in the appropriate context in subsequent chapters.

What You Have Learned

- Sampling is an almost universal part of any act of gathering information.
- The essential problem of sampling is attempting to select a representative sample from a population.
- In order to be able to generalize from a sample, one of a number of types of random sampling techniques has to be used.
- Nonrandom sampling techniques are valuable research tools, even though they do not permit generalizations to be drawn.
- Many factors affect the size of a sample required for a particular study.
- Sampling implies ethical considerations to be taken into account by researchers.

Focus Questions

1. What is the essential problem of sampling?
2. What are the two types of sampling procedures? What distinguishes them?
3. What is a sampling frame and why is it important?
4. What is a stratified random sample and why would you use it?

5. When are cluster sampling and multistage cluster sampling procedures needed?

6. Which kinds of studies use snowball sampling and why?

7. What factors affect the size of a sample?

Review Exercises

The results of research studies are often reported in the popular media. Generally, information about the population, type of sample, size of sample, research method, and, if necessary, the margin of error are provided. Select an article from a newspaper or magazine that reports on a research study and answer the questions below.

1. What are the sources of the information reported in the article?

2. What type of sample was selected?

3. How was it selected?

4. What was the sample size?

5. Discuss the strengths and weaknesses of the study with regard to the sample selected.

6. Is the margin of error indicated? If so, what use can you make of it in interpreting the information provided in the article?

7. What were the general findings of the reported study?

Weblinks

You can find these and other links at the companion website for *First Steps: A Guide to Social Research*, Fifth Edition: **http://www.firststeps5e.nelson.com**.

Sampling Techniques

Sampling Module (International Development Research Centre—Canada)
       http://www.idrc.ca/en/ev-56617-201-1-DO_TOPIC.html

Sampling Methods (Statistics Canada)
       http://www.statcan.gc.ca/edu/power-pouvoir/ch13/
       sample-echantillon/5214900-eng.htm

Survey Sampling Methods
       http://www.statpac.com/surveys/sampling.htm

Types of Sampling Techniques (video)
       http://www.youtube.com/watch?v=rASK8PqakM

Sampling Tools

Research Randomizer: Random Sample Tutorials (Social Psychology Network)
       http://www.randomizer.org

Sample Size Calculator
       http://www.surveysystem.com/sscalc.htm

Social Survey

What You Will Learn

- the aim and purpose of surveys
- how to judge surveys conducted by others
- how to construct a questionnaire
- how to administer a survey

INTRODUCTION

[B]oys worry more about the future, whereas girls worry more about being liked or being overweight.... [Adolescents] who turn to parents for information are less likely to worry about being liked, failure, their future, and their friends than those who turn to other sources (teachers, Internet, and friends).

S. L. Brown, J. A. Teufel, D. A. Birch, & V. Kancherla. (2006). Gender, age, and behavior differences in early adolescent worry. *Journal of School Health, 76,* 430–438. InfoTrac record number: **A152993873**.

Visible minorities are likely to have a lower level of confidence in the police.... Equal confidence in the police is yet to be achieved in Canada as indicated by the different levels of confidence in the police held by visible minority and non-visible minority groups.

L. Cao. (2011). Visible minorities and confidence in the police. *Canadian Journal of Criminology and Criminal Justice, 53,* 1, 1–27. InfoTrac record number: **A252004890.**

[A]lmost three quarters of those bullied in college were also bullied in high school and elementary school, over half of those who were bullies in college were also bullies in high school and elementary school.

M. S. Chapell, S. L. Hasselman, T. Kitchin, S. N. Lomon, K. W. MacIver, & P. L. Sarullo. (2006). Bullying in elementary school, high school, and college. *Adolescence, 41,* 633–649. InfoTrac record number: **A156808955**.

Immigrants in Canada had better access to care than immigrants in the United States; most of these differences were explained by differences in socioeconomic status and insurance coverage across the two countries.

L. A. Lebrun & L. C. Dubay. (2010). Access to primary and preventive care among foreign-born adults in Canada and the United States. *Health Services Research, 45,* 1693–1720. InfoTrac record number: **A243121670.**

As viewers watched more crime-related television, they were more likely to misperceive realities of juvenile crime and juvenile justice.

R. K. Goidel, C. M. Freeman, & S. T. Procopio, (2006). The impact of television viewing on perceptions of juvenile crime. *Journal of Broadcasting and Electronic Media, 50,* 119–140. InfoTrac record number: **A147216162**.

Patient and healthcare provider communication is vital in cancer care and aspects of the patients' experiences provide valuable insight to what constitutes effective cancer communication.

J. Oliffe, S. Thorne, T. G. Hislop, & E. A. Armstrong. (2007). "Truth telling" and cultural assumptions in an era of informed consent. *Family and Community Health, 30,* 5–16. InfoTrac record number: **A157034720**.

The above statements are all examples of results obtained from surveys. It is likely that in your lifetime you have participated either directly or indirectly in a social survey. The survey method has become one of the most popular techniques of collecting data in the social sciences. Surveys are also widely used by government agencies, market researchers, political parties, community groups, and media organizations.

Of the various social science research methods, the survey is the most familiar to the general public; survey data have become part of our everyday sources of information. The media regularly report the results of surveys on a wide range of topics, and TV documentaries, magazine articles, and newspaper editorials often include survey data.

The media regularly fund surveys on a variety of topics. During election campaigns, for example, numerous polls are funded by newspapers and television stations. But though most surveys are properly carried out, sometimes the media report findings that are based on poorly designed surveys, the results of which are not trustworthy.

Unfortunately, despite the popularity of survey research, the public is generally unaware of the criteria that determine whether a survey is meaningful or meaningless. And because of the popularity of survey research, many people may wrongly assume that to conduct a survey requires little thought and effort. There is the added problem that at times they may have been asked to take part in a bogus survey that is part of a sales campaign but disguised as research.

Dial-in polls, which some may wrongly assume are acceptable scientific research, have also been growing in popularity. Television and radio programs invite listeners to call a phone number, or visit a website, to register their votes on a particular issue, usually with a yes or no response to a question. Do you agree with the government's latest budget? Should pornography be illegal? Should the Toronto Maple Leafs get a new coach? And so on. A short time later, the responses are compiled and reported. But although the announcer may state that the poll reflects only the views of those who chose to call, the impression often left is that the results are representative of the public's opinion. Why are these polls not valid as scientific research? Similarly, why should little importance, if any, be given to results of surveys based on questionnaires included in popular magazines?

CASUAL AND SCIENTIFIC INQUIRY

The Casual Inquiry

Suppose you want to find out whether students at your college or university who do not work part-time do better in school than students who do work part-time. What could you do to find out? A quick and easy way, and one with many flaws, is to collect information from students you know who work part-time and those who do not have jobs. Assume you wanted to draw conclusions about all the students at your college or university, but you collected information from only 10 students, five of whom worked part-time and five of whom did not. Also, assume the students were your friends, and while having lunch in the cafeteria you asked them what their overall grade averages were.

For the social science researcher, this approach has numerous weaknesses. Consider a few of the many questions a social scientist would ask about the research:

- Which group were you interested in studying?
- Whom did you interview?
- How did you choose the students to be interviewed?
- How did you conduct the interviews?
- What question(s) did you ask them?
- How did you determine their academic performance?
- How many courses were the students registered in, and how many did they regularly attend?
- Were they day students or evening students?
- How many hours did the students work?
- If they worked, what were the main activities of their jobs? Were the jobs related to their studies?
- Is the information collected sufficient to make generalizations about all the students in the school?

The obvious answer to the last question is no (see Chapter 4). The example illustrates a serious but common error in everyday, personal inquiry. After having spoken to a few students, the tendency might be to look for a general pattern in the relationship between grades and part-time work. Suppose the five students who worked part-time had lower grade averages than the five students who did not work part-time. The tendency is to conclude that all students in the school who work part-time have lower grades than students who do not work part-time, and therefore that working part-time results in poor academic performance. Both conclusions are highly suspect; in particular, they are based on the limited information gathered from casual conversations with 10 students. Moreover, there may be measures of academic performance other than the grade.

The example above, although flawed, contains some of the main ingredients of a **social survey**: asking questions of a selected number of people, usually to draw conclusions about a larger group of people. However, the survey is far more complex and demanding than an everyday, casual inquiry.

The Scientific Inquiry

Let us consider briefly the main components of the social survey. Later, we will explore them in greater detail. Again, suppose you want to find out whether at your college or university, students who work part-time do better in school than students who do not work part-time.

Because of time and cost it may be impossible, as well as unnecessary, to interview all the students. Instead, you can select a subgroup of students (say about 300 students) to represent all 6000 students at your school. You might, for example, select for interview every 20th student from a list (the sampling frame) of all the students. As you will recall from Chapter 4, this is the systematic random sampling procedure, which, if properly carried out, lets you generalize the results from a few hundred students to all the students in the school. In other words, you select a sample to represent the population.

As you may have realized, the grade average is only one determinant of academic performance, and it is not necessarily the appropriate one for your study. Among the many challenges you would face is to state clearly how you would define and measure academic performance. (Can you suggest several possible ways of measuring academic performance? For example, is the number of courses passed or failed a good measure? Or, would having the students assess their overall academic performance be better?).

A **questionnaire** is constructed to elicit the relevant information about the student's work experience, courses, and academic performance, and it is administered to the sample of 300 students. You can choose to give the questionnaire directly to the students and have them complete it with you present, e-mail them the questionnaire, conduct a face-to-face interview in which you read the questions and take down the answers, or hold a telephone interview.

The students' answers are then subjected to analysis. The conclusions reached from the analysis of the data collected on the 300 students (sample), assuming they all participated in the study, are then generalized to the 6000 students (population).

THE AIM AND PURPOSE OF SURVEYS

Social scientists carry out surveys for various reasons, but they use them mainly to describe the characteristics of a population, such as age or income distribution; to study attitudes and opinions, such as people's opinions about products; to examine the relationship between two variables, such as age and voting behaviour; and to test theories, for example, the human capital theory, which explains income differentials as being due primarily to a return on an individual's investment in human resources, such as education.

Thus, surveys are useful for descriptive and explanatory purposes. As discussed in Chapter 2, descriptive research is concerned mainly with the "how" and "who" of the phenomenon, while explanatory research focuses on the "why." For example, you might design a survey to describe as accurately as possible how many students, or which students, at your college or university work part-time. A different survey design could collect information to assess why these students work part-time.

Surveys are also useful for evaluating a social intervention and aiding in decision making. For example, a survey might be used to determine whether there is enough support to justify setting up a community daycare, to assess whether a national campaign against racism is achieving its goal, or to evaluate the usefulness of a new library service.

The purpose of most surveys is to find out only once about the opinions, attitudes, or other conditions in a population. In other words, they are **cross-sectional surveys**. An example is the census carried out by Statistics Canada. Its aim is to collect information on the number of people in Canada—their age, gender, education, income, ethnic–cultural origins, and other details—at a particular time (see Chapter 4). Another example is a survey that is intended to record the employment status of graduates from a certain university at a given time.

However, if the purpose of the survey is to track down the changes in the future employment status of university graduates who terminate their studies this year, then it is necessary to carry out a **longitudinal study**. Two kinds of longitudinal studies are used: the trend study and the panel study. The **trend study** involves carrying out surveys on the same population over a period of time. In other words, a new sample is selected from the same population at two or more points in time. The **panel study** involves carrying out surveys to track changes in the same individuals (or sample) over a period of time. For example, we might select at regular intervals, say each June for the next three years, a new sample from the same population of university graduates and carry out the survey (trend study). The results can tell about the changes in how many are employed or unemployed, but not why these changes occurred. Another possibility is to survey the same sample of graduate students at regular intervals (panel study). As you can imagine, there are many benefits to doing longitudinal studies, but there are also many difficulties. They are especially time-consuming and costly.

As you may have realized by now, undertaking a survey requires much more than asking a few questions. It is part of a broader research process. A key aspect of the process is first to state clearly what you want to find out. Only after you have a clear idea of your research topic can you decide whether a survey is an appropriate method for your study. Stated differently, your research topic should guide you to the research method, not vice versa.

You may begin with a general area of interest, such as sexual harassment in the workplace. You will then want to familiarize yourself with what is already known about your area of interest and take note of the theoretical and political issues relevant to sexual harassment. Doing so will also allow you to identify variables that previous research has proposed or shown to be related, or not related, to the issue you are examining. For example, how have others defined and measured sexual harassment? To gain a broader understanding of the issue you will also want to talk to members of groups and organizations that deal with sexual harassment in the workplace, including your college or university. What you will have done up to now is an essential and critical part of your research design; it can actually influence the quality of your research.

Next you will want to narrow down the research by focusing on a specific aspect of sexual harassment. In other words, what exactly do you want to study? Only once you are clear about what you want to study can you decide whether a survey is the appropriate research method. The steps involved in survey research are presented in Box 5.1.

BOX 5.1 HOW TO CONDUCT A SURVEY

Establish the aim and purpose of your social survey.
What do you want to find out?

Determine your population.
Whom do you want to study?

Select your sample from the population.
Whom will you actually study (i.e., of whom will you ask the questions)?

Determine the technique for administering the questionnaire.
How will you administer the questionnaire?

continued

Box 5.1, *continued*

Develop the questionnaire.
What questions will you ask?

Pretest the questionnaire, as necessary.
What are the strengths and weaknesses of the questionnaire?

Revise the questionnaire, if necessary.
What changes need to be made to the questionnaire?

Administer the questionnaire to the selected sample.
Organize, summarize, and analyze the collected data.

THE QUESTIONNAIRE

After you have decided that the survey is the appropriate method for your research, you need to develop its main instrument—the questionnaire. Although the questionnaire is largely associated with the social survey, it is also used in other research methods, such as experiments and field research. Most of the information presented here can be applied to constructing questionnaires for use in other research methods.

Preparing the questionnaire is more than a simple process of writing down a few questions on paper. Serious effort and consideration have to go into the wording of even the simplest question. Suppose you want to ask first-year high-school students, "Do you own a computer?" Although this may seem to be a simple and straightforward question, it is actually complicated. For example, to whom does "you" refer? What if the computer belongs to the whole family? What does "own" mean? What if the computer is leased? What do you mean by "computer"? Does the student understand a computer to be the same thing you do? What if the student identifies an electronic game package as a computer and you do not?

You will have to sort out the implications of each question before you carry out the survey. In preparing the questionnaire, you should take note of who will be your respondents (persons who answer the questionnaire) and how you will administer the questionnaire (more on this later). You will also want to begin thinking about how you will record and organize the data.

The next two sections should assist you in avoiding some common mistakes in preparing the questionnaire. We will first examine some points to consider in creating questions and then explore some techniques in questionnaire construction.

Guidelines for Asking Questions

A good questionnaire meets at least two basic criteria:

1. Respondents can easily understand the questionnaire.
2. The questionnaire measures what it was designed to measure.

The content of most questions is influenced by the focus of your study and by how you will measure certain concepts.

Suppose you want to find out the age and gender of people who use the facilities of a local community centre. You will obviously have questions regarding the age and gender of the respondents and questions about their use of the centre's facilities. However, in designing the questionnaire you will also have many questions related to your hypothesis or research question. As noted in Chapter 2, in the social sciences a theory attempts to make sense of interrelated generalizations about some kind of phenomenon. To test the theory, social scientists develop hypotheses, which are testable statements derived from a theory about relationships they expect to find. The hypothesis is expressed in terms of the expected relationship between one or more independent variables and a dependent variable; for example, "The unemployed are less likely to vote in federal elections than the employed." In this case, the researcher is expecting to find that employment status (independent variable) influences voting behaviour (dependent variable) in a federal election.

In designing your questionnaire, you will have to define the concepts carefully, in a manner that can be properly measured; that is, you need to give the concepts **operational definitions**, going beyond the dictionary definitions of the words and, instead, stating the procedures that measure the concept.

For example, what would be your definitions of "employed" and "unemployed," and how would you measure these categories? Note that you do not necessarily have to use official government definitions of these terms; the definition you use is for the duration of your study, and you should be able to justify using that particular definition. Assume your concepts of employed and unemployed are for various reasons defined on the basis of whether the respondent works for pay or profit, disregarding all possible reasons for unemployment, such as illness, temporary layoff, and so on. You may therefore decide to ask a **dichotomous question** (or binary question), which offers two options (generally yes/no, true/false, or agree/disagree):

Do you presently work at a job or business, for pay or profit?   Yes ❑   No ❑

Respondents would be classified as employed or unemployed on the basis of their answer to your question. (For an example of a more involved definition and classification of unemployment, see Box 5.2.)

BOX 5.2   WHO ARE THE UNEMPLOYED?

Every month the media report the latest unemployment rate. But how is the information collected? How are the unemployed "counted"?

To find out the number of employed and unemployed, Statistics Canada carries out a complicated monthly household survey known as the Labour Force Survey. The number of households sampled varies over the years. In recent years the sample has involved approximately 56 000 households that are considered to be representative of households in Canada. It is especially large since, among other factors, it is designed to represent the population 15 years of age and over across the various regions of Canada. A household remains in the sample for six months, and is then replaced by another.

continued

Thus, the survey has the characteristics of a longitudinal study that uses a multistage sample procedure; it tracks changes in the Canadian population of working age over time and selects representative subsamples from different areas of Canada. Obviously, it is no easy task to select the sample and subsamples. But a representative sample alone is insufficient; the questionnaire is also important.

The questionnaire ensures that the same questions are posed to all participants in the survey in order to collect the necessary information. Among the various issues examined by the Labour Force Survey is unemployment. But in order to be able to "count" the number of unemployed, researchers need to have a clear definition of whom to count as unemployed.

Hence, the Labour Force Survey uses the following operational definition for unemployed:

Unemployed persons are those who, during the reference week:

a) were on temporary layoff with an expectation of recall, and were available for work,

or

b) were without work, had actively looked for work in the past four weeks, and were available for work,

or

c) had a new job to start within four weeks from the reference week, and were available for work.

Therefore, it is essential that the questionnaire contain questions that aim to find out what the respondent did in the reference week (in the case of the Labour Force Survey, the week before the interview was conducted). One key question that aims to find this out is the following:

In the four weeks ending last Saturday, [date of last day of reference week], did [you] do anything to find work?

Numerous other questions are posed, including the main reason for being away from work, whether the individual has looked for another job in the last four weeks, and so on. Another example is whether workers are employed full-time or part-time. The operational definition of part-time employment for Statistics Canada "consists of persons who usually work less than 30 hours per week at their main or only job." Thus, various questions are in the questionnaire specifically to collect information on the number of working hours, such as the following:

Excluding overtime, how many paid hours [do you] work per week?

If you carry out a study on unemployment or part-time work, your operational definitions may differ from that of Statistics Canada. For example, your own study may simply define those who work for pay as employed and those who do not

continued

work for pay as unemployed. This would be a far broader definition than Statistics Canada's, but it would be an acceptable operational definition, since it is restricted to the length of your specific study. The operational definition would then require you to pose certain questions in the questionnaire to get at the needed information.

For a copy of the Labour Force Survey questionnaire, together with a glossary of the concepts and definitions, population and sample, an explanation of how the questionnaire is administered, and more, see the *Guide to the Labour Force Survey* (2011, March) available free of charge on the Internet at **http://www .statcan.gc.ca/pub/71-543-g/71-543-g2011001-eng.htm**

Each question measures a specific variable, such as gender, age, employment, and schooling. Thanks to the questionnaire, the survey method takes into account numerous variables. We can therefore measure the relationship among variables, such as years of schooling and unemployment, age and unemployment, and so on. Consequently, knowing the relationships that are to be examined is very important when determining what questions to include in the questionnaire. For example, if you want to examine the relationship between schooling and unemployment, as well as age and unemployment, you will want to include various questions that measure the age, schooling, and employment status of the respondents.

Once you have decided on the questions, you need to consider the form and wording of the questions. Below are some guidelines.

AIM FOR CLARITY Gear the vocabulary and grammar of the questionnaire to the sampled population. If possible, take into account your respondents' knowledge of English and level of education, and pay particular attention to the meaning of words as most of your respondents would interpret them. Use slang or jargon only if necessary, and avoid language that they may find offensive. As well, attempt to make your questions or statements clear and concise, and avoid long, complicated questions.

USE OPEN-ENDED AND CLOSED-ENDED QUESTIONS APPROPRIATELY The two options in asking questions are **open-ended questions** and **closed-ended questions** (or what some call multiple-choice questions), in addition to the dichotomous (or binary) questions that seek only one of two answers, such as yes/no. Open-ended questions allow respondents to provide an answer in their own words and to state whatever they consider is most important. The following is an example of an open-ended question:

What do you feel is the most important problem facing Canada today?

The same question can be asked in closed-ended (or multiple-choice) format, whereby respondents select an answer from the list provided, such as the following:

What do you feel is the most important problem facing Canada today?

Unemployment	❑	National unity	❑
Deficit	❑	Other (please specify)	❑ _____
Inflation	❑	No opinion	❑

The choice of whether a question should be closed-ended or open-ended is based mainly on which is more practical for your research project. Closed-ended questions are generally preferred because they are easier to administer and to analyze, and the answers of the respondents are easier to compare. Closed-ended questions list choices of possible responses that the researcher expects. The researcher should ensure that the category items are *exhaustive;* they should include all expected responses. Consequently, researchers include in certain questions the item "Other (please specify)" and provide a space for the respondent to insert details, as shown in the above example. The category items should also be *mutually exclusive,* with no overlaps among the possible responses. For example, consider the category items in the following question:

(Faulty)   How many hours per week do you work?

    10 or less   ❑

    10 to 15   ❑

    15 to 20   ❑

    20 to 25   ❑

If you work 15 hours per week, which category would you select? You could choose at least two, and thus the categories are not mutually exclusive. Now suppose you work 28 hours per week. Which category would you select? There is no category item for respondents who work more than 25 hours, so the categories are not exhaustive. The following question does meet the two criteria:

(Improved)   How many hours per week do you work?

    Less than 10   ❑

    From 10 to less than 15   ❑

    From 15 to less than 20   ❑

    From 20 to less than 25   ❑

    25 or more   ❑

One of the main drawbacks of closed-ended questions is that the few fixed categories to choose from are determined by the researcher. Respondents are unable to clarify their responses, and therefore important information about their beliefs or feelings may be lost. For example, a respondent may choose "national unity" as the "most important problem facing Canada today" from among the suggested categories in the earlier example, but the respondent is unable to elaborate on the answer. Moreover, the respondent might believe that the most important problem facing Canada is the lack of political leadership, but does not tell the researcher since the desired answer is not listed.

Open-ended questions allow the respondent to point to factors other than those that the researcher may have considered. For example, the respondent may say that the most important problem facing Canada is the lack of political leadership and explain why. But open-ended questions are time-consuming and difficult to analyze. There may be many degrees of detail in the answers. Suppose your questionnaire on high-school dropouts in a job training program includes the following open-ended question: "What encouraged you to register in the job training program?" Assume you receive the following response: "I couldn't find a job, and anyway my friend is in the program, plus my parents were bugging me to do something with myself." Which factor led the respondent to register in the program? Was it the lack of a job or the influence of the friend or the parents, or all three? In addition, more literate or articulate respondents may feel less intimidated in providing answers to open-ended questions than less literate or articulate respondents.

AVOID AMBIGUITY   Make the questions precise and unambiguous for the respondent. Avoid confusion and vagueness. For example, in the question "Do you go regularly to the library? Yes ❑ No ❑," it is unclear what is meant by "regularly." Another example is this question:

(Faulty)   What was your total income last year?

Less than $15 000   ❑

$15 000 to $29 999   ❑

$30 000 to $44 999   ❑

$45 000 to $59 999   ❑

$60 000 or more   ❑

The responses will be inconsistent, since it is unclear whether "total income" means income before or after taxes. In addition, people may not recall their exact total income, but rather will have an approximate idea. An improved wording of the question would be the following:

(Improved)   What was your approximate total income last year, before taxes?

Less than $15 000   ❑

$15 000 to $29 999   ❑

$30 000 to $44 999   ❑

$45 000 to $59 999   ❑

$60 000 or more   ❑

AVOID DOUBLE-BARRELLED QUESTIONS   A **double-barrelled question** combines two or more questions into one, but the respondent is expected to provide a single answer. Therefore, the answer is ambiguous. For example, what does it mean if the respondent answers yes or no to the following question?

(Faulty)   Does the community centre have a library and a daycare?

Yes   ❑

No   ❑

If the respondent says yes, does the answer mean the community centre has both a library and a daycare, or does it mean the centre has only one of the two services? The answer would be just as confusing if the respondent said no. Therefore, attempt to ask only one question at a time. For example:

(Improved)   Does the community centre have a library?

Yes   ❑

No   ❑

Does the community centre have a daycare?

Yes   ❑

No   ❑

AVOID LEADING OR BIASED QUESTIONS Avoid questions or sequences of questions that may direct respondents to a particular answer and bias the results of your study. **Leading questions** can arise in various ways.

1. The question may encourage the respondent to give a particular answer.
2. The question may emphasize issues in a way that influences the answer.
3. The choices of responses may emphasize one response over another.
4. The sequence of questions may direct the respondent to an answer.

The following examples of faulty questions illustrate the different kinds of leading questions. How would you improve them? (Note that the questions do not form a complete questionnaire. The letters are included for the sake of referring to each one later.)

(Faulty)   A. You don't approve of the latest tax increase by the federal government, do you?

        Yes ❏

        No   ❏

(Faulty)   B. Should the provincial government spend even more money on the mentally ill than the millions it has already spent on helping them?

        Yes ❏

        No   ❏

(Faulty)   C. Women should have the right to terminate an unwanted pregnancy.

        Agree   ❏

        Disagree   ❏

        Strongly disagree ❏

        Very strongly disagree ❏

All of the above questions have serious flaws; they encourage respondents to provide a particular answer. Question A leads the respondents to say they disapprove of the latest tax increase; question B is loaded with information that might lead the respondents to say no; question (or statement) C provides three disagreement categories and only one agreement category, thereby implying that one of the disagreement categories is a more suitable answer. Below are suggested improvements to the questions:

(Improved)   A. Do you approve or disapprove of the latest tax increase by the federal government?

        Approve   ❏

        Disapprove ❏

        Undecided   ❏

(Improved)   B. Is the provincial government providing enough or not enough services for the mentally ill?

        Enough   ❏

        Not enough ❏

        Don't know ❏

(Improved)   C. Women should have the right to terminate an unwanted pregnancy.

        Strongly agree   ❏

        Agree   ❏

        Disagree   ❏

        Strongly disagree ❏

Another bias to avoid is that of sequencing questions whereby the information in a question can influence the answer to a subsequent question. Consider the two dichotomous questions below and what is wrong with the sequence of the questions.

(Faulty sequence) A. Did you know that the university will have to increase student fees if it sets up a new computer lab?

  Yes ❏

  No ❏

B. Should the university set up a new computer lab for social science students?

  Yes ❏

  No ❏

The problem with the sequence of the two questions above is that the first question highlights a student fee increase if a new computer lab is set up, and then asks whether such a lab is necessary. Knowing that there will be an increase in student fees may influence a respondent's answer about the need for a new computer lab. Consequently, we will not know how many actually support this need.

If the sequence of the two above questions is reversed, respondents can first indicate whether they believe there is a need for a new computer lab, and afterward whether they are aware that student fees would be increased to cover the costs. You could also phrase these questions differently, of course. The wording and sequence of the questions will depend on what you want to find out. For example, as the sequencing of the two dichotomous questions below illustrates, you may want to find out how many think there is a need for a new computer lab, and of those who believe there is a need, how many would agree to an increase in student fees to cover the cost of a new computer lab. (In a later section we provide some general guidelines in constructing the questionnaire.)

Should the university set up a new computer lab for social science students?

Yes ❏

No ❏ (Skip the next question.)

If you answered yes to the above question, should the university set up a new computer lab for social science students if student fees must be increased to cover the cost?

Yes ❏

No ❏

ASK QUESTIONS THAT RESPONDENTS ARE COMPETENT TO ANSWER Avoid asking respondents questions that they are unable to answer reliably. In asking respondents about specific details from their past, consider whether they are likely to be able to answer such questions properly. Likewise, avoid asking questions that expect respondents to provide an answer to a question on a subject about which they probably know little or nothing. Consider the following question in a study on child development:

(Faulty) At what age did you learn to count from one to five? _____

Think about whether you are able to answer the question. It is doubtful that any of us can recall when we first learned to count, and therefore an answer to such a question is meaningless. In other cases you may have to use a shorter time frame to achieve more accurate responses. For example, suppose you were doing a study on leisure activities and asked the question:

(Faulty)   How much did you spend on magazines in the past year? _____

Again, consider whether you are able to answer the question. One possible improvement to the question is to shorten the time period of recall, as in the following question:

(Improved)   Approximately how much do you spend on magazines in a month? _____

From the answer given, you can then roughly estimate how much the respondent spends on magazines in a year by multiplying the given answer by 12.

ASK QUESTIONS THAT ARE RELEVANT TO MOST RESPONDENTS   Avoid asking questions that are irrelevant to the respondents. Note that the respondents may still give you answers to the questions, but their answers would be useless, as illustrated by the following question:

(Faulty)   Should the federal government restrict the number of international banks in Canada?

Yes   ❑

No   ❑

Don't know   ❑

How would you answer this question? It is doubtful that most of us have given this issue any consideration, unless it has been widely debated in the media. Consequently, it would be difficult to arrive at any convincing conclusion regarding the opinions of the respondents.

Note that although respondents will admit that they are ill informed about an issue, they will nevertheless answer questions about it. For example, a few years ago a telephone survey of a representative sample of adult Canadians found that the majority were only a little informed or not at all informed about the details of the proposed Goods and Services Tax (GST). Yet the majority of respondents said they were against the new tax.

Finding out what the public thinks about an issue may in itself be important, however, regardless of how much they know about the issue. For example, the public may have little understanding of the wide implications of a particular policy, but their opinions may still matter for the government in determining whether there is support for the policy.

AVOID NEGATIVE ITEMS   Negations in a question are grammatically incorrect and are open to misinterpretation. For example, how would you answer the following question?

(Faulty)   Do you agree or disagree that Quebec should not separate from Canada?

Agree   ❑

Disagree   ❑

Interpret your answer. Would others understand the question the same way you did? Some respondents might disregard the word "not" in the question, and answer that they agree, although they are opposed to Quebec separating. And others may disagree even though they favour Quebec separating. To achieve clearer answers, drop the "not" in the question:

(Improved)   Do you agree or disagree that Quebec should separate from Canada?

Agree   ❑

Disagree   ❑

Guidelines for Questionnaire Construction

Let us now consider some general guidelines in constructing the questionnaire. Some suggestions will be quite obvious, but it is well worth taking them into account, since errors are easily committed. In preparing the questions, emphasis is placed on the wording. In constructing the questionnaire, emphasis is placed on the sequence of the questions and the design of the questionnaire.

Respondents are more likely to take a well-designed and properly formatted questionnaire seriously than one that is confusing or that looks cluttered.

INTRODUCTORY REMARKS   If you were asked to take part in a survey, what questions would you ask the researchers before agreeing to participate? You would probably want to know what the survey was about, whether your answers would remain confidential, and how long it would take to complete the questionnaire. The introduction to your questionnaire should answer these questions.

An introduction aims to briefly inform respondents about the purpose of the survey, who is carrying out or sponsoring the survey, the confidentiality of the responses, and the approximate time it will take to complete the questionnaire. You may thank the respondents in the introduction or after they have completed the questionnaire, or both. Depending on the type of survey, you may want to include the introduction on the cover page of a questionnaire booklet, in a covering letter, or at the top of the first page of your questionnaire. For example, you may include the following:

> This questionnaire is part of a study on the leisure activities of college students. It is conducted by a group of students in the Recreational Training Program. It will take approximately 15 minutes to complete the questionnaire. Your participation is voluntary, and your answers will be confidential. We thank you for your cooperation in this research.

If your questionnaire is divided into various subsections, you may want to provide a brief introductory statement to each one. Suppose your questionnaire on the leisure activities of college students contains a subsection on television viewing. You may want to briefly introduce the subsection with the following statement: "In this section we would like to focus on the amount of time you spend watching television and the types of programs you enjoy."

INSTRUCTIONS   Following the introduction, you should include basic instructions for completing the questionnaire, if necessary. In self-administered questionnaires, for example, you need to tell respondents what to do. Should they place a check mark or an X in the box that indicates their answer? Do you want them to be brief in their answers to the open-ended questions and to write only in the amount of space provided?

However, some questions may require specific instructions. If you want respondents to select only one answer from a list of categories, you will have to make that clear. You may note this in the question itself by starting with the statement "From the list below select...." Or you may write the question and then direct the respondent, for example, "Which type of television program do you like? (Please check the one best category.)"

QUESTIONNAIRE FORMAT The presentation and the order of the questions should be clear. At first your tendency might be to place as many questions as possible on each page to make the questionnaire seem shorter. You should avoid that tendency and, instead, spread out the questions, disregarding the number of pages. Respondents and interviewers (persons trained by the researcher to do interviews) will be less likely to become confused and skip questions. As for the fear that respondents might think the questionnaire is too long, remember that it is the time it takes to complete a questionnaire that usually concerns the respondent, a piece of information that you will provide in the introduction.

QUESTION FORMAT   There are various ways of presenting the choices for question responses. One of the most popular for closed-ended questions is to include the responses, adequately spaced apart, in a column below the question. For the respondent it is easy to provide an answer, and for the researcher it is easy to code. (Coding involves assigning usually numerical codes to each response to a question for purposes of analyzing the data, generally by computer. See Chapter 9.) Beside each response is a box, a circle, brackets or parentheses, or a designated space in which respondents can place a check mark or an X to indicate the answer. For example, the respondent is instructed to place a check mark in the boxes next to the selected response:

Are you a full-time day student?

Yes   ❑

No   ❑

Another approach is to place numerical coding values beside each response and ask the respondent to circle the number of the appropriate answer. The advantage to this method is that the code numbers are already specified when the answers have to be processed. For example, the respondent may be asked to circle the number of the selected response in the following:

Are you a full-time day student?

1. Yes

2. No

Whatever approach you use, it is essential to make it as easy as possible for the respondents, or interviewers, to indicate the response, and for you to determine which response was selected.

CONTINGENCY (SKIP TO) QUESTIONS   Some of your questions may be relevant only for certain respondents, depending on their answers to earlier questions. You will have to work out instructions for the respondents or interviewers to allow them to bypass the irrelevant questions. For example, in a study on health you would not want to ask nonsmokers how many cigarettes they smoke in a day. The question format must clearly direct respondents to the next question that is relevant to them. This question is known as a **contingency question**; it is appropriate only for respondents who provided a particular answer to an earlier question. The proper use of contingency questions will eliminate confusion and make it easier for respondents to complete the questionnaire. One practical format is to guide the respondent visually by means of an arrow to the contingency question placed off to the side. You might present the smoking contingency question mentioned earlier in the following way:

10. Do you smoke cigarettes?

Yes ❏ - ┐

No   ❏   │
                                                                        ▼

> 10. a. Approximately how many cigarettes per
>   day do you smoke?
>
> Under 5   ❏
> 5 to 10   ❏
> 11 to 15   ❏
> 16 to 20   ❏
> Over 20   ❏

You may want to ask more than one contingency question. For example, you also might want to ask approximately how much the smokers spend on cigarettes in a week. This question might follow the earlier contingency question. Another approach when there is a set of contingency questions is to place instructions after an answer stating which question the respondent is expected to answer next, for example:

10. Do you smoke cigarettes?

Yes ❏ (Please answer questions 11 to 14.)

No   ❏ (Please skip questions 11 to 14. Go directly to question 15 on page 3.)

Whether you use these or other formats for contingency questions, the aim is to clearly direct the respondent to the relevant questions.

MATRIX QUESTIONS   Suppose you want to ask a few questions with similar sets of answer categories, as is common when measuring the respondents' strength of opinions and attitudes. You can save space and make it easier and quicker for respondents and interviewers to note the answers by constructing a **matrix question**, with the questions or statements in rows and the response categories in columns. In a study on sexist attitudes you might include the following matrix question format:

For each family-care activity, please indicate whether you Strongly Agree (SA), Agree (A), Disagree (D), Strongly Disagree (SD), or are Undecided (U).

	SA	A	D	SD	U
Husbands should participate in					
a. grocery shopping	❏	❏	❏	❏	❏
b. housecleaning	❏	❏	❏	❏	❏
c. child care	❏	❏	❏	❏	❏
d. meal preparation	❏	❏	❏	❏	❏

A principal concern with this format is a problem called **response set**: respondents develop a pattern in their answers, tending to check the same column for every question or statement. They may, for example, check "Agree" for every item. The problem can be partly minimized by alternating the orientation of some statements so that consistency in the answers requires respondents to agree with some and disagree with other items. You might, for instance, change the above example to read:

For each family-care activity, please indicate whether you Strongly Agree (SA), Agree (A), Disagree (D), Strongly Disagree (SD), or are Undecided (U).

	SA	A	D	SD	U
a. Husbands should participate in grocery shopping	❑	❑	❑	❑	❑
b. Husbands should participate in housecleaning	❑	❑	❑	❑	❑
c. Husbands should participate in child care	❑	❑	❑	❑	❑
d. Only wives should prepare meals	❑	❑	❑	❑	❑

Thus, respondents who believe that husbands should participate in all family-care activities are unlikely to answer "Agree" to the last statement.

ORDERING OF QUESTIONS   Once the questions or statements are completed, you will have to determine the order in which they will appear in the questionnaire. The challenge you face is to order the questions in a manner that encourages respondents to complete the questionnaire. In addition, you will have to consider whether the order of the questions influences the respondents' answers.

Opening questions are generally easy and nonthreatening and should make respondents feel comfortable about answering the questionnaire. These include questions on the gender, marital status, and type of work activity of the respondent. Sensitive questions, or questions that respondents may be reluctant to answer, such as those that deal with income, are usually placed near the end of the questionnaire. So are open-ended questions, since they take time to complete. The middle questions are normally grouped around common topics; if necessary, sections are created for each topic and are introduced with a brief statement.

PRETESTING THE QUESTIONNAIRE   Once you have a draft of the questionnaire, the next step is to do a **pretest** to detect any problems with it (see Box 5.3). The overall layout and format of the questions in the pretest questionnaire should be identical to those of the actual questionnaire. Whatever approach you use to pretest the questionnaire, the purpose is to determine its weaknesses and strengths in order to improve the questionnaire for the actual study. You need also to note the time it takes for respondents to complete it.

At first you might ask a few friends or fellow students to answer the questions and tell you of any difficulties they noted. Did they have problems with the wording,

---

BOX 5.3   REVIEWING YOUR QUESTIONNAIRE BEFORE A PRETEST

Questionnaires have to be clear and user-friendly whether they are self-administered or interviewer-administered. Clarity of layout as well as content is essential in obtaining good responses and quality data. Here are the important elements to check for *before* you undertake the pretest of your draft questionnaire.

- *Basic readability.* Do the typeface and font size make for ease of reading? Avoid too-small or too-large font size, too-plain or too-fancy typeface, or too great a variety of fonts and typefaces.

continued

- *Clarity of question sequence.* Number questions plainly and simply so that the sequence is clear. Limit the use of subsections of a question—for example, "2a part ii." Avoid using subsections to disguise the length of the questionnaire.
- *Use spacing to reinforce question sequence.* By leaving more space between the questions than between parts of a question, the respondent (or the administering interviewer) sees clearly where each question begins and ends.
- *Use subheadings.* Most questionnaires move through different topics or phases. Each section is best indicated by subheadings in the same font and typeface as the questions but in bold format. Do not capitalize subheadings; this distracts and slows down the reader.
- *Closed-ended questions are clearer with their response options listed vertically rather than horizontally.* This takes up more space, but it helps the respondent or interviewer see all the possible answers more clearly. Matrix questions are an exception, however (see the discussion under "Matrix Questions" above).
- *Closed-ended question response options should be presented as boxes rather than as blank spaces or lines.* Blank spaces invite unserious responses, lines invite sloppy and misaligned check marks that may be unclear. Both of these create more work when coding the survey responses.
- *Open-ended questions need plenty of space to indicate that you want more than a single word or cryptic phrase as an answer.* More space will invite a longer answer.
- *Instructions to the respondent or interviewer should be clearly identified.* They should be clearly separated spatially from the preceding and following questions. They may be italicized and should be bolded (but not capitalized) for further emphasis.
- *Pre-pretest your draft questionnaire on a colleague or friend.* Get them to respond to the entire questionnaire, including the introductory statement, as if they were a respondent. It might save you time and embarrassment. The pretest then becomes even more valuable as a way of checking for more subtle problems.

the instructions, the sequence of the questions, the design of the questionnaire, whether closed-ended questions are mutually exclusive and exhaustive, and so forth? On the basis of the information gathered, you may want to make revisions to the questionnaire.

Next you will want to do a formal pretest on individuals who are part of the target population of your study but not your sample. Suppose your study is on drug and alcohol consumption among first-year health science students. You might ask a few first-year health science students who will not be in your actual study to complete the questionnaire. You should carry out the pretest as if it were the actual study, and not tell respondents that it is a pretest questionnaire until it is completed. Again, you need to note the time it takes to complete the questions and ask the respondents if they had any problems. You should carefully examine the pretest questionnaires for possible patterns in the responses. If respondents provided similar answers to certain questions, you should consider whether this was due to the format or to the wording of the questions.

Following the pretest, you can make changes you deem necessary. You may want to ask friends or fellow students to examine the questionnaire for blatant errors, such as spelling mistakes, missed questions, improper numbering, and so on. Once the questionnaire is used in the actual study, it will be too late to make corrections.

ADMINISTERING THE QUESTIONNAIRE

There are two ways of administering questionnaires: respondents can fill in the questionnaire and give it back to the researcher, or the researcher can read out the questions to respondents and record their responses. How questions are phrased and which are included in a questionnaire will be affected by the method of administering the questionnaire. Obviously, we would not ask respondents to tell us their gender in a face-to-face interview. As well, we would feel uncomfortable asking (and the respondent would probably avoid answering) certain sensitive questions, such as ones regarding their sexual behaviour.

The method used to administer the questionnaire can also affect which population we select to study. For example, suppose we want to do face-to-face interviews with a sample of university graduates. We may quickly realize that many are spread out in various regions of the country and that face-to-face interviews would not be feasible. If we still intend to do face-to-face interviews, we may have to restrict the population to graduates who reside in a specific geographical area, or we can consider the original population, but mail the questionnaire to the selected sample.

Before we explore the main methods of administering the questionnaire, we must consider why you would or would not agree to participate in someone else's survey. "It's (not) too much trouble?" "It will (not) take too much time?" "The interviewer is (dis)courteous or seems (in)competent?" Whatever the reasons, it is clear that if you agreed to complete the questionnaire you made a special effort. In agreeing to complete a questionnaire, respondents are doing the researcher a favour, and not vice versa; they are giving up a portion of their precious time. Therefore, it goes without saying that you should treat all individuals in the selected sample with respect, whether they accept or refuse to complete the questionnaire.

There are two main categories of surveys: self-administered questionnaire surveys (administered to individuals or groups and directly filled in by the respondents) and interview surveys (administered face to face or by telephone and filled in by the researcher). Which approach you choose will depend on many factors, not the least of which are the objective of the research and the time and resources available. For example, if your questionnaire contains many sensitive questions regarding sexual behaviour, you will want to consider how best to ensure anonymity and perhaps avoid face-to-face interviews. If you need to reach out to individuals in a wide geographical area, it may be more convenient to do a telephone survey, a mail survey, or an online survey.

Self-Administered Questionnaires

There are three main types of **self-administered questionnaires**.

1. Individuals complete the questionnaire and hand it back to the researcher.
2. The researcher administers the questionnaire in a group setting.
3. The questionnaire is mailed to individuals and returned when completed.

In general, the **response rate**—the percentage of those in the sample who complete the questionnaire—is higher when the researcher completes the questionnaire directly with the respondents. The completion rate is lower when respondents have to mail in the questionnaire, return it to researchers at a later date, or drop it into a box left in a certain area.

INDIVIDUALLY AND GROUP-ADMINISTERED QUESTIONNAIRES Because of lack of time and resources and the type of sampling procedure selected, you may decide to do an individually or group-administered questionnaire. In both cases the completion rates are potentially high. Whichever approach you use, however, it is essential to be courteous and to avoid comments that may influence responses.

In an **individually administered questionnaire** the respondent usually completes the questionnaire with the researcher present. One advantage is that respondents can ask the researcher for clarification, which can ensure a higher rate of completed questionnaires. A disadvantage is that this method is time-consuming, as the researcher has to monitor one respondent at a time. It can also be costly in terms of travel and time, and it can be difficult to arrange schedules convenient for the researcher and respondents. If the researcher cannot be flexible, some respondents may be unable to participate in the survey.

Another possibility, if it meets the objective of the research, is to administer the questionnaires to a group, say a class. An obvious advantage of a **group-administered questionnaire** is that it saves time and money. But there can also be disadvantages: someone may make an amusing comment that suggests to others that the research is trivial, respondents may talk to each other about the questions, and some may participate only to be part of the group and therefore provide incomplete answers. Thus, a challenge for the researcher administering a group questionnaire is to simultaneously make the participants comfortable and to create a mood that signals that the research is serious.

You should avoid having others administer the questionnaires unless you have trained them in what to do. It is best not to ask professors, parents, friends, or acquaintances to administer questionnaires for you in their classes, workplaces, sports clubs, or other places. If possible, it is also wise to avoid asking people in authority to request subordinates to complete the questionnaires. Many problems may arise when someone else administers the questionnaire: your friend "forgot" or "will do it next week," respondents lied because the supervisor might see their answers, respondents were allowed to take the questionnaires home and never returned them, and so forth.

If you need permission to carry out a survey in a particular place, including a class, you should obtain it as early as possible. Many unforeseen obstacles may arise that can delay the survey. A school may require a proposal for a committee of parents, who then will get back to you; at a business organization only the owner, who is on a two-week holiday, can give the permission; your favourite professor, who would certainly approve, has exams planned for the very week you want to survey the class; and so forth.

MAIL SURVEY A method of administering the questionnaire is through the **mail survey**. In this approach, questionnaires are mailed to the selected sample with a covering letter explaining the purpose of the research and a self-addressed, stamped envelope for returning the questionnaire. If it is necessary to maintain anonymity, identification codes are placed on the questionnaire and matched to the list of names in the sample. The covering letter usually explains that the identification code serves only to enable follow-up requests to those who do not return the questionnaire. The identification code is destroyed once the questionnaire is returned, and it is then impossible to identify the respondent.

A major advantage of a mail survey is that you can reach out to a large sample spread across a wide geographical area. A mail survey is relatively inexpensive when compared with handing out individual questionnaires to a wide area. But you should not underestimate the expense of a mail survey, such as the cost of stamps for first-class mail and envelopes of different sizes—one for sending the questionnaire and the other for returning it. Moreover, you need to consider this question: If you received a questionnaire in the mail, would you complete it? Likely most of us would not. The major weakness with the mail survey is its low response rate.

To improve the response rate, you can follow up the initial mailing with requests to complete the questionnaire. The week following the initial mailing, you might send postcards thanking respondents for participating and reminding them to return the completed questionnaire if they have not yet done so. About three weeks after the initial mailing, you can do a second mailing to respondents who have not returned the questionnaire, again sending them the questionnaire, a covering letter, and a self-addressed stamped envelope for returning the questionnaire. You can, of course, continue to do more follow-up requests, including phone calls, to increase the response rate. Researchers typically do one to three follow-ups.

Internet-Based Surveys

With the escalating popularity of the Internet has come an increase in web-based surveys and e-mail surveys, or what some identify as online surveys. Each of these types of surveys has certain advantages over other types, mainly that they are quick to administer. But they also pose various difficulties, such as sampling limitations, and each offers its own challenges to the researcher.

E-MAIL SURVEYS   The increasing use of e-mail as a means of communicating has also led to a rise in using it to administer questionnaires. E-mail surveys can save time and cost and may be easy to administer. Once it has been set up, the e-mail questionnaire can be distributed to a large number of potential respondents across large distances. The questionnaire can be either part of, or attached to, the e-mail message. Participants can e-mail back the completed questionnaire or complete a printed copy to mail back by surface mail. It is generally assumed that respondents may also be more willing to answer sensitive questions, such as ones about their health, drug use, or sexual behaviour, since they are not responding with an interviewer present.

E-mail surveys face various obstacles, however. The target population is restricted to people who can be reached by e-mail and for whom you must have correct e-mail addresses. Further, potential respondents may have different e-mail and word-processing software; may have different levels of expertise with e-mail usage and computers; may be concerned with privacy over the Internet; and may think your e-mail is "junk mail" or have security software that classifies it as spam. Any of these, and more, can influence their participation. Moreover, the person who receives the questionnaire may not necessarily be the one who completes it. But perhaps the most important drawback is that many people refuse to respond to unsolicited e-mail and may not even read it. This in turn can raise further concerns about who the respondents are and how reflective the sample is of the target population. Consequently, researchers face various challenges in using e-mails as a mode to administer questionnaires.

In general, given the various constraints, researchers have largely limited their use of e-mail surveys to specific target populations, such as a student population or members of an organization for whom they have the e-mail addresses. For example, a team of researchers wanted to find out whether physicians would allow children and adolescents with single, normal kidneys to participate in contact/collision sports. They chose to do their study with members of the American Society of Paediatric Nephrology and obtained all the necessary approvals, including from the ethics committee of their university. They sent each member an e-mail questionnaire, of which nearly 31 percent responded. The e-mail survey had the advantage of reaching potential respondents over a wide geographic area. Moreover, it focused on a clearly identified population that could be contacted by e-mail. A main result of the study was that the majority of respondents would not allow contact/collision sports participation by patients with single, normal kidneys (Grinsell, Showalter, Gordo, & Norwood, 2006).

Briefly, if you need to carry out an e-mail survey, you should be sure the target population has access to e-mail; that you have, to the best of your knowledge, their correct e-mail addresses; and that the questionnaire is in an electronic format that all of your potential respondents can easily open and handle. Further, we suggest that you take into account some of the steps used in administering a mail survey and apply them to the e-mail survey. For example, as with mail surveys, one key concern should be with improving response rates. Therefore, first send an e-mail that acts as a covering letter informing them of the survey and inviting them to participate and, either then or later, depending on the study, provide them with the questionnaire. Perhaps you could think of a creative way of offering an incentive for their participation in the survey. A short time after e-mailing them the questionnaire, you might want to follow up with an e-mail kindly asking them to return the completed questionnaire, if they have not already done so. You should again include your questionnaire with the e-mail message in case they discarded your earlier e-mail. You must of course be concerned with the tone of your reminders, and be reasonable in how often you send them. Otherwise, some may attach little importance to the questionnaire and respond only to stop you from sending more e-mail reminders.

WEB-BASED SURVEYS The popularity of the World Wide Web has given rise to web-based surveys, mainly by businesses, but also increasingly by academic researchers. In the case of a web-based survey, people are informed about the survey, invited to complete a questionnaire, or asked if they are interested in completing the questionnaire by linking to another website. Web-based surveys have various advantages, not the least of which is that after they have been set up they are relatively fast to administer and data-compile. Moreover, if necessary, there can be graphics, videos, and audios that can be part of the questionnaire. Web-based questionnaires may be presumably easier and quicker to complete than those on paper, since respondents are promptly directed to appropriate questions, such as contingency questions.

Web-based surveys, however, have various drawbacks. An obvious one is that people have different levels of experience and comfort in using the Internet. Researchers will have to consider how they will reach their target population and how the sample will be selected. Further, people may be reluctant to complete them, not the least because of concerns with privacy. Respondents may have different technologies for accessing and communicating over the Internet and use different browsers, which, together with other factors, may influence their participation. They are also more likely to stop completing a web questionnaire than one administered by an interviewer. Consequently, researchers need to take into account various issues, including measures to ensure that only the selected sample can have access and complete the questionnaire.

Researchers also face the challenge of having the necessary skills or knowledge with the needed technology and software to administer web-based surveys. They need to be concerned with the presentations of the web page and questionnaire, such as the layout. For example, is the graphic design of the web page and questionnaire too complex? Should there be pull-down instructions? Should respondents see one question or several at a time? Will the questionnaire appear the same in different browsers? Clearly, all of these may influence the response rate. There is the added concern with privacy, particularly since it is relatively easy to collect information without the respondent necessarily knowing. This also raises ethical concerns. For example, is it right for researchers to take note of the respondent's Internet protocol address, the time and date the questionnaire was completed, and how long it took to complete, without telling respondents they will collect such information?

If you need to carry out a web-based survey, it is probably best for now to restrict it to a target population that can easily access the specific web page with a password. An example would be administering the survey at a college or university, or an organization such as a place of employment. As with the e-mail survey, we suggest you take into account some of the steps used in administering a mail survey. Consider various means of improving response rates, such as contacting potential respondents by surface mail, e-mail, or telephone to invite them to participate in the survey and inform them how to access the web-based survey. After a brief period, contact them again to gently remind them to complete the questionnaire and thank them for their participation. You should be reasonable in how often you remind them or they may carelessly complete the questionnaire to stop you from sending them more reminders. These various steps and more, for example, were taken by researchers who carried out a web-based survey to find out about the influence parents may have on college drinking.

The researchers carried out a web-based survey among first-year university students between 18 and 19 years old living on campus. They received approval for the research by the appropriate university committee and selected a random sample of 477 students out of a list of 1933 eligible students. Each was sent a letter through campus mail explaining the purpose of the study, a copy of a consent form, and, as an incentive, a free pen with the university's name on it. They later received an e-mail with a web link to the consent form and the web-based survey that would take about 15 minutes to complete. Students were contacted by phone and e-mail up to three more times to request their participation. When students linked to the indicated site, they had to first read and agree to the electronic informed-consent notice and then were connected to the survey. As a further incentive, students who completed the survey were eligible to win a $100 gift card at a bookstore. The data were collected over four weeks and the responses were then downloaded into statistical software for analysis. Of the selected sample, 265 students participated. Briefly, the researchers found that over two-thirds responded as having experienced at least one drinking problem; more than one-third believed their parents would approve of their drinking occasionally; and those who perceived more parental approval were more likely to report at least one drinking problem (Boyle & Boekeloo, 2006).

Web-based and e-mail surveys require researchers to consider many of the same issues as when using other survey techniques. But, while new technology such as the Internet offers researchers different possibilities in administering questionnaires, it also presents added challenges in meeting the requirements of a scientific inquiry as opposed to a casual inquiry. To better assess and improve the use of web-based and e-mail surveys, researchers have been carrying out methodological assessments on response rates, sampling, privacy, and other related issues. Clearly, with the increasing popularity of the World Wide Web and e-mail, their use as modes of questionnaire delivery is expected to grow.

Interview Surveys

FACE-TO-FACE INTERVIEWS   Another approach to administering a questionnaire is the **face-to-face interview**, wherein researchers (or interviewers) read out the questions and record the answers. One definite advantage is the high response rate. In addition, the interviewer can make certain that all the questions are answered. Interviewers can also clarify questionnaire items that are confusing to the respondent. If necessary, interviews can be carried out in different languages. (However, special attention must be given to the translation.) Interviews can be carried out with respondents who are illiterate or who feel uneasy about writing their answers, especially to open-ended questions. Interviewers can also take into account difficulties that were not foreseen, and gain insights that may help in interpreting the data. But there are also various disadvantages, mainly the time and cost of doing the interviews and the unintentional influence the interviewer might have on the respondents, through appearance and tone of voice.

The manner in which interviews are carried out largely depends on the sample selected. When designing the questionnaire, you properly took into account the sample in phrasing the questions. It will now be necessary to take into account how you communicate verbally and visually with the respondents. It is important to consider the many factors that might influence how respondents feel about the research and the interview. Your clothes and grooming, for example, may suggest to the respondent how serious you are about the interview. They may also influence how comfortable the respondent will feel in being interviewed by you.

What would be your reaction if you were interviewed by someone who kept stumbling over words and phrases and who seemed unfamiliar with the questionnaire? Not only would the interview take longer than necessary, but you would probably become irritated. Thus, for the respondent to feel comfortable and take the interview seriously, the interviewer must be familiar with the questionnaire and read the question items without errors.

TELEPHONE INTERVIEWS   The telephone has become an increasingly popular means of carrying out survey interviews. Many advantages account for the popularity of **telephone surveys**, especially the money and time saved compared with face-to-face interviews. You can easily reach respondents by telephone across wide areas and do many more interviews in an afternoon. (And you can dress as you please.) Among the various disadvantages of this approach is the obvious one that respondents without telephones cannot be reached. In addition, some of the phone numbers you have to work with might be incorrect. Another drawback is that the anonymity of respondents is reduced, which in turn can affect their answers. Respondents might answer hastily because they were interrupted in the middle of another activity. They might hang up with only part of the questionnaire completed. In addition, many people do not like to answer questions over the phone.

Recent technological developments have created new challenges for survey researchers who want to use telephone interviews. The proliferation of such technologies as answering machines and telephones with call display allow people to screen calls and may make it difficult for the researcher to contact the respondents. The growing use of cellphones has created sampling difficulties, because there are no directories of cellphone exchanges. Consequently, an increasing number of people who rely exclusively on cellphones may not be included in a sampling frame. This may be a serious weakness in carrying out certain studies, such a national opinion survey on the political views of the voting-age population. One way to circumvent this problem is to select the sample from a list of randomly created phone numbers, say across the different area codes. This technique, known as random-digit dialling (RDD), would generate both landline and cellphone numbers. (However, other issues may arise in using RDD that are beyond the scope of this book.)

While certain technology has offered obstacles for researchers, other technology has simplified and contributed to the popularity of the telephone survey. For instance, because of advancements in computer technology and the falling costs of the technology, interviews are usually carried out using **computer-assisted telephone interviewing (CATI)**. In general, CATI involves interviewers with telephone headsets sitting in front of a computer. After the introduction, the interviewer reads the first question that appears on the screen. The interviewer types in the respondent's answer, often by pressing a single key, and the response is automatically stored in a central computer. The interviewer then asks the next question, and the process continues until the end of the questionnaire. This technique cuts down on the time needed for the interview, reduces the possibility of interview errors, and speeds up the data analysis.

In face-to-face and telephone interviews, the interviewer has to record the answers as given, especially answers to open-ended questions. However, for some questions it may be necessary to **probe**, or elicit or encourage a response, when the answer is inappropriate or vague. For example, a respondent might say "maybe" to a closed-ended question with the choices "strongly agree," "somewhat agree," "somewhat disagree," or "strongly disagree." The interviewer can then probe the respondent to give an appropriate answer by asking, for instance, "Do you mean you agree somewhat or disagree somewhat?"

Probing is especially useful in eliciting clearer responses to open-ended questions. There are various ways to do this. For example, a respondent might say the economy is the most important issue facing Canada. The interviewer may quietly wait as if expecting more information. If the respondent does not elaborate, the interviewer asks, "In what way?" or "Can you tell me more about it?" Thus, probes are neutral requests intended to complete, clarify, or elaborate on an answer. Of course, interviewers must be careful not to influence the nature of the answer.

BOX 5.4 TECHNIQUES OF SURVEY ADMINISTRATION: MAIN ADVANTAGES AND DISADVANTAGES

Technique	Description	Main Advantages	Main Disadvantages
Self-Administered Questionnaire			
a. Individual	Respondent reads and completes questionnaire.	• High response rate • Respondent can ask for clarification	• Time-consuming to monitor • Possibly costly if travelling is involved
b. Group	Questionnaire is administered to a group.	• High response rate • Saves time and money by reaching out to a large number of respondents	• Respondents may talk to each other and influence responses • Individual may feel compelled to participate
c. Mail	Questionnaire is mailed to a selected sample.	• Can reach out to respondents spread out across wide geographical area • May be able to maintain anonymity	• Low response rate • May be costly and time-consuming
d. Internet-based surveys	Questionnaire is administered over the Internet.	• Quick to administer and relatively inexpensive • Easily accessible to a large population and across wider distances	• Target population must have access to the Internet or an intranet • Sampling problem
Interview Questionnaire			
a. Face to face	Researcher reads out questionnaire to respondent.	• High response rate • Interviewer can clarify questionnaire item, as well as reach out to respondents less willing to participate if they had to read the questions and write the answers	• Costly and time-consuming • Characteristics of the interviewer may unintentionally influence the respondent's answers
b. Telephone	Interview is carried out over the telephone.	• Saves time and money • Can reach out to a large number of respondents across a wide geographical area	• Sampling problems; unable to reach persons without telephones and maybe those with cellphones • Respondents can easily discontinue the interview

If several interviewers are involved in the research, as in a group project, all have to be clear about how to probe. One way of ensuring that all use similar probes is to include an example of a probe next to questions in which probing may be necessary.

QUESTIONS TO ASK ABOUT A SURVEY

Virtually every day, the media provide the results of public opinion polls: surveys on opinions of public interest. Subjects can range from immigration policy to favourite sitcom actors to which party would win if an election was held. But before you accept the results of such surveys, it is well worth considering various questions. (See also the Weblinks at the end of the chapter.)

Who Sponsored the Survey (Poll)?

Organizations sponsor surveys to gain certain information. What was the motive for carrying out the survey? For example, the government might want to find out about the public perception of the educational system; a newspaper might want to collect more information about a news story; a special interest group might want to know if there is support for banning a certain product; or a business firm might want to know if it should promote a new product.

The media usually inform us if they sponsored the survey and when it was carried out. However, they may provide little information when reporting results of surveys sponsored by others. In such cases you will have to use your judgment. Are we told who actually carried out the survey? The organization that sponsors the survey is usually not the same organization that conducts it. Besides Statistics Canada, numerous private agencies in Canada specialize in conducting surveys.

Knowing who sponsored the survey can help you to judge how much confidence you should place in the results. If during an election campaign the survey was sponsored by a political party, you might hesitate to accept the results. The political party may have sponsored it to gain information for a political campaign. But regardless of who sponsored the survey, the validity of a survey depends mainly on the sampling technique and on the quality of the questions.

How Was the Sample Selected?

A popular assumption is that the greater the number of people interviewed, the more accurate the survey. But this is not necessarily so; accuracy depends on various factors, including the sampling method (see Chapter 4). It is essential to know how the people interviewed were selected. A study that relies on people simply deciding to participate in a survey can have many biases. To warrant generalizations about the opinions of a larger group, the sample should be selected using appropriate sampling techniques. Was the study based on a random sample or a nonrandom sample?

When and How Was the Study Carried Out?

A particular event that occurred around the time the public opinion survey was carried out may influence the results. If an oil tanker spills millions of gallons of oil off the coast of Canada, then the event is likely to influence the results of a survey on environmental pollution.

In addition, how were the respondents interviewed? What was the technique used to administer the questionnaire? What was the response rate? Is any explanation given of why some may have refused to participate? Dial-in polls or visiting a website, both of which techniques allow everyone to express their opinion, are increasingly popular, especially among the media, but these have no validity, no matter how many people express their views. The respondents chose to phone or access the site, and some may have even responded more than once. It is best to disregard the results of such surveys, as well as the results of surveys in magazines based on questionnaires that readers send in.

What Was Asked?

Unfortunately, when the media report the results of a survey other than one they sponsored, they provide little information on the wording of the questions. But the questions are key in any survey; there may be unintentional—or intentional—biases in the questionnaire. If you do not know the questions, you can use other criteria to judge the survey. Are the results reported in a respectable publication? Was the survey carried out by a reliable organization? If all we are given is general results because they have entertainment value, it is best to take the results lightly.

ETHICAL CONSIDERATIONS

Is it wrong to tell respondents that a study is on housing, when in fact it is on whether they are willing to pay more taxes for community services? Is it wrong for an interviewer, without the knowledge of respondents, to observe and note down information about their dwellings? Is it wrong to tell respondents it will take about 10 minutes to complete the questionnaire, even though it usually takes about 20 minutes? Is it wrong for the researchers to collect other than the stated information about respondents accessing a web-based questionnaire? Is it wrong for researchers to hold back information from a national survey because the results were not to their liking?

While it is easy to agree that a study that causes bodily harm is unethical, it is often more difficult to see the ethical issues in other circumstances or with the survey method. The survey might be perceived as ethically neutral, since it involves merely asking questions. But there are many ethical considerations to take into account in the survey, including the nature of the questions; after all, questionnaires are often aimed at obtaining personal information about respondents.

Suppose you were carrying out a survey on the sexual attitudes of university students. A few professors agreed to allow you to ask their students to fill out the questionnaire, but they all asked you to do it at the beginning of their classes and to take as little time as possible. In one of the larger classes, the professor introduced you and made a favourable remark about your research project, without explaining what it was about. Since your time was limited and the class was large, you decided to hand out the questionnaire and simultaneously explain that the survey was on sexual attitudes. In a short time you had a large number of completed questionnaires, and off you went, happy that little effort was required. But is what you did ethical? Imagine you are a student in that class. How free would you feel to refuse to fill out the questionnaire? Would you be concerned about how your professor might perceive your lack of compliance, especially if marks were assigned for participation? Moreover, how would you feel if a questionnaire was handed to you, and only after you received it were you told that it was about a sensitive subject?

It is essential that researchers consider what is proper and improper with regard to both why and how a social survey is carried out. It is also imperative that researchers consider the impact of their research on the respondents. You need to put yourself in the place of the respondent, rather than the researcher, and then ponder whether what you are doing, or how you are doing it, is right or wrong. In other words, you must give importance to ethical considerations throughout the research process. However, certain ethical considerations may not necessarily be evident to you. Therefore, it is helpful to discuss your research with others to determine if there are any ethical issues you may have failed to notice.

While each survey research project involves specific ethical issues, general agreement exists about certain ethical considerations. A key consideration is that participation in the social survey should be voluntary, with some exceptions, as some would argue for the census of population carried out by Statistics Canada. In the earlier example, students should have been told that their participation was voluntary. It would also have been preferable to ask the professor to leave the room during the survey.

Survey questions sometimes aim to reveal information that the respondent might consider sensitive or threatening. For example, how would the respondents feel if they were asked to reveal some deviant behaviour or information about some unpleasant past experience? Each question in the survey should be given ethical consideration in deciding whether to include it or how to phrase it.

Two other issues to consider, especially when the survey includes certain sensitive questions, relate to assuring the respondent that the questionnaire is anonymous or that the survey is confidential. For a questionnaire to be anonymous, it must be impossible to identify the respondents. Therefore, it is wrong to promise **anonymity** in face-to-face interviews, since you can identify the respondent. Anonymity is possible only in surveys in which you cannot link the questionnaire with the respondent. An example is a mail survey, in which the questionnaires are not identified in any way with the respondents. But, as noted earlier, preserving anonymity is very difficult even in mail surveys, especially if you want to do follow-ups.

Confidentiality is more common with surveys. Confidentiality means you are able to identify the responses of a respondent, but you are committed to not revealing the information publicly. For example, in a face-to-face interview you might ask the respondents whether they take illegal drugs, and thus you would be in a position to identify them. But you are committed not to reveal the responses of a respondent publicly. Researchers use various techniques to ensure that the identity of the respondent of a questionnaire is kept confidential.

There are, of course, various other issues that should be considered when participants might experience stress and discomfort as a result of the research technique, and in the following chapters we will point to some of these. But a first and essential step is to recognize that ethical considerations are not restricted to only one method; rather, they must be of concern to all researchers, regardless of the method they use, and the survey is no exception.

BOX 5.5 ADVANTAGES AND DISADVANTAGES OF THE SURVEY METHOD

ADVANTAGES

- The method allows the researcher to use a sample to describe the characteristics of large populations (e.g., a random sample of students' opinions in a college or university, which in turn reflect the opinions of all students in the college or university).
- The researcher can collect information on many issues (e.g., the questionnaire can include questions on the respondent's gender, age, marital status, religion, yearly income, occupation, voting intentions, and so on).
- The survey is reliable, since all the respondents are presented with the same questions.
- It allows the researcher to formulate operational definitions on the basis of answers to certain questions (e.g., the operational definition of part-time workers may depend on the answers to questions regarding work for pay and the number of hours worked in a week).
- The researcher can find out quickly what the public thinks about certain issues (e.g., carrying out a survey within hours of an event to find out the public's reaction).

DISADVANTAGES

- The questionnaire might fail to touch on issues important to the respondents.
- The information collected might be artificial, giving no true "feeling" for what the respondent is experiencing (e.g., there may be lots of information on the difficulties the unemployed face, but no feeling of what it is like to be unemployed).
- Certain closed-ended questions assume respondents view issues in degrees, such as "Strongly Agree, Agree, Disagree, Strongly Disagree," but this is not necessarily the way most think of the issue.
- Once the questionnaire is distributed, it is no longer possible to change the design of the study.
- The questionnaire may encourage respondents to express opinions about issues to which they have given little consideration (e.g., the respondent may have been unaware that pornography existed on the Internet but might still express an opinion about the issue).

What You Have Learned

- Social surveys are used to help us understand certain aspects of a social phenomenon.
- The main instrument of the social survey is the questionnaire.
- Much careful effort goes into framing the questions and constructing the questionnaire.
- Two main categories of surveys are self-administered questionnaire surveys, including Internet-based surveys and interview surveys.

- Internet-based surveys are an expanding area of survey research.

- The objective of the research and the time and resources available, among other factors, will determine which approach is chosen to carry out the survey.

- There are various ethical considerations that the researcher should take into account in conducting a social survey.

Focus Questions

1. What are the main differences between a casual survey and a scientific social survey?

2. What are the main steps in developing a social survey research project?

3. Why would a researcher use the social survey method?

4. What are some mistakes to avoid in wording the questions?

5. What are some general guidelines to consider in constructing the questionnaire?

6. What are the main ways that questionnaires are administered?

7. What has been the impact of the Internet on survey research?

8. What are the major issues to look for when reading accounts of survey research?

9. What ethical considerations are involved in carrying out a social survey?

Review Exercises

1. Select one of the articles cited in this chapter's introduction. Discuss the objective of the study and the method with particular emphasis on how the questionnaire was administered.

2. Examine one of the questionnaires available at the Statistics Canada site noted in the Weblinks section below, and describe the design of the questionnaire and, if possible, how it is administered.

3. Search for a media report on a recent poll and answer the questions indicated in the above section "Questions to Ask about a Survey."

Weblinks

The various sites provide further information on the survey method, including planning the survey, designing the questionnaire, collecting the data, and more. One of the sites provides a booklet, *What Is a Survey*, intended for nonspecialists. The other sites offer advice on designing and administering the questionnaire, and one offers numerous examples of surveys conducted by Statistics Canada.

You can find these and other links at the companion website for *First Steps: A Guide to Social Research*, Fifth Edition: **http://www.firststeps5e.nelson.com**.

What Is a Survey? (Booklet)
> **www.amstat.org/sections/srms/pamphlet.pdf**

Questionnaire Design (Statistics Canada)
http://www.statcan.ca/english/edu/power/ch2/questionnaires/
questionnaires.htm

Definitions, Data Sources and Methods (Statistics Canada)
http://www.statcan.ca/english/concepts/index.htm

Survey Design Tutorial (also available as pdf)
http://www.statpac.com/surveys

Guide to the Design of Questionnaires
http://iss.leeds.ac.uk/info/312/surveys/217/guide_to_the_design_
of_questionnaires

How to Write a Survey or Questionnaire (video)
http://www.howcast.com/videos/383549-How-To-Write-a-Survey-
or-Questionnaire

20 Questions a Journalist Should Ask About Poll Results
http://www.ncpp.org/?q=node/4

Experimental Research

What You Will Learn

- the aim and purpose of the experimental method
- the logic behind experiments
- the steps involved in setting up an experiment
- variations of the experimental design

INTRODUCTION

Relaxing on the steps of your school on a warm spring day, you and two friends embark on a debate about which is the better-tasting cola drink: brand A or brand B. One insists it is A and the other, B. "How do you know that?" you ask. Both quickly respond that they tried the two brands and preferred one over the other. And ever since, they quench their thirst with their favourite brands. "But if you usually drink only one brand, how can you know for sure that one is better-tasting than the other?" Without hesitation, both insist they can easily tell the difference. But you note an inconsistency: both said that their preferred brand is sweeter. This makes you wonder, "Can you really distinguish between the taste of the two brands?"

They tell you to taste the two and see for yourself. One friend then pours brand A cola in one glass and brand B in another. She tastes the two and tells you she can definitely tell which is her preferred brand. They now turn to you to do the same. You rarely drink cola, but you try anyway. You cannot distinguish a difference in taste between the two. They suggest that you should drink more cola, and that if you carried out a survey you would find that people can tell the difference. "But that would tell me only whether people think they can distinguish between the two, not whether they can actually distinguish between the taste of the two brands," you object.

So how can you find out? You realize your friends knew which glass their preferred brand was in. Their ability to distinguish between the two brands may have depended on knowing which brand they were drinking. What if they did not know? You redo the test, but this time your friends do not know which glass their preferred brand is in. Do you think the results will be the same?

Although it contains several flaws, your simple taste test would have the basic features of an **experimental study**, whereby the researcher attempts to control aspects of the situation and then observes the change that results. You would first observe your friends' ability to distinguish between the tastes of the two brands when they knew which glass held their brand. You would then deliberately hide this information from them and see if they could still distinguish between the two. Of course, a properly conducted experiment is more demanding, as we shall see in this chapter.

Experimental research is a technique that aims at demonstrating and specifying or clarifying cause-and-effect relationships. In the above example, the experiment would be set up to discover whether people's preference for a brand of cola resulted from, or was caused by, their ability to distinguish between the tastes of the different brands. In chemistry, experimentation will determine how much a mixture will need to be heated before its components will react and form a new compound. You have probably undertaken physics experiments in high school, such as altering the placement of weights in order to balance a lever.

While the experiment is viewed as the prime technique in scientific research, the basic logic of experimental research is also very much a part of everyday problem solving. For example, on a cold, damp day your car will not start. You do not want to spend money on a tow truck, so you open the hood. You look for any loose electrical leads, you check the battery by turning on the lights, and you may clean the spark plugs or dry off the distributor cap. After checking each problem area, you try to start the motor. What you are doing is eliminating alternative causes of the problem. Similarly, when we learn to cook, each meal is an "experiment." With every repetition of the meal we learn more precisely how much seasoning to use, exactly how long to cook things, and so on.

Finding out why your car or other appliances will not work and learning how to cook better-tasting food involve a *search for causes,* which defines the experimental method. What is causing the trouble with the car or the vacuum cleaner? Why does the cake not rise or the tomato sauce taste funny? What causes these problems?

In scientific research a simple experiment would be one in which we manipulate a variable to assess its effect on another variable. (Review the discussion of variables in Chapter 2 if you need to refresh your memory.) But while experiments in certain natural sciences, such as chemistry, can be relatively tidy and simple to manage, experiments in social research face various constraints.

Most importantly, social research involves human beings. As we will see later, there are ethical considerations and legal restrictions in using people in experiments. Even when people volunteer, other factors in their lives can occur that researchers have no control over and that can influence the results of the experiment. Furthermore, human beings are likely to change their behaviour simply as a result of knowing they are part of an experiment.

Social scientists have attempted to overcome some of the obstacles that researchers face in doing experiments by developing different types of experiments. One type is the **laboratory experiment**, which is an artificial situation that the researchers create and control. Another type is the **field experiment**, which is conducted in real-life settings where researchers give up some of their control over the situation. In both laboratory and field experiments, however, researchers retain some control over the situation and manipulate some variable(s). In contrast, the **natural experiment** involves no such manipulation. Researchers observe changes in social behaviour as they occur after natural events, such as floods or earthquakes, or social events, such as plant closures or riots.

Each type of experiment is covered in this chapter. However, the focus is largely on the laboratory experiment, since the other types are seen as approximations of the laboratory experiment. In addition, an advantage of the laboratory experiment is that it is generally on a small scale and is controlled by the researchers. Field experiments are generally large, expensive, and long-lasting. Natural experiments involve waiting for a suitable event to occur. Regardless of the type of experiment, the underlying logic is the same.

THE EXPERIMENT IN SOCIAL SCIENCE RESEARCH

Experimentation in the social sciences is used by different disciplines but is most associated with psychology. Until the late 19th century, psychology was seen as part of the study of the "mind" and was treated as part of philosophy. Over time, psychologists became increasingly focused on understanding causal factors in human behaviour, mainly through experimental research. Below are three key figures who have in various ways helped develop and spark discussions on experimental research in psychology.

WILHELM WUNDT (1832–1920)

Wilhelm Wundt, a professor of philosophy, was an early pioneer of experimental psychology at the University of Leipzig, Germany. He is credited with establishing in 1879 the first experimental psychology laboratory. His approach was quite controversial, but it has played a key role in the subsequent development of psychology as an experimental discipline. The approach Wundt employed was *introspection*—the study of mental phenomena. He set up tests, puzzles, and problems for his subjects to perform and asked them to describe what went through their minds as they worked on these tasks. His subjects were fellow scholars, professionals, and upper-class university students.

IVAN PETROVICH PAVLOV (1849–1936)

Pavlov was a Russian physiologist interested in the biology of digestive processes. He noticed that the dogs he used in his experiments learned to anticipate feeding times by the sounds made during food preparation. This finding led him to explore *conditioning*—a learning process in which repeated exposure to a stimulus (a ringing bell) followed by the appearance of food led the dogs to interpret the stimulus as a sign of food, even outside feeding times. Today's research design of attempting to control other influences has in many ways originated from Pavlov's research strategy.

B. F. SKINNER (1904–1990)

One of the best-known psychological researchers, Skinner explored conditioning in great detail. He was a leading advocate of *behaviourism*—a perspective in psychology that emphasizes the study of only what is observable. Behaviourists believe that we should not assume the existence of internal mental processes, but rather should study actions that are clearly observable. Some also credit Skinner with having encouraged methodological precision in psychology research. He developed various types of "Skinner boxes"—devices that set up a task for the experimental subject and rewarded the subject when the task was successfully accomplished.

THE LOGIC BEHIND EXPERIMENTS: CAUSATION

We noted in Chapter 2 that the idea of causal explanation flows from the basic premise of all science that there is order or patterning in nature and social reality. For many researchers, the research process is a sequence that moves from discovering and describing patterns to explaining why and how the patterns work. In the social sciences there is some debate over the possibility and importance of causal explanation and the kinds of explanation that might be appropriate. As we saw in Chapter 2 and shall see in Chapter 7, many social science researchers argue that the social world is so complex that adequate description is at least as important as and possibly more important than causal explanation.

Experimental research is mainly explanatory. It is designed to ascertain the degree to which one thing affects or causes another. In this way the researchers aim to build knowledge about cause-and-effect relationships, or to account for why and how things happen. Suppose you tried the simple taste test on a few friends who all had a preference for one of the two brands of cola. Assume that the results showed they were unable to distinguish between the tastes of the two brands of cola. You examined the relationship between preference for a brand of cola and the ability to distinguish between the brands of cola. But your inquiry will probably not stop here; you would want to know why the relationship exists or does not exist. Asking *why* is looking for a cause, and the experimental method provides a means of searching for the cause.

However, neither nature nor social life comes already labelled as sets of causes and effects. Consequently, it is difficult to discover and measure causal relationships. The search for causes, as we saw in Chapter 2, combines logical analysis and empirical research. Here, we shall focus on three conditions to establish causality: temporal order, consistent association, and the elimination of plausible alternatives.

Condition 1: Temporal Order

There must be a *clear time sequence*, such that the cause consistently appears before the effect. For example, say you want to find out whether the time spent studying for an exam affects the grade obtained. Obviously, the cause (time spent studying) has to come before the effect (grade on an exam). Or, suppose you want to see the effect of a film on the dangers of smoking on a group of students who smoke. You could first give them a questionnaire to find out what they know about the dangers of smoking, then show them the film, and then give them a post-film questionnaire. The film may provide them with information that they did not know before seeing it. Obviously, the students would have to see the film before you can determine the effect of the film on them.

In actual research it often can be quite difficult to establish a clear temporal order of cause and effect. For example, social scientists have explored and described the way of life of some urban poor as a "culture of poverty." But research in this area has been plagued by a chicken-and-egg controversy. This culture is characterized by, among other things, negative attitudes toward regular work and undisciplined work behaviour. Some researchers argue that the culture of poverty is a consequence of economic conditions. In other words, low wages and insecure jobs discourage people from being hardworking and reliable. Therefore, bad working conditions (independent variable) cause a cultural pattern of negative worker attitudes and behaviour (dependent variable). Other researchers argue that the culture of poverty is learned. It is transmitted from generation to generation and moulds people so that they are incapable from the outset of fitting into stable routines and disciplining themselves. Consequently, there is a supply of willing casual workers owing to a cultural tradition (independent variable), and these workers take up jobs that others will avoid (dependent variable).

Thus, experimental research requires an assumption of a clear sequence that can be put to the test. But what is cause and what is effect also depends on how you analyze the situation.

Condition 2: Consistent Association

There must be a *consistent pattern of association* between cause and effect. In other words, whenever or wherever you see the effect, you will (at least in most cases) find that the cause is also present. In the above example on the effect of a film about the dangers of smoking, what the students learned was presumably from the film. There was an association between the film (cause), which appeared before their change in knowledge of the dangers of smoking (effect).

However, the association is rarely perfect, and this poses problems in causal interpretation. There may have been a sharp increase in the students' knowledge of the dangers of smoking, or a very slight increase, no increase, or even a decrease. To cope with this ambiguity, social scientists use statistical interpretation to see whether there is enough of a relationship to warrant a causal interpretation.

Condition 3: Elimination of Plausible Alternatives

Before accepting the idea that there is a causal connection, researchers have to satisfy themselves that the connection is the *best possible explanation*. They have to be able to rule out alternative explanations. For example, the students' knowledge of the dangers of smoking may not have been affected simply by the film. The students may have taken part in other events during the period between the time they saw the film and the time they were given the post-film questionnaire to determine the change in their knowledge of the dangers of smoking. Or, possibly, the pre-film questionnaire got the students thinking about the dangers of smoking, and this in itself increased their knowledge. (Consider how you could set up the experiment differently to avoid this problem. Later, we will suggest how to improve the experiment.)

To rule out other possible explanations, social scientists look for alternative explanations. The experiment allows for this by isolating the variables involved so that no other cause can affect the results. As we will see later in this chapter, social scientists have developed different experimental designs to help them try to eliminate other possible influences. Moreover, an experiment can be repeated many times to see if the association between possible cause and effect is consistently found.

BOX 6.2   HOW TO CONDUCT AN EXPERIMENT

1. Select a research topic and develop a clear causal hypothesis focused on a few key variables.
2. Operationalize:
   - Isolate the key elements of the causal connection.
   - Decide how to establish contrasting situations or treatments for the experimental group and the control group.
   - Develop valid and reliable measures for the initial conditions and the effects of the independent variable, and pretest these where necessary.

continued

3. Design the study:
   - Review the experimental design and check alternatives to eliminate potential sources of error and ethical problems.
   - Ensure that only the independent variable can be responsible for any changes over the course of the experiment.
   - Consider running a pilot test to ensure that the experiment will run smoothly.
4. Select the subjects:
   - Consider the population the hypothesis applies to and set up a method of sampling from that population.
   - Consider how the subjects (participants) are going to be allocated to the experimental and control groups.
   - If the experiment is long-term, consider using larger numbers to compensate for any dropouts.
5. Consider the issues of internal and external validity:
   - Try to anticipate any threats to internal validity, such as biases in dividing subjects into experimental and control groups.
   - Do likewise for external validity, by considering problems in generalizing from your experiment(s).
6. Perform the experiment:
   - If necessary, measure the initial conditions or characteristics of both experimental and control groups (pretest).
   - Expose the experimental group to the independent variable.
   - If necessary, measure the final conditions or characteristics of both groups (post-test).
7. Debrief the participants:
   - Explain why any deception was necessary.
   - Ensure that any stress arising from the experimental experience is relieved.
8. Analyze and interpret the data, and prepare a report.

STEPS IN EXPERIMENTAL RESEARCH: THE LABORATORY EXPERIMENT

Essentially, a laboratory experiment involves manipulating one variable while attempting to either control the influence of other variables or hold them constant. The focus is then on the effect the manipulated variable (the independent variable) has on another variable (the dependent variable). In our smoking example, showing the film is the manipulated variable. The effect of the film is measured by comparing students' responses to the questionnaire before and after seeing the film.

The laboratory experiment is considered the basic approach to experimentation. Other approaches are seen as approximations to the laboratory experiment. Therefore, our discussion of the major steps involved in experimental research focuses mainly on the laboratory experiment.

Step 1: Formulating the Experimental Hypothesis

The point of an experiment is to expose or measure causation at work. Therefore, it follows that you have to be clear about the relationship you expect to find between two variables in a setting over which you have much control. For example, what did you expect to find in showing the film on the dangers of smoking to a group of students? You expected the change in their knowledge of smoking to be dependent on their viewing the film. Stated differently, viewing the film was the independent variable, or the variable that was manipulated. And the dependent variable was the change in their knowledge of the dangers of smoking after viewing the film, or the effect of the manipulated variable. Hence, the first step in an experiment is to clarify the causal connection between what are assumed to be the causal or independent variables and the effects or dependent variables. Without a clear hypothesis, no well-designed experiment can emerge.

An example of this process is Stanley Schachter's (1959) development of an experiment on anxiety or fear and affiliation. (There is much more to Schachter's work than is presented here, but this summary is sufficient for our purposes.) Schachter had read a number of books and articles on religious hermits, convicts, and prisoners of war who had been socially isolated for long periods of time. He noticed that all the accounts of their experiences reflected high levels of anxiety or even fear. He began to wonder whether isolation from others, for whatever reason, was itself the cause of these distressing reactions. In other words, was there a causal connection between isolation and anxiety or fear? This question led him to consider how this connection might operate under more general conditions and led him to formulate the hypothesis that people who are not isolated turn to others as a result of anxiety. Notice the drastic simplification that begins at this stage. Schachter was not interested in the different kinds of stresses that religious hermits, as opposed to prisoners of war, may experience; his interest was in stressful experiences in general. What emerges from this review of the literature is a clear, uncomplicated hypothesis: anxiety or fear leads people to affiliate or to turn to the company of others.

Thus, the first step in experimental research is to assert the relationship that exists between two or more variables—that is, to state your hypothesis. Your next challenge, however, will be to devise a way to measure the relationship between the variables—in the study by Schachter, to measure the relationship between anxiety-provoking experience (independent variable) and affiliation (dependent variable).

Research Info 6.1   *Repeat Viewing of a Movie and Comprehension among Young Children*

Does repeat viewing of a movie improve comprehension among four- to six-year-old children? Does the presence of a mother facilitate the comprehension?

Research on young children's exposure to television programs and videos suggests that these are complex phenomena. One-shot exposure to a TV program is different from watching a video several times. As well, studies have shown that repeated viewing of curriculum-based programs improves comprehension. And the little research on repeat viewing of movies that were not curriculum-based has also found some improvement in comprehension.

Consequently, given what was known from previous research, Skouteris and Kelly (2006) wanted to further examine the effect of repeat viewing of non-curriculum-based movies. In addition, they wanted to find out if there was a relationship between the time spent looking at the screen and comprehension, and whether the children's visual attention changed with repeat viewing. To find out, they carried out an experiment among 77 four- to six-year-olds, whose first language was English, most of whom were from middle-class families, and all of whom resided in suburbs of Melbourne, Australia.

The children were randomly assigned to one of four experimental groups. Half would watch the animated movie once and the other half would watch the movie five times. Each group was further divided into two subgroups, one in which the child watched the movie alone and another in which the child watched it together with his or her mother. They all watched a 70-minute movie that had not yet been released for public viewing. The researchers implemented various procedures to retain, as much as possible, control over the experimental process, including what mothers should or should not say to the children, the comprehensive questions for the children, and when and how to carry out the repeat viewing. Essentially the researchers found that repeat viewers had higher comprehension scores, and that the presence of a mother did not improve comprehension.

Any experimental study needs careful design. Given the age of the participants, designing this experiment was a major challenge. Note also that the sample was selected to minimize the potential effects of class, ethnic, and other external social factors.

Source: Helen Skouteris & Leanne Kelly. "Repeated-viewing and co-viewing of an animated video: An examination of factors that impact on young children's comprehension of video content." *Australian Journal of Early Childhood.* Issue 31 (2006), Pg. 22–31.

Step 2: Operationalizing

Designing an experiment requires having considerable imagination and creativity. An experiment involving human stress immediately raises both ethical and practical problems. As we will discuss later, social researchers need to follow strict guidelines that ensure human participants in experiments are not subjected to unnecessary or excessively upsetting experiences. Researchers have to carefully consider such limitations when designing the experiment.

Designing experiments is also a process of simplifying "reality" so that only the key elements in the causal relationship are involved. Schachter ultimately developed an elegant, straightforward experimental design. Two groups were to be set up: the **experimental group** was to be subjected to a strong anxiety-provoking experience and its reactions were to be measured; the **control group** was to be placed in an identical situation or environment as the experimental group but without being subjected to such a strong anxiety-provoking experience. (Strictly speaking, since both groups were subjected to specially manipulated experiences, they were both experimental groups. It was impossible to create a strict control group situation in this experiment.) The disturbing or anxiety-provoking experience is the independent variable. If anxiety-provoking experience does cause people to turn to each other (affiliation), then the two groups should be clearly different in their reactions by the end of the experiment.

In designing the experiment we also need to consider whether the procedure actually measures what it is supposed to measure. If it does this, it has **validity**. We are also concerned about the extent to which the experiment would yield the same results if repeated. Experiments and other research techniques that can be consistently repeated are considered to have the desirable quality of **reliability**.

The anxiety experience Schachter set up involved an assistant pretending to be a medical researcher whose task was to administer electric shocks using a formidable-looking machine. The experimental group was to be subjected to a clearly stressful situation. When meeting with the experimental group, the assistant acted in a cold, distant way and said that the shocks would be quite uncomfortable, although not physically harmful. The control group was not placed in a clearly stressful situation. When meeting with the control group, the assistant was friendly and informal and went out of his way to emphasize the mild nature of the shocks.

Before the shocks were supposedly to be administered, the assistant told each group that there would be a delay in the proceedings while the machinery was set up. He then told them that they could wait either together or individually in different rooms. Before leaving to do so, they were asked to fill out a questionnaire. The questionnaire probed how anxious they were and asked whether they intended to wait in the company of others or individually. When all of the questionnaires were completed, both groups were then told that the experiment was at an end and no shocks were to be given.

Clearly, the two groups experienced very different situations: the experimental group experienced high anxiety-provoking stress (being threatened with severe shocks), and the control group experienced low stress (being threatened with mild shocks). As very few of us are masochistic, the threat of physical discomfort is almost universally stressful. In this sense, the experimental or independent variable has validity—it is a true or real stressor for most people. It also has reliability, because the threat of physical discomfort will produce stress in most people most of the time and so can be used when the experiment is repeated with different groups by different experimenters. (The questionnaire used in the study was previously shown to be reliable and valid; it was used before over a wide range of applications.)

Step 3: Designing the Study

In setting up the experiment, the researcher attempts to rule out possible explanations other than the one intended. Schachter, for example, selected two samples that were as similar as possible and created situations in which he could measure the changes. The experimental group was treated to a stressful situation and then compared with the control group, which did not face the stressful situation. The use of a control group is common in the experimental design.

However, other experimental designs exist. Sometimes there is no control group in the experiment; instead, the researcher compares a pre-intervention condition with a post-intervention condition. This is the design of the earlier example about the effect a film has on students' knowledge of the dangers of smoking. The design of this experiment aimed to measure the effect of the film on a single group. But, as we noted earlier, the pre-film questionnaire in itself may have encouraged students to think about the dangers of smoking and influenced their knowledge, irrespective of the film. Could we improve this experimental design?

One approach is to combine the two mentioned designs and have both an experimental group and a control group. Each group is given the same pre-film questionnaire but only the experimental group is shown the film. After a delay, possibly the time it takes to view the film, the experimental group and the control group are given the post-film questionnaire. Then the knowledge of the two groups on the dangers of smoking is compared. Presumably, if the experimental group shows a sharp change in knowledge, it was due to the film.

Figure 6.1 illustrates these three possible designs for testing the effect of the film. As you move from the simple to the more complicated design, consider how each design improves on the previous one. Ultimately, the only limit to the number of designs is the ingenuity of the researchers in determining the various ways to carry out their experiments given various constraints of the research (see, e.g., Research Info 6.1).

For example, to find out the limitations of gathering information from children between the ages of two and five by asking them yes–no questions, Fritzley and Lee (2003) conducted four experiments in which young children were asked comprehensible and incomprehensible yes–no questions. Each child participated in only one experiment. In the first experiment children were asked to identify familiar objects (e.g., a cup) and unfamiliar objects (e.g., an electrical fuse) and were asked the object's colour and use. The children's responses were then analyzed according to their age group (two years, three years, four years, five years) and the patterns for each age group were then compared. In the second experiment, two nonsense questions (Is this for nurking? Is this made of doow?) were included to check for the effect of comprehension on response bias. The responses were then checked for the different age groups.

The third experiment replicated the second experimental test to check the validity or generalizability of the procedure. The fourth experiment repeated the straightforward questions of the first experiment, with a repetition of the questions after a one-week interval to check for the impact of memory.

This experimental design allowed Fritzley and Lee (2003) to determine that the youngest children answered questions with a consistent yes bias. At ages four and five, children showed no response bias toward questions they understood, but a no-saying bias in responding to the meaningless questions. The responses of three-year-old children were mixed and inconsistent, suggesting to the authors that this was a period of transition in their cognitive development.

What is important to note is that social scientists design experiments to arrive at generalizable conclusions. Fritzley and Lee (2003) argue that the results have a broader significance for social science. Five-year-old children tended to answer negatively to questions they did not understand (including the nonsensical questions) even when they were given the option to say "I don't know." The researchers suggest that saying no implies that you actually know the answer to a question, whereas saying "I don't know" is a clear admission of ignorance. They note that survey research shows that about 20 percent of adults attempt to answer unanswerable or meaningless closed-ended questions. This suggests

Figure 6.1 Experimental Designs

Pretest—Post-Test Design: **Only Experimental Group**

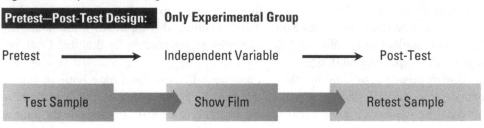

Pretest ⟶ Independent Variable ⟶ Post-Test

| Test Sample | Show Film | Retest Sample |

Control Group Design: **Experimental and Control Group**

Experimental Group:

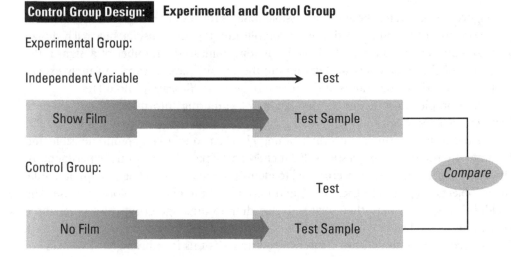

Independent Variable ⟶ Test

| Show Film | Test Sample |

Control Group:

Test

| No Film | Test Sample |

Compare

Pretest—Post-Test Control Design

Experimental Group:

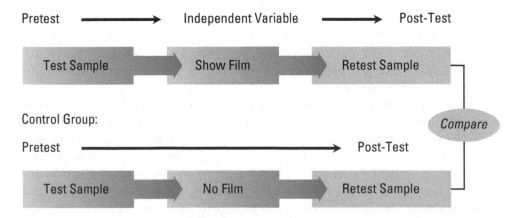

Pretest ⟶ Independent Variable ⟶ Post-Test

| Test Sample | Show Film | Retest Sample |

Control Group:

Pretest ⟶ Post-Test

| Test Sample | No Film | Retest Sample |

Compare

that responding to questions is a social process in which we try to present ourselves in a favourable light—as competent, knowledgeable people—a response that is learned early in life. It is important, then, that we learn as much as we can about the ways we respond to questions, partly because it is so basic to social life but also because it is a fundamental part of social research.

Step 4: Selecting Subjects

The key to a successful experiment, such as the one by Schachter, is that any differences between the experimental and control groups are a result of the experimental manipulation—the introduction of the independent variable into the situation—and not the result of other factors. It is important to ensure that both groups are identical and that the only difference between the two groups is that one experiences the experimental stimulus, while the other does not. Thus, the room in which the experiment occurs, the apparatus, and the person presenting the experiment to the group are identical in Schachter's experiment. The only difference is the way the experiment is presented to the two groups.

To ensure that the experimental and control groups are identical, the researchers assign the subjects (the participants in the experiment) to each group in a manner that ensures that the subjects are as equal as possible in terms of relevant characteristics. The most common technique is that of **random assignment**, whereby the subjects are randomly allocated into two subgroups. It uses the same basic mechanism and assumptions as the selection of random samples discussed in Chapter 4: every subject should have an equal chance to be placed in one group or another. Thus, subjects could be randomly assigned to the different groups by numbering the list of subjects and selecting numbers by means of a random number table. The first name drawn would go into the experimental group, the next into the control group, and so on. However, unlike random sampling, which determines who takes part in your research, random assignment determines whether a subject will be in the control group or the experimental group.

Step 5: Determining Internal and External Validity

The purpose of random assignment is to make certain that all subjects have an equal chance of being selected to either group. Their ultimate placement into the two groups is not dependent on a choice made by either the researchers or the subjects. If researchers could choose, then their conscious or unconscious desire to see the experiment succeed might lead them to stack the experimental group with certain kinds of people. For example, it could lead them to place friendly, sociable people in the experimental group, so the group would be made up mainly of subjects who would be more likely to seek the company of others. In a study on stress, the personality characteristics of those in the experimental group, rather than the stressful situation, could account for their reactions in the experiment.

A similar problem might occur if the participants were permitted to choose for themselves which of the two groups they would go into. Perhaps the more nervous individuals would volunteer to be part of the control group to try to avoid stress. Again, any reactions on the part of the experimental group could now be interpreted as resulting from the types of people in the groups, rather than from anything that was controlled in the experiment.

Therefore, in assigning the subjects randomly to the different groups and controlling other factors of the experiment, the aim is to design the experiment in such a way that changes are due to the independent factor. If the researcher is confident that the experimental results (the measured or observed differences between the experimental and control groups at the end of the experiment) are due to the experimental or independent variable, the experiment is said to have **internal validity**. Internal validity means that alternative reasons for the results have been eliminated by the way the experiment was organized and carried out.

It is also important in checking the experiment's design to consider its **generalizability**. If this experimental setup or procedure is repeated many times with different groups by different researchers, will the same relationship between the variables involved be found? If an experiment is supposed to tell us something about people in general, as was Schachter's experiment, then we will have to check it out with men and women, different age groups, and so on. If experimental conclusions can be generalized to a broad range of human populations, the experiment has **external validity**. Schachter's first experiments, for example, were initially conducted in England with young female students, but researchers could repeat the experiment elsewhere with samples of other population groups randomly assigned to experimental and control groups and expect the same results.

FIELD EXPERIMENTS

Until now, our discussion has focused on the laboratory experiment. Its advantage is that researchers have much control over the situation for discovering and measuring cause-and-effect relationships. They artificially isolate a limited range of variables and focus as precisely as possible on the key variables in a causal relationship. By controlling the way the variables interact, they may be able to measure the impact or the degree of causation involved. A disadvantage of such experiments is that they do not create a real-life setting. Indeed, many real-life situations are difficult, if not impossible, to create artificially. Social life involves complex, interrelated systems of elements, not discrete, easily separable components of independent and dependent variables. Moreover, as we noted and will say more about later, the use of experiments in social science raises many practical and ethical problems. Consequently, researchers have often been forced to compromise and to look for research designs that provide some of the advantages of the laboratory experiment without incorporating all its features.

Unlike the laboratory experiment, the field experiment is conducted in a real-life social situation, such as a subway or a library. Suppose your study involved finding out whether students were willing to help a drunk student. You might set up a situation in which an apparently drunk student sits near the entrance to the library and shows signs of needing help standing up. You could then record whether he received help, who came to his assistance, and how much time it took. An advantage of such an experiment is that it studies social behaviour in a more realistic situation than in a laboratory, but you would have far less control over factors that could interfere with the situation. Perhaps certain students did not approach the drunk student because they expected a nearby professor to help him. Since you are unable to control the many possible factors that could have interfered, it is difficult to establish a clear causal relationship. Thus, a challenge in setting up a field experiment is to have as much control as possible while manipulating the key causal factor.

An interesting example of a field experiment is one conducted by Ginzberg and Henry (1984/1985), two anthropologists who wanted to test the degree of racial discrimination in Toronto's labour market. The researchers attempted to discover if there were differences in the number of job offers obtained by equally qualified White and Black applicants. They also wanted to see if applicants from different racial and ethnic backgrounds were treated differently in job interviews.

The researchers used two field experiment procedures. First, similarly qualified Black and White applicants applied in person for the same job. Second, applicants using five different ethnic accents (White-majority Canadian, Slavic, Italian, Pakistani, and Jamaican) made phone inquiries about advertised job openings. In both cases, employers were confronted with apparently normal people searching for jobs in the regular fashion. In fact, while one set of applicants—those for semiskilled and unskilled jobs—were high-school and university students who would normally be seeking these jobs as temporary work, another set applying for more skilled work were professional actors trained to participate convincingly in the more demanding job interviews required for this level of work. By matching applicants in terms of their qualifications, having them apply for the same jobs, and having them apply by phone and in person, the researchers controlled for many real-life variables, and the results are convincing because of the realistic nature of the experiment. What do you think the results showed?

An important difference between field experiments and laboratory experiments is that the former are often used in applied research. Field experiments are undertaken to examine social policies and institutions, such as schools, welfare organizations, and government training programs. When considering changes in policy, government agencies often introduce pilot studies or test cases where the new arrangement is introduced on a small scale and the results are compared with the old ways of doing things. Such studies have the advantage of uncovering problems with the new policies before the government has gone to the expense of changing the entire system.

Since the 1970s there has been increasing debate over welfare payments to very poor, chronically unemployed people in our society. In the United States this debate has been accompanied by a number of evaluation experiments—field experiments aimed at discovering whether welfare payments discourage people from searching for work, whether guaranteed sums are better than amounts that fluctuate according to recipients' job-search activities, whether payments tied to training are better incentives to seeking work than no-strings-attached payments, and so on. Such experiments have generally involved random samples of very poor welfare families, assigning some to a special treatment, such as payments tied to work training or a guaranteed minimum payment (the experimental group), and assigning the rest to continue with the existing arrangement.

The two groups are then followed up regularly through interviews and reports that keep track of their family incomes, expenditures, job-search activities, and perhaps other aspects of their lives, such as marital stability. Contrary to popular opinion, these studies find that no welfare payment schemes are clear disincentives to seeking work.

This research technique has all the components of a laboratory experiment except that society itself becomes the laboratory. Because the experiment occurs in the field, the researchers have no control over factors that might affect the situation but that have not been considered in the research hypothesis. For example, surges of economic growth encourage the long-term unemployed to seek work, while economic recessions discourage people without skills from actively looking for jobs.

The researcher also may not fully control the application of an independent variable in a study set up by a government agency. For example, the government agency might insist on screening recipients instead of randomly assigning them to a treatment group, or the agency might severely limit the length of time permitted for the study. As well, selecting and dividing groups or areas into those to whom the new policy is applied and those who remain with the old policy may be influenced by political or administrative concerns, which could undermine the validity of the outcome. However, the major advantage of this and other kinds of field experiments is that they involve real issues in realistic settings.

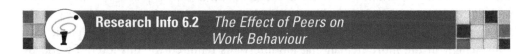

Research Info 6.2 *The Effect of Peers on Work Behaviour*

WHAT IS THE EFFECT OF PEERS ON WORK BEHAVIOUR?

Studies have examined this issue but have had difficulty estimating the extent that the behaviour was modified because of peer effect. In other words, through observation one may be able to see an effect but cannot necessarily separate the peer effect from other confounding factors. Falk and Ichino (2006) carried out an experiment in which they set up two conditions in which subjects did a simple task in a controlled work environment. In one condition, which served as the control group, subjects did a task alone in a room. In another condition, the same type of task was done by two subjects in the room.

The subjects were high-school students from schools in a city in the canton of Zurich (Switzerland). Announcements were placed in the schools offering students a four-hour simple job that provided an attractive payment. The researchers employed 24 subjects and randomly assigned eight in the single treatment and 16 (eight pairs) in the pair treatment. The work task mainly involved stuffing letters into envelopes. It required no prior knowledge and was easy to measure. The work was performed in a high-school building in a natural work environment. Subjects were told that they had to work for four hours without a break. Payment was independent of output.

The results showed a positive peer effect, in that output within pairs was largely similar but differed considerably between pairs. The results also showed that average output in the pair-treatment condition was higher than in the single-treatment condition, indicating that the peer effect increased output.

Unlike in laboratory experiments on work behaviour in which subjects carry out tasks that resemble work, in the field experiment by Falk and Ichino (2006) subjects performed a regular task for which they were paid and that was carried out in a natural work environment. However, as the authors note, they were unable to address the peer effect over time, and their study involved doing a simple task for which subjects were paid a fixed amount. Would the same results be obtained in a more typical setting in which the pay is based on group performance or on individual performance?

Source: Armin Falk & Andrea Ichino. "Clean evidence on peer effects." *Journal of Labor Economics.* Issue 24 (2006), Pg. 39–58.

NATURAL EXPERIMENTS

Laboratory and field experiments are invaluable research tools, but their precise focus and high levels of control are major obstacles to their use in many fields of social research. Ethical considerations further inhibit the use of laboratory and field experiments. However, the *logic* of experimentation often guides much nonexperimental social research. This logic is applied to a great variety of situations where various kinds of comparisons may be made that approximate in some way differences between control and experimental groups, or a pretest versus post-test situation.

Consider the impact that natural disasters, such as floods, fires, and earthquakes, or negative social events, such as plant closures, accidents, or terrorist attacks, can have on social behaviour. These phenomena cannot be manipulated as if they were independent variables. Instead, researchers interested in the influence of these phenomena on social life would carry out a "natural experiment" that, unlike the field experiment, uses naturally occurring independent variable(s), and participants are not randomly assigned. For example, say you want to study the recovery of flood victims. Obviously, you can carry out the experiment only after the flood has occurred. It is likely that you neither anticipated the flood nor considered the study before the flood happened. Therefore, you would have no baseline information on the victims before the event. You cannot be sure whether the changes among the victims were due to the flood or to other factors, or whether they would have occurred without the flood. The information gained by your study, however, can provide important insights and even suggest ways of helping flood victims. Since it is difficult to test a cause-and-effect relationship, a natural experiment often serves as a source of hypotheses that need further evidence.

In some cases events occur from which data are available and can be examined as if an experiment were carried out and be used to test a theory, even though the event and source of data were not intended for that purpose. For example, the economic theory of crime predicts that a rise in policing resources results in a drop in crime rates. But for various reasons, the magnitude of this effect has been difficult to demonstrate. While an increase in policing resources may enable more monitoring of criminal activity and result in a rise in arrests, it is also possible that potential criminals realize that with more monitoring they are more likely to be arrested. Heckelman and Yates (2003) wanted to further understand the monitoring and deterrent effects, and tested the theory by examining data on National Hockey League games in the 1999–2000 seasons. (As we will see in Chapter 8, researchers often use existing data to test theories.)

Although Heckelman and Yates (2003) did not create the situation or assign referees to the games, the available NHL data allowed for a natural experimental analysis. They compared games that had one referee with games that had two referees, and found that the latter had more penalties called, suggesting there was a monitoring effect. They noted that the number of referees in the game was not an important deterrent for committing infractions. Infractions occur during the heat of competition and are not necessarily planned. The results of the natural experiment led Heckelman and Yates (2003) to conclude that "the act of committing a sports infraction may be more analogous to a crime of passion rather than a calculated benefit–cost analysis performed by a rational criminal" (p. 713).

Often a natural experiment is a completely unanticipated event that provides the opportunity for either exploratory or hypothesis testing research. A chance sighting of an advertisement by a flying saucer cult led psychologists to infiltrate the group in order to observe their reactions when their prediction of the end of the world failed. This allowed them to test certain theories of cognition in a "real life" situation! Another classic example in social psychology was Hadley Cantril's (1940) study of panic. In 1938 a Boston radio station broadcast a dramatization of the H. G. Wells science fiction novel *War of the Worlds*. The radio drama began unannounced, with what sounded like an official government emergency newscast announcing the invasion of New England by Martians and the efforts of the U.S. armed forces to contain the invasion. Many thousands of listeners enjoyed the play, but a significant minority fled for the hills on hearing the news bulletin. Cantril interviewed a sample of those who had panicked and those who had remained at home. He then compared the two groups. What do you think he discovered?

In many other applications of experimental logic, natural experimentation involves more exploratory investigation. For example, the historian Natalie Davis was asked to be a consultant for a television drama based on her research on peasant life in 17th-century France. This research was based on documents that had very little direct evidence about the thoughts and feelings of the peasants themselves. By interviewing actors who had to make the history come alive by portraying the characters' thoughts and emotions, she obtained new insights into the documentary evidence itself. These insights were later incorporated into a new edition of her work.

This more open-ended, exploratory approach is characteristic of social science research into the "natural experiments of history." For example, anthropologists, economists, geographers, historians, political scientists, and sociologists interested in postcolonial, developing societies, often undertake systematic comparisons in order to assess the importance of complex factors that shape these societies. Many African countries share a history of being sources of supply of slaves to the Americas, but vary in the extent or intensity of the slave trade. What long-term consequences of the slave trade can be seen, and how does the variation in the extent of the trade affect these consequences? Other factors such as differing policies of English versus French colonialism, or the nature of anti-colonial movements in different ex-colonies, offer opportunities for careful contrast and comparison, providing a kind of retroactive natural experiment that can be usefully analyzed.

To conclude this comparison of the various types of experimental research, there are three components of a true experiment. First, at least two groups or conditions are compared. Second, the researcher can control, predict, or evaluate the experimental stimulus or treatment. Third, the subjects are randomly assigned to the experimental and control groups. There are many obstacles to applying this research design rigorously, so social researchers have developed approximations or substitutes to use when one or more of the key conditions are relaxed. In sum, while the application of classical laboratory experimentation is restricted in social research, its logic influences many areas of social science.

DOES PERSONALIZATION INCREASE THE RESPONSE RATE?

Some theories of social interaction suggest that, for different reasons, the more personalized a request for help or for a favour, the more likely a positive response will occur. Consequently, the theories would suggest that an e-mail request that uses a person's first name in the salutation rather than a generic salutation such as "Dear student" should increase response rate.

Heerweigh (2005) carried out an experiment over the Internet to find out if the personalization of an e-mail to participate in a web survey would increase response rates. From a list of Belgian university students that contained their names and e-mail addresses he randomly assigned half into a "nonpersonalization" condition and the other half into the "personalization" condition. They were all sent an e-mail with a salutation, which for the former read "Dear student" and for the latter used the student's first and family name. They were all invited to complete a web survey on attitudes toward marriage and divorce. A week after sending the e-mail, reminders were sent to those who had not yet responded. The survey was closed after a few weeks.

The results found that the personalized e-mail salutation increased response rate and web survey log-in, giving support to several broad theories of social interaction. However, the author cautions about extrapolating the findings since the population consisted of students. He suggests future research should be conducted with other groups.

Source: Dirk Heerweigh. "Effects of personal salutations in e-mail invitations to participate in a Web survey." *Public Opinion Quarterly.* Issue 69 (2005), Pg. 588–599.

INTERNET-BASED EXPERIMENTS

In recent years an increasing number of experiments have been carried out over the Internet, and have been referred to by various terms such as online experiments, web-based experiments, or Internet experiments. As with other research methods, the Internet has opened up other possibilities for experimental research and the means for collecting and analyzing data. But the decision to use the Internet in experimental research, as for other methods, is determined by what you want to study.

Basically, for certain Internet-based experiments, people are invited to take part or unknowingly participate in an experiment by e-mail, as in the above study (see Research Info 6.3). Another approach might be to instruct people with certain characteristics to link to a specific website, where they would be given instructions on what to do and possibly even offered an incentive for participating. Or certain people could be invited to participate and given a password that would allow them to access the website; researchers then would have some control over the choice of participants.

In other cases, researchers may impose no restrictions on who participates and allow access to the website to all Internet users. While they would have no control over who would participate, they could choose to include in the results only those who match certain criteria, perhaps related to gender, age, and schooling. Again, how the experiment would be designed and carried out would in large part depend on what the researchers want to study. For example, research on helping behaviour that has focused on the similarities between the solicitor and the helper has generally relied on face-to-face or phone interaction. Research on the self has found that the first name is seen as part of one's identity. Guegue (2003) wanted to find out if a similarity effect could also be obtained if no verbal communication occurred. He designed an Internet-based experiment to determine if a solicitor and a subject with the same first name communicating by e-mail would enhance helping behaviour. He sent 50 undergraduate students a similar e-mail message from a hypothetical student, with a request to complete a questionnaire on their diet habits. Students were randomly assigned to an experimental or a control group. In the case of the experimental group, the first name used by the hypothetical student was similar to that of the student receiving the e-mail. In the control group the name used was quite different from the first name of the student receiving the e-mail. Guegue (2003) found that "compliance to the request was significantly higher in the same first-name condition. No difference between the two conditions was found according to the delay of response." However, aside from other limitations of the study noted by the researcher, Internet-based experiments have certain constraints that in themselves suggest that the findings should be interpreted cautiously.

Many of the issues to consider when carrying out an Internet-based experiment are common to those of the survey method discussed in Chapter 5. Clearly, Internet-based experiments may have various advantages over the laboratory experiments. Researchers are able to reach out to a larger number of participants across wide areas and to samples that are culturally and demographically diverse (Reips, 2002). Consequently, Internet-based experiments may be more reflective of the population than laboratory experiments, which generally have small samples and are restricted to a certain demographic group, such as university students. The study can be cost-effective in both time and money, since you do not need the physical space, equipment, and hours to conduct the experiment as in a laboratory. Participants do not come to you; you take the experiment to them by means of the Internet. And their participation is completely voluntary, since they can easily stop taking part at any time. Moreover, for certain experiments, the Internet allows researchers to be relatively flexible about when participants need to take part.

Needless to say, these advantages, and others, largely depend on the type of experiment. In the example above, the researcher's experiment on the similarity effect was appropriate for using the Internet. But not so for certain research, such as experiments that require much control over the situation or in which participants carry out an act that needs to be measured directly. In addition, using the Internet involves certain limitations and disadvantages.

A major disadvantage of Internet-based experiments is the potential lack of control over key elements of the experimental conditions. There might be multiple submissions and participation by others than those expected. Further, the researcher is not present and may be unable to communicate with the participants in a timely manner to answer questions or concerns that participants may have about the experiment. In addition, possible participants may not take part in the experiment because of their worry about privacy over the Internet. Or your efforts to contact them may fail because your message is believed to be spam and is ignored. There are of course many other factors to consider. The material sent over the Internet, whether as an attachment or on a website, for example, is also dependent on the potential participant's computer hardware and software with regard to how the material's graphic presentation will be viewed and its download time. Since the researchers are not present, some participants are more likely to quit than if the experiment was carried out in a laboratory. Moreover, there is the ethical consideration of what people participating in an Internet-based experiment are told, such as when they assumed they were involved in another activity than an experiment. For example, in the study noted above, students were unaware that they were taking part in an experiment. Instead they assumed they were responding to a hypothetical student, with a request to complete a questionnaire on their diet habits.

In recent years researchers have developed many creative ways to conduct Internet-based experiments. But whatever the creative approach, in conducting an Internet-based experiment, it is imperative that you take into account the different steps involved in designing an experiment, consider what obstacles you will face, and determine the level of control and technical skills required. Briefly, consider the possible disadvantages and take appropriate and necessary measures to avoid or alleviate them. (For more examples of Internet-based experiments see the Weblinks at the end of the chapter.)

METHODOLOGICAL PROBLEMS IN EXPERIMENTATION

Experiments are most successful when the researcher has the most control over all the elements involved. But in social research such control is constrained by ethical considerations and undermined by the very nature of the material being experimented on. In addition, when human beings know they are part of an experiment, they are likely to behave differently than they usually would. This tendency is known as the **Hawthorne effect**. Studies carried out at the Hawthorne plant of the Western Electric Company in the late 1920s and early 1930s found that whether physical conditions were improved or made worse, the workers being studied increased their productivity. Later, researchers concluded that the unintended influence had been the workers' response to the interest shown in them. According to these researchers, the workers ignored the changes in physical conditions and produced more simply *because* they were part of the experiment.

Social scientists have since discovered that their experimental "subjects" are not passive, reacting only to the instructions and conditions consciously set up by the experimenter; instead, they are "participants," bringing their full range of sensibilities and social abilities to the experiment. As participants they actively cooperate by trying to guess the purpose of the experiment and then "helping" the researcher by modifying their behaviours to achieve that purpose. They pick up subtle, unintended cues from experimenters and react to these, even though the cues may be irrelevant or counter to the purposes of the experiment.

Experiments in social science often take lengthy periods of time, during which the participants are released and returned to the laboratory several times. During this time the participants may encounter other participants and share their experiences. This interaction may produce group dynamics, such as deliberate attempts to help with the experiment. If the encounter is between individuals who are in different groups in the experiment (experimental versus control groups), however, an awareness of the differences in treatment or experience may develop and lead to rivalry or resentment between the groups.

Even if these communication processes do not take place, many other things may affect the lives of the participants because of the length of time involved, undermining the internal validity of the experiment. People, especially young people, grow and change over time, and the experimental experience may reinforce or interact with these maturation processes. The passage of time also permits the occurrence of events that may have an effect on the experiment. For example, if an experiment is focused on prejudice, then political events, an outburst of ethnic rivalry, or a newspaper story may affect the participants' attitudes and interact with the experimental conditions in ways that are difficult to identify.

If the experiment involves repeated testing, such as performing skill tests, quizzes, or problem-solving tasks, the sheer repetition of testing will lead to improvement, regardless of any experimental manipulation. Furthermore, the longer the time taken up by the experiment, the more likely it is that fatigue and boredom will contaminate people's responses. In the case of very lengthy experiments, there may be dropouts, and researchers will have to check that the number of dropouts does not significantly change the balance between the experimental and the control group.

ETHICAL CONSIDERATIONS

Experiments in social science raise serious ethical questions. Many social science experiments involve some degree of *deception*—as was the case with Schachter's experiment, in which the purpose and the procedure of the experiment were misrepresented to the participants. The false pattern is itself the independent variable. In field experiments, professional actors often are hired to simulate muggings, to break social conventions, or to pretend to need help in some public location. Often, these deceptions are harmless in the sense that no physical threat is involved, but very often some element of stress is created as an unavoidable part of the deceptive experimental procedure. In some cases this stress may be quite severe, as was the case with the notorious obedience experiments conducted by Stanley Milgram (1974). (See the Weblinks at the end of the chapter for more on this and other experiments.)

Milgram set up situations in which naive subjects were urged to give successively more intense electric shocks to a victim (in reality, a collaborator with the researcher) whenever he made mistakes in reading. Despite the fact that the task involved shocking the victim to an apparently dangerous level and despite the realistic pleadings and screams of the victims, about two-thirds of the naive subjects followed the directions of the researcher.

The publication of Milgram's experimental research ushered in a period of intense debate among social scientists, especially psychologists, over the nature, appropriateness, and ethical problems of experiments. The results of that debate have been the tightening of ethical regulation of human research across the social sciences and the introduction of a new procedure in human experimentation: **debriefing**. The term originated in the armed forces, where it meant a review of a military operation after it was carried out in order to learn all that could be useful for future operations. In psychology the term was used by Milgram as a synonym for what he called "dehoaxing"—making clear to the participants of an experiment that they had been hoodwinked and explaining why. In Milgram's case it was also found necessary to counsel some of the participants who had been seriously distressed at the thought of inflicting physical harm on others. Since Milgram's experiments, psychologists have generally tried to avoid intensely stressful forms of experimentation, which had been a distinct trend in the 1950s.

Table 6.1 Advantages and Disadvantages of Experimental Research Techniques

	Advantages	Disadvantages
Laboratory Experiments	Allow for much control over the situation.	Artificial (i.e., one has little confidence that the results of the experiments would be the same in real life).
	Possible to manipulate one variable while controlling or keeping constant the influence of other variables.	Many aspects of human behaviour are difficult to examine in the laboratory (such as intimate dating relationships or family violence).
	Strong at measuring cause-and-effect relationships.	Difficulty in measuring cause-and-effect relationships.
Field Experiments	Allow manipulation of variables in real-life situations. Subjects are likely to behave as they normally would in such a situation. Often useful in applied research, such as measuring the effect of a government policy.	Researchers lack some control over the situation. Unexpected developments can influence the situation. Generally longer-lasting and thus more expensive and difficult to conduct than laboratory experiments.
Natural Experiments	Studies social behaviour in its natural setting. May be beneficial in suggesting how agencies might help people affected by natural disasters, such as floods, and by negative social events, such as riots.	Weak in providing proof of cause-and-effect relationships. No control over independent variable. May have to wait for the event to occur.
Internet-Based Experiments	Easy access to large, demographically diverse population. Less time constraints as to when to participate. Lower costs as to material required and to administer.	Risk of the participation of others than the expected subjects, multiple submissions, and high dropouts. Might be weak in experimental control. Dependent on the use of computers, Internet, and related factors.

1. What is the hypothesis that is being tested? What are the variables in the hypothesis?
2. What kind of experimental design has been selected in this study? For example: Post-test only? Both pretest and post-test? And so on.
3. What is the independent or experimental variable? How is this variable measured in the experimental design?
4. Have control and experimental groups been set up? How were subjects selected for the experiment, and how have they been allocated to control and experimental groups?
5. Are there any possible alternative causes apart from the experimental stimulus to account for changes in the experimental group?
6. Could the findings of this experiment be applied to real-life situations and experiences?
7. Are there any aspects of this experiment that raise ethical questions?

What You Have Learned

- The experiment is a technique designed to uncover causal relationships.
- Causal interpretation involves showing a clear time sequence, ensuring consistent association, and eliminating alternative explanations.
- Laboratory experiments allow researchers much control over the situation and require the use of carefully developed measures; random assignment of participants into experimental and control groups might also be required.
- Field and natural experiments provide greater realism but less experimental control.
- The Internet is providing a new area for conducting experimental research.
- Experiments have been valuable in social research, but they face serious challenges in terms of internal and external validity.
- Experiments with human participants often involve deception and stress, which raises difficult ethical issues.

Focus Questions

1. What are the main steps in a laboratory experiment?
2. What distinguishes a control group from an experimental group?
3. Why would researchers assign subjects randomly to the experimental and control groups?
4. What distinguishes a laboratory experiment from a field experiment?
5. What distinguishes a laboratory experiment from a natural experiment?
6. What are the advantages and disadvantages of field and natural experiments?

7. What are the advantages and disadvantages of Internet-based experiments?

8. Why is debriefing an essential part of experiments with human participants?

9. What are some important issues to consider when evaluating experimental research?

Review Exercises

Search for an article on an online database that is based on experimental research, or select one in the Research Info boxes in this chapter, and answer the following questions. Note that it is possible that not all these questions are answerable because of lack of information in the article, lack of clarity, or, in some cases, technical complexity, as in statistical exploration of the data.

1. What is the hypothesis tested by the experiment?

2. Why is this hypothesis of interest or concern to the researcher(s)?

3. What kind of experimental design has been set up for this research? That is, is it a pretest/post-test design? A control group design? A pretest/post-test control group design? Or some other design?

4. What is the independent or experimental variable? What is the dependent variable?

5. How have the variables been operationalized and measured in the experimental design? That is, is a video shown or a treatment given or tasks set as a stimulus? Are questions asked, reactions observed, or tests applied to study reactions?

6. How many participants are involved? How were they recruited, selected, or otherwise involved in the experiment?

7. Were the participants divided into experimental and control groups? How were they divided or allocated to the groups?

8. What were the research results? That is, what patterns and relationships are found in or can be inferred from the resulting data?

9. What conclusions do the researchers draw from the results? Do the results support the hypothesis? What do the results suggest for further research?

10. Do the researchers identify any limitations or problems with their research?

11. Do the researchers identify any ethical issues or problems with their research? Do they debrief the participants?

Weblinks

You can find these and other links at the companion website for *First Steps: A Guide to Social Research*, Fifth Edition: **http://www.firststeps5e.nelson.com.**

Experimental Method

The following websites highlight the importance of experimental design and the necessary steps to conduct an experiment.

Experiment–Resources.com

> **http://www.experiment-resources.com/index.html**

Research Methods Knowledge Base—Experimental Design

> http://www.socialresearchmethods.net/kb/desexper.php

Research Methods—Experimental Design

> http://allpsych.com/researchmethods/experimentaldesign.html

Random Assignment: Tools

The sites offer information on the importance of random assignment and online programs that can be used to assign participants randomly to different experimental conditions.

Research Randomizer

> http://www.randomizer.org

QuickCalcs: Random Assignment

> http://www.graphpad.com/quickcalcs/randomize1.cfm

Experiments (Ethical Issues: Texts and Videos)

Below are three famous experiments: Asch's experiment on group pressure, Milgram's obedience experiment, and Zimbardo's Stanford Prison Experiment. Each was influential in providing further insight into our understanding of human behaviour and how we change our behaviour in certain situations. But ethical concerns were also raised, especially of the latter two experiments.

The Asch experiment on group pressure
Article:

> http://www.experiment-resources.com/asch-experiment.html

Video:

> http://www.youtube.com/watch?v=iRh5qy09nNw&feature=related

The Stanley Milgram obedience experiment
Articles:

> http://www.experiment-resources.com/stanley-milgram-experiment.html
>
> http://www.experiment-resources.com/milgram-experiment-ethics.html

The Stanford Prison experiment (documents of original experiment) articles, slide show, and video:

> http://www.prisonexp.org

Video:

> http://www.youtube.com/watch?v=Z0jYx8nwjFQ&NR=1

Web-Based Experiments

The websites below provide information on how to conduct an experiment over the Internet, the benefits and challenges of conducting such research, and numerous links to mainly web-based psychology experiments.

A Guide to Running Surveys and Experiments on the World-Wide Web
    http://psychlaw.fiu.edu/guide/guide.asp

Psychological Research Online: Opportunities and Challenges
    http://www.apa.org/science/leadership/bsa/internet/internet-report.pdf

Psychological Research on the Net
    http://psych.hanover.edu/research/exponnet.html

Online Psychology Research
    http://www.onlinepsychresearch.co.uk/

Online Social Psychology Studies
    http://www.socialpsychology.org/expts.htm

7

Field Research

What You Will Learn

- the purpose of field research
- how to conduct nonparticipant observation research
- how to conduct semiparticipant and participant observation research
- the ethical issues associated with field research

INTRODUCTION

One day, several weeks after the semester has started, you walk into class to find a class-mate sitting at the desk you normally occupy. How would you feel? Until this disruption occurred you had not noticed that it was "your" desk. But you and your classmates quickly establish your territories in every classroom with very little thought. Some of you habitu-ally seek the back seats, some the front, others sit by a window, and so on. Once seated you mark your territory by various signs of ownership—relaxing your body posture, perhaps even stretching and sprawling out; putting your bag and coat on the desk; scattering pens, paper, and folders around.

Of course, in addition to such territorial activities, a lot of other social processes occur in the classroom. Social connections from outside are brought in, as indicated by groups of friends sitting together. Social connections are also made in the classroom as gangs, cliques, and partnerships form over the course of the term. As the course proceeds, you and your classmates will perform different roles in the class. You may be a "keener," coop-erating with the teacher by responding to questions and actively engaging in discussions. Or you may be the "class clown," mocking the lectures and discussions by pretending to be extremely stupid or making off-the-wall comments. Other roles you may recognize (or even play at) are the aloof, disconnected, silent type, or the disruptive, "don't give a damn" type.

Teachers represent a variety of types, as well: the very strict sergeant major types who bully a class into submission, the magnetic personalities who can charm and beguile, the jokers, and the storytellers. Finally, there is the complicated process of classroom dynamics. How do the students and teachers play out their roles or act out their moves? How do these roles and actions change as the term wears on?

You probably recognize a lot of your own classroom experience in the above descrip-tions. Thinking about it will probably lead to modifications and extensions of these ideas, and soon your classroom experience will never be the same again. It will no longer be a comfortable, semiconscious set of taken-for-granted routine activities, but rather some-thing that when analyzed shows a lot of complicated social activity. It is the aim of most fieldwork to describe social activities in depth so that these complications can be appreci-ated and understood.

When you begin to think about the social organization and actions in a classroom, it quickly becomes apparent that a great deal is going on. Observing all this is no simple task. It takes time and concentration. Observing a class while being in class will probably lead your attention away from what you are supposed to be studying. Being a student may be the only way to get inside the class, but observing at the same time as you are acting out the student role will be very challenging.

Field research requires researchers to go to the groups they wish to study. Instead of bringing human subjects to specially prepared experimental settings or taking carefully developed and pretested questionnaires to a select sample of respondents, the researcher goes into the field and observes the flow of social life as it naturally takes place. Unlike the field experiments discussed in Chapter 6, the field research we discuss here does not try to provoke reactions in order to test hypotheses. On the contrary, field researchers try to minimize the impact of their presence on those they observe, and let hypotheses or interpretations develop from their observations. Field research involves a wide range of research techniques, including observing individual and group behaviour; interviewing people; collecting documents and artifacts; and photographing, sketching, and mapping physical aspects of group life. (For the history of field research see Box 7.1.)

BOX 7.1 HISTORY OF FIELD RESEARCH

THE PRE-SCIENTIFIC ERA

Explorers, travellers, missionaries, and conquerors described the foreign lands they travel to. For example, Thucydides, a Greek general in the 5th century BCE, described the Persians and other peoples in the lands where his army fought, and Marco Polo described his travels to China in the 14th century. The aim of these writings was mixed: to inform, to entertain, and often to draw moral conclusions about appropriate and inappropriate ways of living.

The scientific revolution in 17th-century Europe led to the belief that social problems could be solved by actions based on truthful information. For example, John Howard travelled throughout Europe to study prison conditions in the late 18th century. His work led to the foundation of the John Howard Society, which was dedicated to the humane treatment of the jailed. Charles Booth lived with the London poor as part of a study of the lives of working people in the late 19th century. The aim of these writers was to apply the knowledge they obtained toward solving pressing social problems.

THE ERA OF SOCIAL SCIENTIFIC FIELD RESEARCH

By the turn of the century, the social sciences had emerged as academic disciplines; that is, they became concerned with knowledge for its own sake, rather than for its ability to solve moral and social issues. This change was reflected by individuals who promoted field research as a set of procedures for gaining knowledge. Bronislaw Malinowski was the first anthropologist to live with a group of people for a long period of time in order to study them. His work established the British school of social anthropology. Franz Boas was an American anthropologist

continued

concerned with the disappearance of Native people's culture as a result of the settlement and industrialization of North America. He trained his students in Native languages so that they could interview Native informants in order to build up a picture of their culture prior to the coming of White society. At the same time, the Chicago school of sociology developed under the leadership of Robert Park, who was trained as a journalist as well as a sociologist. He urged his students to go out into the streets and neighbourhoods and write about social life on the basis of direct observation and investigation. There was little systematic attention to the methods of field research.

This changed in the 1950s, when other social researchers developed strong arguments for the use of surveys and experimental logic. In response, those influenced by the Chicago school began to lay out their ideas in a more systematic way—a perspective that became known as *symbolic interactionism*. This approach emphasized the importance of understanding social activities and social settings in terms of the meanings they have for the participants. This method continued the Chicago tradition of creating descriptive, qualitative studies of neighbourhoods, marginal occupations, deviants, and social movements.

From the 1960s on, qualitative research became increasingly popular and controversial. Historians developed "oral history" techniques, such as in-depth interviews of survivors of significant recent events, and became increasingly sensitive to the elite bias of much historical documentation. Anthropologists were now studying non-European societies, which were urbanizing rapidly, and turned to sociological writings on fieldwork in urban settings to guide them. But symbolic interactionism came under increasing criticism from feminists and others for being too much concerned with defending itself from the scientific emphasis of other social researchers and too little concerned with developing a systematic, qualitative research perspective that could include an understanding of the importance of the emotional and irrational aspects of social life.

Field research has many applications. One use of this method is to study social behaviour so familiar that it consists of unconscious routines and activities. Another use of fieldwork is to study small-scale, intimate social settings, such as bars, community groups, and gangs. The social organization of the classroom is an example of a setting where both of these applications come together.

The Aims of Field Research

Field research can vary considerably in both its focus and its aims (see Box 7.2). Social anthropologists attempt to describe the way of life of an entire group of people; to do so in terms of the people's own understanding, experience, and values; and to capture the way of life as a complete whole. Sociologists do the same when they try to study an entire community, such as a small town, an ethnic group, or a neighbourhood in a big city.

BOX 7.2 WHAT DO FIELD RESEARCHERS DO?

THE AIMS OF FIELD RESEARCH

1. To discover and describe patterns of social life as they occur naturally.
2. To understand a group member's own point of view.
3. To get a complete picture of a group's way of life, beliefs, customs, etc.
4. To describe and account for patterns in social interaction from the perspectives of the participants.

PURPOSE OF THE FIELD RESEARCH PROCESS

1. To be "on the spot," recording social life as it occurs.
2. To be sufficiently involved with the group to be an eyewitness and/or to directly experience group life.
3. To observe group life without upsetting or disturbing it.
4. To understand the insider's point of view without losing the analytic perspective of an outsider.
5. To notice both explicit (recognized, conscious, spoken) and tacit (unacknowledged, implicit, unspoken) aspects of culture.
6. To notice both ordinary and unusual events and activities.
7. To see activities, events, and relationships as parts of a complete whole, not as unrelated items.
8. To use a variety of techniques and social skills in a flexible manner as the situation demands.
9. To produce data in the form of extensive written notes, maps, diagrams, and pictures, and where appropriate to record and collect data in the form of photos, video- and audiotapes, and artifacts of group life.

Other social researchers focus on a smaller field of study, such as a bar, a workplace, or a youth gang. As well, they often study a specific kind of experience or activity, such as a visit to a tattoo parlour or a day in the life of an addict, and they focus on the details of social interaction and communication that create and organize these "micro" social worlds.

As we pointed out in Chapter 2, there is a division within social research concerning the appropriate philosophy and general approach to social analysis. Field research is often associated with a "humanistic" approach emphasizing qualitative aspects of social phenomena. Many field researchers argue that the explanation of social behaviour has to be sought in the reasons and motives people operate with when they interact socially. Reasons and motives are culturally and historically determined, passed on through socialization, and modified through social interaction and experience. Consequently, understanding social life requires an understanding of underlying values and perspectives, and observation of the ways these are carried through and modified in social situations. Researchers carefully and systematically describe the social life of a group with the aim of understanding that way of life from the viewpoints of group members. When combined with a systematic description of physical and technological aspects of life, this approach is called **ethnography**—"ethno" meaning people or folk, and "graphy" meaning to describe or study.

Field research not only studies many different social phenomena, but it also uses various techniques or research tools, unlike the experiment or the questionnaire-based survey. The key to field research is its flexibility in making use of whatever techniques are most appropriate and fruitful, given the characteristics of the research setting or field.

The Varieties of Field Research

Unlike much survey and experimental research, most field research is the work of a single researcher, although team investigations are occasionally done. Therefore, the scope of the research is limited to intimate and small-scale institutions and groupings that have some clear separation or boundary in relation to the larger society. Social anthropologists typically study small-scale tribal societies or peasant communities that are geographically concentrated as well as partly isolated from other groups. Sociologists study small-scale groups, such as gangs, cults, occupations, or neighbourhoods, or places where people come together repeatedly, such as workplaces, bars, hospital wards, schools, jails, or shopping malls. Historians study groups of people with some experience in common, such as a war, an economic depression, or a political event. In addition, social researchers study events, such as riots, social movements, or accidents, in a "field research style," bringing together eyewitness accounts and experiences to discover the overall pattern or meaning of what was often a rapidly changing complex of events. This approach—the intensely focused, usually multi-methods study of a restricted range of phenomena in great depth—is often referred to as the **case study method**, which we shall discuss later in this chapter.

Field research can be used to study a variety of social groups and situations. It employs various techniques, including three basic patterns of or approaches to fieldwork, each defined by a specific type of relationship between the observer and the observed. (See Research Info 7.1.) In **nonparticipant observation** the observer goes "where the action is" and observes without interacting with the group under observation. The observer tries to blend into the background to watch and eavesdrop on those around, who carry on their actions unaware that they are being observed.

 Research Info 7.1 *Markets and Traditions*

DOES A MARKET ECONOMY DESTROY TRADITIONAL CUSTOMS?

There have been many debates about the impact of economic markets on other areas of social life since the industrial revolution. Today these debates centre on the idea of globalization. Is the world becoming socially homogenized as a result of the penetration of corporate enterprise? Curry (2003) reviews the current debates and argues that his research suggests that while customs may adjust to market forces, people do not abandon their cultural traditions as their economy becomes incorporated into the world market.

Curry focuses on a group of Papua–New Guinea islanders who had traditionally supported themselves by subsistence horticulture and hunting. They had been resettled and now were cash-crop cultivators of palm oil nuts, which they sold to a large offshore processing company. Curry had done research on this community before and used his contacts to set up a new study. He devoted seven months to ethnographic fieldwork, during which time he undertook many in-depth interviews and observed daily life, including community and ceremonial events. The interviews usually involved several members of a smallholder's family, frequently with refreshments and a great deal of free-flowing conversation not always directly related to the topic. The informality of the interviews allowed people to get comfortable with the researcher and to raise issues of concern to them. "Throwaway" comments were often useful leads that would not have emerged in structured interviews. Such leads would be followed up later in the interview and in interviewing others to check the validity and generalizability of a particular viewpoint.

The smallholders had tribal traditions of extensive gift-giving and resource sharing for ceremonials at weddings, births, funerals, and religious occasions. In the past, these gifts would be produced from gardens or the hunt, and many gift-giving occasions were tied to harvest and hunting seasonal cycles. The move to a different region and cash-crop agriculture had not eliminated the gift-giving, but the ceremonies were increasingly tied to the palm oil harvest and payment schedules. As well, traditional patterns of mutual aid were now extended to new situations, such as borrowing money. Bank loans were not matters of individual negotiation but were obligations assumed by an entire group of families in support of a loan for a single family. All these arrangements were bound together by a persisting framework of magical, religious, and kinship beliefs inherited from the past.

Source: G.N. Curry. "Moving beyond postdevelopment: Facilitating indigenous alternatives for 'development'." *Economic Geography*. Issue 79, (4) 2003, Pg. 405–424. InfoTrac record number: **A111532917.**

The anthropologist Edward Hall (1959) carried out such observations in Asia and the Middle East and noticed how social status is expressed by behaviour. He found that when families walk in public, the members walk in an obvious order. In the Middle East, husbands are preceded by their wives and children, whereas in the Far East, husbands walk ahead. Both patterns separate husbands from the rest of the family and clearly establish the wives' duty of looking after the children.

In North America these kinds of observational studies have looked at the "highway code" underlying pedestrian movement on city sidewalks, behaviour in subways, and the hidden rules of sociable touching. In each of these studies, direct observation of large numbers of naturally acting citizens, unaware that they were being observed, showed consistent, repeated patterns of behaviour. From these patterns the researchers inferred cultural rules and values that organized the behaviour. For many fieldwork researchers this is a major problem with nonparticipant observation. How can you be sure the observers' interpretation of cultural rules makes sense to the people whose behaviour is observed? Nonparticipant observation gets at people's *behaviour,* but it does not get at people's experience and understanding—the *meaning* of the behaviour in the social group where it occurs.

In order to get at the meaning of social life, field researchers argue that it is necessary to *participate* as well as to observe, although researchers may try to limit their participation and be "hangers-on." This practice is referred to as **semiparticipant observation**: the researchers join in social activities as a way of establishing trust but generally maintain some distance between themselves and the group.

A classic example of this approach was Erving Goffman's *Asylums* (1962), a study of mental hospitals and other closed institutions. Goffman realized that being too closely connected with one of the groups in an asylum would limit his access to the other groups. If he came in as a patient he could not easily mingle with the medical staff, the orderlies, or the nonprofessional staff. As a make-believe medic, he would not be totally trusted by patients, and nonmedical staff would probably treat him with reserve. To solve this problem, he obtained the hospital administration's agreement to create an entirely fictitious position with responsibilities that were so vague that he was not identified with any of the key subgroups in the mental hospital. Consequently, he was able to be everywhere and anywhere and speak to anyone at any time.

Other situations where semiparticipant observation is necessary include those where gender, age, or ethnicity clearly identify the observer as an outsider. Most anthropological studies are of this type, because the researcher is so clearly from another society. But the outsider status also applies to studies of youth by adults, of people with disabilities by the able-bodied, of slum dwellers by middle-class academics, and so on. This is the most common type of field research. Of course, semiparticipant observation does involve some loss of information, because some parts of the social group or subculture remain closed off to the researcher, who is an outsider.

To overcome such access problems, a few researchers have become full-fledged participants in the groups they have observed. A dramatic example of this **participant observation** approach comes from the world of journalism, rather than social science. John Howard Griffin, a White journalist, decided to explore American society as it was experienced by Blacks in the early 1960s. He did this by undergoing a painful series of skin coloration treatments (later, he developed skin cancer as a result of these treatments), which enabled him to pass as an African-American. He published a book, *Black Like Me* (1976), about his harrowing and painful experiences as a Black in an often hostile White society. This book had a big impact on popular opinion, because it provided White Americans with a sense of what it was like to be Black in America.

These three approaches to field research can be used alone or in combination with each other and with other research techniques. Each method has particular advantages and applications, as well as certain limitations, so let us look at each one separately.

NONPARTICIPANT OBSERVATION RESEARCH

Why Use Nonparticipant Observation?

Anthropologists, sociologists, and psychologists say that human beings are largely products of socialization into a culture. Not only are our values, beliefs, and attitudes shaped by learning how to think, believe, and act in social groups but also our emotions and physical responses are socially learned. Much of this learning takes place at a very early age, and we become so well trained that our conformity to social requirements becomes a habit and takes the form of routines that we are unaware of. It is only when we move into other social groups or see others with different rules and routines that we may notice the many rules we follow.

Because we are unaware of so many of our cultural rules, it is difficult to discover them through interviews. The interview could be confusing and uncomfortable and might lead to some resentment, or be broken off by the subject before it was finished. Observation, which does not require conversation, is a better way to discover the everyday, commonplace, nonverbal behaviours by which we unconsciously express cultural rules. Observation is also useful in studying young children, who may be operating according to cultural rules but are unable to explain what these rules are. Researchers also depend a great deal on observation in the early stages of conducting fieldwork in unfamiliar settings in order to pick up nonverbal clues to help them identify regular patterns of interaction and to distinguish the important from the unimportant in the flow of social life. Thus, observation is both a technique essential for the study of phenomena that are difficult to get at in other ways and an exploratory technique that prepares the way for, and may accompany, other techniques.

Choosing a Question

As with all research, the first step in an observational study is deciding what to study (see Box 7.3). For instance, social life is full of unconscious, nonverbal routines. Some are simply not open to study through observation by outsiders because they occur in private settings: undressing and dressing routines at bedtime or on waking, for example, or nonverbal patterns in sexual seduction. However, a vast array of nonverbal action takes place in accessible public places, such as parks, shopping malls, sidewalks, bars, cafés, schoolyards, and public transit vehicles. These locations and their associated social behaviour patterns offer many interesting possibilities for observational research. For example:

How do people avoid collisions or unintended intrusions?

How do people preserve spheres of privacy in the midst of crowds?

What are the differences between men and women in their public gestures?

What happens when there is a breakdown of social orderliness?

What adjustments to behaviour are made when places become crowded or when crowds thin out?

Do people change their behaviour when certain other groups appear, such as a group of nuns, a detachment of police or security guards, or a gang of children or teens?

Any of these questions, or others like them, would be an appropriate focus for an observational study.

BOX 7.3 STEPS IN NONPARTICIPANT OBSERVATION RESEARCH

1. Choosing a research question: an appropriate topic of study using this technique.
2. Identifying the location: where the actions of interest are likely to be seen.
3. Preparing to make observations: exploring the field, selecting sites, checking appropriate times.
4. Developing observational skills.
5. Doing a trial run.
6. Observing: watching, listening, collecting, and recording data.
7. Reanalyzing data and writing the research report.

Identifying the Location

Defining the focus will determine *where* the observations should take place. At one extreme, defining the study as an exploration of the rules of elevator behaviour will lead to a focus on a single type of setting: elevators. At the other extreme, investigating how people maintain their identity as loners in public places would require a very broad range of locations: bars, sidewalks, malls, parks, and so on.

Observation depends on being at the right place at the right time in an appropriate state of readiness. All this requires some preparation; you cannot just turn up and do it. How long can you stay in one spot without arousing people's suspicions? Are there police or security checks that might turn out to be a problem? Since observation takes time you have to consider some basic comforts and supports. Can you get to food and drink easily without interrupting the observation for too long? Where are the washrooms? Are there places you can go to write field notes quickly if the field is far from your home or place of study?

Clearly, locations have to be checked out to identify the best places for people-watching. Where can you place yourself so that you are inconspicuous (and therefore unlikely to interfere with what is going on) but you can see quite easily the kinds of behaviour you want to observe? Several careful inspections of the observation sites are necessary, then, to find out about these basic features.

Such explorations are also required for other reasons. Most public places have a rhythm of activity, which has to be discovered in order to get the best results. Cafés near warehouses and truck terminals may be intense hives of activity when most of us are asleep or just dragging ourselves out of bed. Bars and clubs may be really jumping only long after parents have turned in. Shopping malls are quiet in the mornings and populated by young mothers and senior citizens, become crowded with all sorts of people during the midday lunch period, and then quiet down, only to become more populated again as children and teenagers come out of school and people leave their places of work. Public places are populated by different types of people depending on the hour of the day, the day of the week, or—as in the case of parks, amusement parks, and public beaches—the season of the year. It is essential to know or to discover these cycles of activity before you draw any conclusions from your observations.

Getting a sense of the type of people who will populate the place or places where you are observing is also essential to finding out how you have to be dressed and behave in order to blend in and avoid attracting attention.

Developing Observational Skills

Another important reason for checking observation locations is the need to experience the difficulties and requirements of observation—in short, to train yourself to overcome the pitfalls and develop your observational skills. Even sparsely populated parks, malls, or school playgrounds can be full of distractions—people, behaviour, and events that are interesting but irrelevant to your study. Nothing much may happen for long periods of time and then along comes a flood of interesting activities. You may find yourself feeling very self-conscious, possibly even embarrassed, like an intruder who should not be there. These feelings can easily become a preoccupation and draw your attention away from the actions of others. You may be asked for directions, approached by a panhandler, or eyed by an attractive member of the opposite sex. Any of these events can take your mind off the task at hand. Experiencing these diversions and discomforts before you begin your real period of observation is useful.

You should include in these trial runs some training of your ability to observe with ease and alertness. You should not try to push yourself to the limits if you feel tired. Your observations will be less reliable and the quality of your study reduced. You need to learn to pace yourself and take breaks. Breaks can be used to bring your notes up to date or to review them to make sure they are legible, understandable, and do not leave out some important items that may still be fresh in your memory.

You can also develop your observational capacities during these pretests. You need to look for characteristic, repetitive, commonplace behaviours and populations; listen to the characteristic sounds of the setting, the levels and modes of communication (shouts, relaxed talk, formal talk, hushed whispers, nonverbal gestures, grimaces, etc.); get a sense of the *ambience* of the place or places (friendly or not, relaxed or not, casual or business-like and formal, etc.). Through all these observational acts you build up a picture of the settings or environment, the populations, and the actions and interactions of these populations.

Taking Notes

Finally, the trial observations will bring you face to face with the major problem of observation, namely, *recording your observations*. Scribbling notes in public places often sets you apart from the majority of the people there. Maybe you are a struggling writer at work on the next Great Canadian Novel. Perhaps you are a journalist or an undercover police officer. Possibly you are a student working on some essay notes, or someone writing a shopping list to replace the one left at home. Some of these identities (journalist or police officer) might be disruptive of people's natural behaviour; others (student or forgetful writer of lists) are unthreatening. You have to judge for yourself how open you can be in combining watching with recording.

You will find that recording can take away from watching and that it is often better to break up periods of observation by retreating to a convenient café to write your notes in the field in a full, clear, legible way while the details are still fresh in your mind. Not only do you give yourself a break from the pressures of observation, but writing your observations can focus your mind on what to look for when you return to observe.

Problems with Nonparticipant Observation

TIME AND EFFORT Unless you decide to videotape behaviour in public places (potentially a much more intrusive observation technique), observation is inexpensive but very time-consuming. To get good results, considerable preparation—checking out locations and pretesting—is necessary. Even then you may end up spending a long time with little to show for it. Unpleasant weather keeps people at home; an economic recession empties the bars, malls, and pleasure parks. Observation also makes great demands on the observer. We all differ considerably in our ability to keep focused in crowded, active situations; to be alert to subtle nuances of behaviour and action; to remain sensitive to visual and other stimuli; to witness and recall; and to act unobtrusively.

RELIABILITY Observation and related techniques (such as participant observation and case study analysis) are research methods where the *personal equation* (see Chapter 2) is very significant. Because observation depends on the observer's ability, checks on the reliability of the observations are necessary. Several observers in the same or a similar location can have their data carefully compared. During the training or trial observation period, observation can be done in tandem with another or several other more experienced observers. Ultimately, the replication of observation studies is an important check on the reliability of each study.

VALIDITY AND SAMPLING A single observer, or even a small team of observers, is also limited in ways that raise questions concerning the validity of observations. Observation is time-consuming and energy-draining for those who do it. Consequently, the amount of firsthand observation is quite limited and has to be concentrated on a few carefully chosen locations. Previewing sites before this selection is made is also probably restricted to places near the researcher's home or place of study. There is no random sampling from a carefully developed sampling frame. Locations are chosen for their suitability and accessibility, resulting in a convenience sample. As a result, observation studies may be uncertain bases for generalization.

THE KIND OF DATA COLLECTED A final problem associated with observation studies, one they share with qualitative techniques in general, arises from the kind of data they produce. All three variants of observation—nonparticipant, semiparticipant, and participant—produce what anthropologist Clifford Geertz (1973) has termed "thick description." Thick description is description that conveys the depth, richness, or complexity of social life. Thick description is lengthy, elaborate, and "literary," rather than concise and summarizing (as statistical description is). Field researchers tend to end up with a mass of detailed descriptions of settings, behaviours, gestures, spatial positioning, and so on. Such material is difficult to break down into typologies or to code. As a result, data analysis is often complicated and time-consuming.

SEMIPARTICIPANT AND PARTICIPANT OBSERVATION

Fieldwork observation may be viewed as a continuum ranging from pure observation to semiparticipant observation to fully participant observation. A participant observer "gathers data by participating in the daily life of the group or organization he studies" (Becker, 1958, p. 652). How much participation the observer engages in varies greatly, depending on the nature of the group and the commitment and perhaps even the foolhardiness or bravery of the observer. John Howard Griffin's act of changing his physical appearance is probably the most extreme case of becoming as far as possible indistinguishable from those you study.

The boundary between semiparticipant and participant observation is a matter of degree rather than a clear divide. Even fully participant observers are always partly outside the group because of their concern with describing and analyzing what is going on as it is happening. While ordinary people just get on with ordinary life, the observers continually pull back to take notes, to remember, to probe more deeply, so that they are always, however slightly, mentally hanging back from full participation.

In contrast to pure nonparticipant observation, a participant observer speaks to the people being observed. By conversing with the subjects of observation and participating in their lives, the researcher tries to discover the group's own experiences and interpretations of their activities, to find out the history of the group, and to understand aspects of group life that may not be accessible to direct observation. Yet many of the techniques of direct observation are still essential in participant observation.

A participant observer tries to gather whatever kinds of information are available in the situation, including observing behaviour directly; looking at written records, letters, diaries, and such; or taking note of physical aspects of the group (buildings, tools, clothes, arts, environment, resources, and so on). To record and analyze these different items of information, researchers often use diverse research techniques. Consequently, participant observation is much more complicated to describe and to undertake than nonparticipant observation. The participant observer is gathering varied types of data at the same time as he is caught up in the life of the group. He is interpreting what happens and what is spoken about and is continually making decisions about what events or actions have to be followed up, whom to speak to next, and so on. These tasks require a great deal of social alertness and the ability to be flexible and to improvise in the field of study.

Fieldwork is not easy to describe, because field situations vary enormously and are apt to change from day to day as the researchers encounter new aspects of group life, make new contacts, and so on. Not only do the groups studied vary and change during the fieldwork, but fieldworkers also differ in their styles of involvement and in their data-gathering observational techniques. Some people are more gregarious and at ease mixing in with the flow of interaction, talking to and questioning the people they meet. Others may be very good at self-effacement, at tagging along for the ride in less obtrusive ways, and eavesdropping. Both approaches can get good results, even though they employ very different personal approaches to social interaction. In this respect, fieldwork research is more like a craft or an art than a science. But this does not mean "Anything goes," as the following discussion of fieldwork procedures should make clear.

Defining the Focus

Participant or semiparticipant observation research is essentially a learning experience. Field researchers go into the field with a general sense of what they are looking for, but new questions, issues for analysis, or hypotheses often emerge as the fieldwork progresses.

The first step in a participant observation study is selecting a subsociety or subculture that is of interest and that you can gain access to. Potential subjects might be deviant or unconventional groups, such as delinquents or cult members; culturally distinct groups, such as ethnic and religious minorities; geographically limited locations where intense, repeated social activity occurs, such as in bars or on factory shop floors; or ongoing subcultures whose members share common concerns, values, and perspectives without definable territories, such as Trekkies or computer hackers. The key characteristic of all such groups is their distinctiveness, which sets them apart from wider society. This distinctiveness may set up barriers and accessibility problems, which you will have to deal with in preparing and setting up your research and often in conducting the fieldwork itself.

Preparing for the Study

A good field study requires a great deal of careful preparation. It requires good relations between the observer and the observed, and it depends on extensive access to the group in general and to individuals willing to act as guides, interpreters, and informers. The broader the range of cultural events, artifacts, records, and perspectives you have access to, the deeper will be your understanding of the subculture. Participant observation is therefore a wide-ranging, time-consuming process. Because of the time and the "people skills" involved, participant observation makes great demands on the ability of the researcher to be open and flexible. The researcher must make the best use of new opportunities for observation as they arise from the flow of life, and must cope with times when group life becomes less hospitable to outsiders.

The more you have studied the group and prepared the way for the field research—by carefully approaching potential informants or people with appropriate connections and by reviewing what is known about the group—the smoother your entrance into the group and the more at ease you will be once in the field. The more background you have obtained before entering the field, the less time you will need to become familiar with the setting. At the same time, you have to keep an open mind about the group and avoid letting this background information colour or bias your experiences and observations in the field.

Such background preparation is also necessary in order to decide the probable role that you will adopt in the field. The advantages and disadvantages of various types of participation during fieldwork are set out in Table 7.1. In some groups, such as outlaw motorcycle gangs and religious cults, you will have no choice but to participate fully or almost fully. In many other cases you have some choice in your degree of participation. Are you going to participate as fully as you can, or will you just tag along? How do you intend to identify yourself to the people you are going to study? Are you an independent researcher, or are you identified with a specific project, institution, or sponsor?

Table 7.1 Advantages and Disadvantages of Field Research Techniques

	Advantages	Disadvantages
Nonparticipant Observation	Limits chance of observer disturbing normal behaviour. Noninvolvement of observer encourages objectivity. Good for studying everyday behaviour in public places.	Provides access to behaviour patterns only. May involve problems interpreting what is happening without other data.
Semiparticipant Observation	Necessary when researcher is obvious outsider. Gives researcher direct access to group, and thus to members' beliefs and ideas. May be useful to be an outsider if group is divided by conflicts. Outsider role reinforces objectivity.	Observer's presence may alter group activities. Some aspects of group life may still be inaccessible.
Participant Observation	Necessary for the study of "closed groups," such as cults and criminal groups. Greater involvement of researcher leads to empathy and understanding of group. Greater involvement opens up range of group life to observation.	Can be very psychologically demanding. May involve personal danger or participation in illegal acts. Researcher may be trapped in a particular role. Requires a long period of preparation. Researcher may lose objectivity.

The more fully you participate, the greater the range of observations and experiences you can take in. But your involvement might itself influence the group. Extensive participation also increases the risk of becoming trapped on one side of social divisions existing within the group, and such participation also influences the researcher. The more involved in the group you become, the greater the risk of losing your objectivity and seeing things exclusively from the group's viewpoint.

In general, it is easier to be open about who you are and what you are doing than to conceal your identity and your research aims. The greater the gap between your account of yourself and reality, the greater the potential embarrassment and damage that accidental disclosure will bring. Going in openly as an observer makes interviewing informants, asking questions or being generally inquisitive, and recording data much easier. Recording observations in the field will be especially difficult if you have to take notes about everything while you are away from the events and people themselves. By being open about observing, you allow yourself some leeway to take the occasional note, perhaps even photos and videos, and to ask questions about things as they happen.

Entering the Field

As with nonparticipant observation, it can be extremely useful to visit the site of your field research before actually beginning the research. This visit allows you to become familiar with the sights, sounds, social activities, and other basic characteristics of the scene and gives you a sense of what to expect. You also become a familiar part of the scene yourself the more you visit it, so that you are partly established there when your fieldwork actually begins. These visits and the process of gradually fitting in may also be useful in making contacts. In fact, this technique may be the only way to make contact if you are studying very loosely organized, highly informal groups, such as youth gangs, addicts, prostitutes, or regulars at a bar.

It is simpler to gain entry into more organized settings, such as communities, hospitals, and workplaces. You might join a community organization; get a job at a workplace; or do volunteer work in hospitals, prisons, homes for the elderly, and so on. Being a researcher here is an extension of being a somewhat inquisitive joiner, volunteer, or new employee.

The major problem in the loosely organized field setting is, perhaps, the suspicion that such groups have of outsiders and the slow and sensitive process of working your way in and establishing trust. The major problem with the organized field setting is the risk of getting trapped by a particular role at a specific level of the organization, which limits your access to other people and parts of the field. Of course, entry into the field or group you have chosen to study is no guarantee of access to information. Trusting relationships have to be set up and maintained throughout the fieldwork. Field researchers have to be on their toes and alert to keep up this trust.

Doing Fieldwork

The initial period of fieldwork is spent getting used to the processes of observing and participating—learning how to fit into the group. While you should take a lot of notes, quickly learn how to "interview on the run," and observe as much as you can, you will need to reinterpret many of your early observations as you learn more about the group's life. Angry and aggressive gestures that at first seem to indicate tension and conflict may later be revealed as just a style of "coming on." Conversely, pleasant or polite behaviour may mask a lot of hostility, and politeness itself may be a very subtle way of expressing anger and resentment. As you participate more, you will observe new behaviours that will lead you to reinterpret what you have already seen, as well as repetitions of events that firm up some conclusions or interpretations you have already formulated.

As for recording what you see, note taking is the least intrusive technique—it is much less obvious than tape recording and videotaping, which require accompanying written description and presentation in a final report anyway. The best recording technique is to jot down the essentials in a pocket-sized notepad and to develop the ability to recall what people said, who was there, how they looked and acted, and where the incident took place.

Field interviews often have to be very brief, as they occur in lulls between periods of activity. Field researchers often have to interview and observe the setting simultaneously, which poses great difficulties if you have to take notes on the interview as well. Abbreviating the notes by writing down the topics of the conversation as they take place allows you to observe at the same time. You can fill in the details of the conversation later.

Whatever notes you have been able to take in the field should be written up in full every evening or in breaks during fieldwork. Because only brief notes are possible in the field itself, field researchers have to rely on memorizing what they see and hear in the field. The longer the period between the initial encounter and the full writing-up, the more unreliable the record because of probable lapses of memory. Since the point of fieldwork is to understand the way of life of a group from the group's own perspective, it is particularly important to accurately report conversations, recreating people's statements verbatim and preserving their terminology and speech patterns as closely as possible.

Collecting and Analyzing the Data

In addition to writing up field notes on a daily basis, you will soon find it necessary to review these notes; earlier misunderstandings can be cleared up in light of later knowledge. You should sort through the accumulating record to look for repeated patterns of events, activities, communications, and interrelationships that may provide working hypotheses and a frame of reference for further observations in the field. What connections between events and people do there seem to be? Are there alliances and conflicts that repeat themselves (or shift) from one situation to another? Do there appear to be hidden lines of communication, alliances, or relationships that explain otherwise puzzling actions and events? Do you need to look for some specific items of activity or see events, groups, or individuals you have not yet concentrated on? Are areas of observation still confusing, or do you have ideas that have too little support?

This process of sorting, sifting, and analyzing your field notes means that data analysis and **data collection** become much more closely connected than is the case with other research methods. (See the section in Chapter 9 on qualitative data analysis.) With other methods, you collect your data first, and *then* you analyze them. In field research the movement back and forth between data analysis and data collection can sometimes make it difficult to distinguish what actually happened or what something means to the group under observation; your best recollection and description of what happened or its meaning to the group; your selective interpretation of the event or meaning as you recorded it in your field notes on location; and your shifts in understanding or perception as you later transcribed, reanalyzed, and reinterpreted your field notes.

Although there is no absolute or perfect resolution of these problems, some precautions may be taken that allow the initial researchers and others to check on possible sources of unreliability:

1. The initial field notes (on table napkins or backs of envelopes, or in torn, crumpled, and dirty pocket notebooks) should be kept at least until the full report has been written and accepted by the appropriate reviewing groups.
2. The write-ups from these field notes should be kept and should be clearly identified as write-ups.
3. All notes should be clearly dated.
4. Field note reviews, analyses, and interpretations should either be written separately from the field note write-ups or clearly identified as secondary material. Writing these interpretations is best done in a clearly labelled, separate notebook, with appropriate dating of all entries.
5. It is often worth the effort of keeping a **personal journal** to document your own feelings, anxieties, perceptions, and responses to day-to-day events as the fieldwork occurs.

These entries too should be carefully dated and reviewed regularly to check for possible sources of bias, negative or positive emotional responses, blind spots, and so forth.

The procedures of data analysis in fieldwork are covered in Chapter 9.

Leaving the Field

As the fieldwork phase approaches completion, ethnographic researchers need to consider the process of leaving the field. Have you already made clear through your introduction to or involvement with the group that you will be moving on? If not, you need to give some thought to how to leave. Fieldwork may be an intimate research technique that involves building close, personal ties between the observer and the observed. Many fieldworkers report feelings of guilt as the time for leaving approaches. Sometimes the group has used the fieldworker's services or has expected the researcher to provide some service or benefit, and a bargain or implicit contract has been set up. "We'll tell you about our lives if you stand up for us in the courts" (in studies of deviants), or "publicize the injustices committed against us" (in studies of ethnic and religious minorities). Have the researchers delivered on their end of the bargain? Are they planning to?

If you are leaving with boxes full of field notes and other data, you are clearly indebted to the community or network you have studied. At the very least, you owe the group a courteous farewell by thanking everyone and acknowledging this debt. Apart from being good manners, leaving the field "in good standing" may allow you to keep contacts and to do some follow-up observation and interviewing that might improve your study or add to it at a later date.

THE ADVANTAGES AND DISADVANTAGES OF FIELDWORK

Advantages

Supporters of field research argue that it has four major advantages over any other kind of research technique: *holism, depth, appreciation of complexity,* and *meaningful understanding of subjects.* The importance attached to these advantages depends on the aim and focus of the researchers.

Social anthropologists often stress the first advantage, *holism,* which means that researchers can see the way of life of a group as a whole and see how its various parts fit together. Researchers can see how attitudes and actions fit or do not fit, how public and private spheres of life are separated and how people pass from one to another, how different parts of a group interrelate, and so on.

Depth refers to the fieldworker's ability to get behind the surface appearances of group life. Fieldwork makes possible access to parts of life that are normally closed to outsiders. In modern societies, with their great variety of ethnic and religious groups, subcultures, and countercultures, this special access to their inner life is very important. These societies are also permeated by stereotypes generated by the mass media, which makes it more important to develop techniques of social research that go beyond public opinion.

By *complexity,* field researchers mean that there is always more than meets the eye in social life. Social relationships are often ambiguous and ambivalent. Groups that are in long-term conflict tend to work out patterns of interdependence and tacit understandings that set limits to their conflicts. Teachers, for example, have techniques for handling or controlling class clowns. However, they may on occasion welcome the class clown as someone who breaks up normal routines and introduces an element of liveliness and alertness to the classroom. In addition, the class clown momentarily takes the spotlight away from the teacher and provides some comic relief for everyone in the class.

Researchers who focus on very specific and limited elements of social interactions—greetings, joke-telling, being waited on at a bar, and so on—find that these interactions involve many little rules, patterns, cues, and signs. These very commonplace, everyday pieces of social behaviour have great depth and complexity, which are best revealed by some ethnographic technique. As well, there are rhythms, cycles, or seasons to social life that occur over long time periods and that researchers, even participant observers, sometimes do not easily see because of their limited stay in the field. But ethnographers argue that these things have even less chance of being grasped by surveys, experiments, and limited forms of direct observation techniques.

Perhaps the most important advantage of fieldwork for many researchers is that it is a way of studying *how people understand their own lives.* Fieldwork is very much aimed at understanding the point of view of the people studied. What do their lives *mean* to them? What are their *reasons* for the way they do things? Getting at the meaning of a way of life is a crucial step beyond mere description: to understand meaning is to understand *why* people act the way they do. In this qualitative approach to social research, then, to understand meaning is to explain a way of life in its appropriate, human terms.

Disadvantages

The field researcher has to deal with five major problems. First, the very presence of an observer, however participatory you are, may alter the way the group operates. All field research requires a settling-in period, during which you become accepted as a member of the group or as just part of the scenery. Only after the group has accepted your presence will it settle back into its normal ways of behaving.

The second concern is that not only does the group have to get used to you but also you have to get used to the group. You will have to abandon your own culture and learn the group's ways of believing, thinking, and acting. If this resocialization does not occur, you risk **ethnocentrism**, imposing your own cultural interpretations on the lives of those you are observing.

The third concern is that you need to ensure that the methods used to reveal social life do not distort or create false pictures of that life. Interviews, however in-depth, may reveal how people think about their lives, but they do not necessarily show how they *act.* Interviews and conversations may produce a picture of ideals, beliefs, and thoughts but will leave out many other aspects of group life. In addition to observing people talking, you have to see how people actually behave, communicate with each other, express their emotions, break rules, and so on. Consequently, you have to use a variety of research techniques, each of which will reveal some part of social activity—a procedure known as **triangulation**. The information from each method can be used as a cross-check on the other data to provide a more reliable picture of the entire group's way of life or an in-depth look at a particular segment of life.

The fourth problem is that fieldwork often depends on special relations with specific individuals within a group or social scene. You will most likely have made contact with one or more of the group members prior to going into the field. These individuals will pave your way into the group and will probably continue to be important guides and **infor-mants** as the fieldwork proceeds. It is very important, then, that these contacts are indeed well informed and knowledgeable about the group and its culture. If they are high or low in the social "pecking order" or are marginal or deviant in relation to the group, their information will have certain biases, which you need to take into account.

The fifth problem is that field research can be a demanding and time-consuming process, and its success in many ways hinges on the particular abilities of the field researcher. This renders fieldwork one of the least *replicable* techniques of study. The long period of preparation and the time taken to properly prepare for and pursue the fieldwork means that repeated observations of the same community tend to be separated by very long time periods. Even replications using different communities involve considerable time lags. Because of these long time lapses, so much is likely to have changed that later researchers are no longer really repeating the study, but instead are investigating a much changed and, therefore, different group. As a result, it is difficult to tell the degree to which differences between earlier and later studies are due to different interpretations, researcher bias, and so forth, and how much is due to real social change. Lack of replicability means that we depend on one researcher's or one research team's interpretation of group life. Consequently, some social scientists argue that fieldwork is valuable as the first stage of a longer research program that should move from pure description toward more focused data gathering designed to test hypotheses.

BOX 7.5 QUESTIONS TO ASK ABOUT FIELD RESEARCH

1. What is the purpose of this study? Is it designed to describe the way of life of an entire community or era? Is it designed to describe the culture or subculture of a group or part of a group? Is it designed to identify the characteristics of a specific activity or experience?
2. What was the design of the field study? How much direct participation was involved? How deep and extensive was the participation? Did the researchers depend mainly on their observations alone, or did they gather other kinds of evidence? What was the length of time of the field research?
3. How much background information do the researchers provide about the interests and perspectives that led to the study? What do they tell you about their prefield preparation? What do they tell you about their own reactions to their experiences in the field?
4. Did the researchers depend on one or a few contacts, guides, or interpreters in the field? What was the status of these people in relation to the rest of the group(s) being studied?
5. Were the researchers excluded from any group activities? Did they obtain alternative sources of information on these activities?
6. Are there any indications that the researchers' social or cultural backgrounds affected their field research, their interpretations, or the reactions of those being studied?
7. Are there any aspects of the research, such as dependence on covert observation, that raise ethical questions?

INTERNET-BASED FIELD RESEARCH

The rise of the Internet and the rapid spread of computer ownership and use have opened up new opportunities for field researchers, in terms of both *what* they study and *how* they study. Web communication has reinforced preexisting social ties among certain groups, such as political opposition movements in authoritarian nations and ethnic and religious groups with a widely dispersed membership. Web communication has permitted ties to develop among individuals who may not have communicated before, such as sufferers from a particular health problem who now can form support and information groups on the Web. Internet and computer technology has created an entirely new set of subcultures and countercultures such as hacker "communities."

All of these kinds of groups are of interest to ethnographic researchers and may be accessed through Internet contact, in addition to or instead of traditional field research in which the researcher travels to the group's physical location and observes, interviews, and participates in group life. In addition, websites and textual documentation of discussion groups, bulletin boards, and so forth, are now a significant part of modern culture and deserve to be studied as significant contemporary artifacts in social life.

For example, before a serious, degenerative, nervous system disease limited his ability to travel, Gold (2003) had been an ethnographically oriented researcher on dispersed communities—particular ethnic or religious groups who live in widely scattered settlements around the globe. His research had shown that such communities are often held together by "weak ties" of nostalgia and sporadic communications, without ongoing, regular contacts common to other communities. But, because of his illness, Gold could no longer carry out research that involved extensive travel, and turned to an Internet-based support network of other people who were similarly afflicted. His experiences with this group led him to reflect on the parallels between the "virtual community" of Internet groups and the geographically dispersed communities of his earlier research.

Gold suggests that the Internet opens up for study widely dispersed groups, transcending distance and international boundaries. Internet groups tend to be open, with few controls over eavesdropping and participation. The lack of face-to-face involvement seems to render people more available to each other and to be more generally open in their communications. The ease of asking people to communicate offline via private e-mail also allows more in-depth communication without violating confidentiality and anonymity online. Mutual aid groups such as those Gold studied tend to be supportive of researchers, but he recognizes that this is not universal and that a major ethical dilemma exists concerning identifying oneself as a researcher.

A major shift in skills required of field researchers participating on the Internet is identified by Gold. Traditional field research involved the observation of physical dimensions of the field: gestures, actions, tone of voice, and so on. Internet research involves textual information and the researcher is forced to use literary and documentary analytical techniques to the exclusion of others or to a far greater extent than before.

Internet technology is helping researchers by making it easier than before to gain access to groups of people having a common interest or experience but having no other links to each other. For example, researchers interested in the subculture of Tolkien *(The*

Lord of the Rings), *Harry Potter* or *Star Trek* fans once had to wait for a convention to occur in order to study them. Now these fans are to be found in continuous communication on the web. At the same time, the Internet makes things easier for people to participate in research at home, at times of their own choosing.

For example, Moloney, Dietrich, Strickland, and Myerburg (2002) set up four discussion groups on the Internet to create a forum through which the experiences associated with the onset of midlife migraines could be explored. The aim of the research was exploratory, since little research on midlife migraines had been published. Twenty-two women between the ages of 40 and 55 were recruited by word of mouth, flyers, and Internet postings. Background information on their health and their migraine history were obtained through structured questionnaires, and four Internet discussion groups were set up. Participants used pseudonyms so that they would be anonymous both to fellow participants and to the discussion moderator. The moderators were primarily discussion facilitators rather than discussion "steerers." During the research each participant was interviewed in depth to obtain detailed case study data in addition to the Internet discussion records. At the end of the discussion group sessions, each participant was asked to evaluate the study by phone or in person through a semistructured interview, and a small payment was made for participation.

Moloney et al. found that using the Internet saved time and costs once the technology was set up. However, they strongly recommend that Internet contact with participants be supplemented by traditional, direct research tools, such as interviews and surveys. Such methods provide additional data that can be used to triangulate with the Internet data. Such methods also provide social contacts among the participants that motivate them to participate in the Internet activities.

Studying Internet-based groups, websites, and documents opens up new challenges for social research. Does the researcher reveal her identity and purpose to a web group on the first encounter or later on? Either move risks upsetting the group and potentially ending the research. Is "lurking" on the Internet the equivalent of covert observation, or is the openness of discussion groups and the like the equivalent of behaviour in public spaces and therefore open to public and research observation? These questions suggest that the issue of informed consent may be difficult to deal with in Internet research.

Issues of sampling are significant since computers and Internet access are expensive and use tends to be restricted to those with higher incomes and education. In addition, participants can easily disguise their identity so that the researcher may not know for sure the social background of all or many of the group under study. As well, participants may not be consistently involved in the discussion group, and the factors accounting for different levels of involvement may be important but difficult to identify.

Another set of issues arises from the fact that interaction on the Internet is primarily textual. The physical aspects of face-to-face interaction—gestures, facial expressions, voice tone, and so on—are unavailable to contextualize and deepen the meaning of the text. Without these physical signs much meaning may be lost or the wrong meaning imposed. The importance of textual data requires that Internet researchers acquire or deepen their skills in textual analysis techniques. In light of these problems, most researchers develop a multi-method approach to their research, combining Internet communication with phone or face-to-face interviews, mail surveys, and so on.

THE CASE STUDY METHOD

Fieldwork-based studies tend to be in-depth investigations of small communities, groups, networks, or organizations. Such research often involves a combination of data-gathering techniques, including different forms of observation, structured and unstructured interviews, biographical data, private and public documents, artifacts, and both quantitative and qualitative data interpretation. These features are characteristic of the case study approach or method. A case study refers to research based only on one or a few examples of interest that are studied in great depth and detail.

The **case study method** was established in medical practice before the rise of social science. Scientific medicine developed as medical practitioners learned to document symptoms carefully and systematically for each and every patient who consulted them. By accumulating accurate and detailed descriptions of many individual cases, medical practitioners might ultimately discover the various patterns of diseases. Sigmund Freud, trained in medical and biological research, used the data accumulated in the treatment of many people suffering from nervous disorders to draw conclusions about normal patterns of psychological functioning. Jean Piaget systematically observed his own children, using these observations as part of the empirical foundation for his theory of cognitive development. In medicine, psychiatry, and psychology, then, there is a tendency to use the case study as an initial phase in the development of a theory from which testable hypotheses may be generated and later tested by other means, such as experiments or large-scale, quantitative surveys.

However, in many areas of social science case studies are seen as necessary and valuable for reasons other than generating hypotheses. Historical studies involve intense, detailed analyses of widely varying documentary material (private, public, biographical, institutional, quantitative, qualitative), as well as artifacts and archaeological material requiring various kinds of research methods and interpretative techniques. Much of this research is aimed at emphasizing temporal and regional variations and differences, to highlight uniqueness or specificity between different cases. Historical studies in anthropology, geography, political science, and sociology often compare and contrast cases and attempt to identify elements of *both* difference and similarity. Economics often uses case studies to analyze the impact of contrasting economic policies, or to compare firms with different entrepreneurial strategies. Even where a case study has been undertaken by a researcher emphasizing its uniqueness, later researchers may well incorporate that "case" as a variant of the cases that they have studied, framing it in terms of common or general features found in other cases.

Because case studies differ widely in their methods and their aims, it is difficult to speak of a case study *method* outside of medicine and psychiatry. In the social sciences case study research involves intensely focused research on one or a few cases using a variety of methods, which depend on the discipline of the researcher. In terms of their aims, case studies may be ranged between two contrasting emphases. First, cases are studied as instances of a wider pattern of phenomena about which generalizations may be developed and tested. Second, cases are studied for their ability to reveal features of specific phenomena of interest on their own terms rather than as bases for generalization. In both of these situations selection of cases involves purposive sampling—that is, deliberately including cases that appear to fit the study's specific focus of interest.

ETHICAL CONSIDERATIONS

In fieldwork-based studies, two ethical issues are particularly marked: *invasion of privacy* and *deception*. Is the observation of your classmates and teachers an invasion of privacy? The case might be made that because you are looking at behaviour in public places, rather than intimate or private activities, it is not. However, your teachers and classmates do not know that they are being watched. They have not been given a choice to either accept or reject becoming subjects of your research. In this sense a basic ethical principle of all social research—**informed consent**—has been violated. Of course, asking for permission to observe classroom processes and revealing your intention to observe the behaviour of students and teachers may well lead to modifications of those behaviour patterns. You will no longer see the social dynamics of the classroom in their natural state, and thus the validity of your observations may be suspect.

Does the researcher's right to do research, then, override the general public's right to privacy and informed consent? There is no clear, rigid rule here. Responsible research involves carefully weighing the balance of rights and responsibilities.

Field research often places researchers in positions where they must make difficult ethical choices. Your observations of teachers and students might reveal some negative features of classroom life. Some students may have developed an effective technique for cheating on tests. A teacher may slant lessons in ways suggesting bigotry and prejudice against some ethnic or religious minority group. Do you take your findings to the student newspaper? Do you confront the cheating students or challenge the teacher? Do you go to a dean or academic ombudsperson? Or do you accept the unsavoury findings, submit them in your course essay, and leave your research methods teacher with these moral dilemmas? Once again there is no easy, standard answer to these kinds of questions.

The issue of deception is equally complex. Certain social groups, such as religious cults or extremist political sects, require **covert observation**. Covert observation occurs when researchers join or observe a group without revealing their true identities or purposes. Such concealed participant observation is a very controversial technique in the social sciences. Most researchers see covert observation as a last resort, to be used only when open observation is impossible and in situations in which significant new knowledge might be obtained.

Because participant observation research involves a high level of contact between the researchers and their research subjects, the issues of confidentiality and anonymity, or of protecting the research subjects from harm arising from publication of research results, are also important. Superficial changes of informants' names or place names are often insufficient to protect the anonymity of the group and its members. The very depth, richness, and holism of fieldwork studies often make it easy to work out which particular community and individuals are being described. Even if the researchers do their best to maintain anonymity, readers, reviewers, and journalists are often able to work out where the study occurred and who some of the subjects are. In the case of deviant groups, such breaches of anonymity or confidentiality may be especially damaging.

The exact nature of the ethical problems associated with fieldwork will vary from study to study. Studies of deviant groups or of public figures' private lives; studies of cults or other groups requiring covert observation techniques; and studies of controversial and divisive aspects of social life, such as abortion or racism, are likely to involve the greatest difficulties. There are no easy answers to the ethical problems associated with research. You will have to answer these basic ethical questions to your own satisfaction at every stage of the research process.

What You Have Learned

- Field research is an approach to social research that attempts to discover patterns in social life through careful observation of social activities in their normal or natural settings.
- There are three basic types of field research: nonparticipant observation, semiparticipant observation, and participant observation. Each has its own advantages and limitations.
- Field research often combines the three types of observation with each other and with other research techniques, such as documentary research or in-depth interviews.
- Field research involves qualitative data analysis techniques that are similar to those used in historical research but are different from the data analysis used in most experimental and survey research.
- Certain kinds of fieldwork, especially covert observation, raise difficult ethical questions.
- The Internet has altered both what field researchers study and the methods they use in fieldwork.

Focus Questions

1. What are the aims of field research? How does field research differ from survey and experimental research?
2. Why is nonparticipant observation useful? Which kinds of social phenomena is it best suited to study?
3. Why are reliability and validity significant problems in field research?
4. Why is semiparticipant observation necessary?
5. What are the advantages of openness in fieldwork?
6. Why is replication a significant issue in field research?
7. What special ethical problems are associated with fieldwork?

Review Exercises

1. Read the article cited in Research Info 7.1 and answer the following questions. Note that it is possible that not all of these questions are answerable because of lack of information or lack of clarity in the article.
   a. What is the purpose of this study? Is it purely descriptive or exploratory, or is it aimed at explanation or hypothesis testing?
   b. Is it designed to describe the way of life of an entire community? Is it designed to describe the culture or subculture of a group or part of a group within a community?

c. Is it designed to explore social change or is it cross-sectional or static in its approach?

d. What was the design of the field study? How much direct participation was involved? How deep and extensive was the participation? Did the researcher depend mainly on observations alone, or did he or she gather other kinds of evidence? What was the length of time of the field research?

e. Were arguments presented in defence of a qualitative approach? If so, what were these arguments?

f. How much background information does the researcher provide about the interests and perspectives that led to the study? What does the researcher tell you about his prefield preparation? What does he tell you about his or her own reactions to experiences in the field?

g. Did the researcher depend on one or a few contacts, guides, or interpreters in the field? What was the status of these people in relation to the rest of the group(s) being studied?

h. Was the researcher excluded from any group activities? Did he obtain alternative sources of information about these activities?

i. Are there any indications that the researcher's social or cultural background affected his or her field research, his or her interpretations, or the reactions of those being studied?

j. Do any aspects of the research, such as dependence on covert observation, raise ethical questions?

Weblinks

You can find these and other links at the companion website for *First Steps: A Guide to Social Research*, Fifth Edition: **http://www.firststeps5e.nelson.com**.

General Resources

The following sites have useful resources on conducting qualitative research, as well as other links to websites on related topics and electronic journals.

QualPage: Resources for Qualitative Research
> **http://www.qualitativeresearch.uga.edu/QualPage/**

Qualitative Research Resources on the Internet
> **http://www.nova.edu/ssss/QR/qualres.html**

Forum: Qualitative Social Research
> **http://www.qualitative-research.net/fqs/fqs-eng.htm**

The International Journal of Qualitative Methods
> **http://ejournals.library.ualberta.ca/index.php/IJQM/index**

The Qualitative Research Web Page (videos and book)
> **http://qualitativeresearch.ratcliffs.net**

Qualitative Methods Workbook

http://www.ship.edu/~cgboeree/qualmeth.html

Participant and Non-Participant Observation—Making the Right Choice (videos)

http://www.uea.ac.uk/ltg/activities/rlo/index.html

Case Studies

Writing Guide: Case Studies

http://writing.colostate.edu/guides/research/casestudy/index.cfm

Ethnographic Research

How to Do Ethnographic Research: A Simplified Guide

http://www.sas.upenn.edu/anthro/anthro/cpiamethods

A Synthesis of Ethnographic Research

http://www-bcf.usc.edu/~genzuk/Ethnographic_Research.html

Ethnographic Research

http://faculty.chass.ncsu.edu/garson/PA765/ethno.htm

Oral History

Oral History Society

http://www.ohs.org.uk/advice/

Oral History Primer

http://library.ucsc.edu/reg-hist/oral-history-primer

Ethical Conduct in Participant Observation

Guidelines for Ethical Conduct in Participant Observation

http://www.research.utoronto.ca/ethics/pdf/human/nonspecific/
Participant%20Observation%20Guidelines.pdf

Indirect or Nonreactive Methods

What You Will Learn

- why indirect research methods are necessary
- when and how to use unobtrusive measures
- when and how to use content analysis
- when and how to use available data
- when and how to use historical documents

INTRODUCTION

Suppose you have become interested in your schoolmates' level of health awareness and health-related lifestyles. You may have developed a questionnaire that included a number of questions on dietary or eating habits. On pretesting the questionnaire, you find that there is an apparently high level of awareness of healthy eating. Intuitively, this does not seem quite right to you. You are aware that you and your friends consume large amounts of pop, fries, and pastries in the cafeteria. The responses to some of your open-ended questions seem too pat, sounding like the lectures on healthy living you got in high school. Around the same time that this questionnaire is on your mind, your grandmother asks what you are doing at college, and you tell her about your survey on eating habits. She is not very impressed with the "health craze" and says that in her youth, people had so little choice they felt lucky to get three square meals a day and "no one worried about fat and fibre." This starts you wondering not only about your questionnaire but also about dietary history and modern-day obsessions with health.

Thinking about your study further, you begin to see that there are different ways of checking on students' eating habits apart from directly asking questions in a survey. You could become an observer in the cafeteria, or work with others and try to record the variety of food and drinks consumed. Perhaps you could persuade the cafeteria management to show you the sales records to find out which foods are most popular. If you have a really strong stomach (and some good work gloves), you could even go through the cafeteria trash to get a sense of food consumption patterns from the leftovers in the garbage!

Your rethinking might also lead you to become interested in the influence of high-school education on health awareness. You begin to dig out your old high-school folders and discover quite a lot of material, in the form of classroom handouts, pamphlets, and the like. But you now realize that you would have to read and analyze this material and carefully compare it with the statements in your survey if you are going to show the influence of schooling on health attitudes.

Your grandmother's comments have also sparked your curiosity, and you have now begun to pick up old paperbacks on dietary matters and health from garage sales. These books are both plentiful and cheap, and you begin to realize your grandmother was right: there is a "health craze." You begin to think that it would be interesting to put current dietary thinking into the context of this longer-term ebb and flow of dietary trends.

Your research has now grown more ambitious and probably more interesting. It has also become more complicated. You are moving away from relying solely on the direct research method of the survey—you have now added the three main **indirect or nonreactive methods** of getting information on the topic that interests you:

1. Using physical evidence of human activity (sorting through the cafeteria trash) is known as **unobtrusive measurement** of behaviour. It has long been used in archaeology and geography but has only recently come to be used as a technique by anthropologists, historians, and sociologists interested in contemporary social life and culture.
2. Analysis of ideas, themes, and topics in documents (school class materials, teachers' guides, etc.) is known as **content analysis** and has been used most prominently by political scientists.
3. Tracing changes in human behaviour and ideas by reading available documents (old paperbacks from garage sales) is also, of course, the primary method in historical research.

The research methods we looked at in Chapters 5, 6, and 7 are **direct research methods**. With the exception of nonparticipant observation, the subjects in these methods of research are aware that they are being studied. When researchers set up an experimental situation and recruit participants to be tested in some way, when they interview or mail questionnaires to selected respondents, or when they observe and perhaps participate in a group, participants, respondents, or actors will become aware that they are being studied.

As we have seen, this awareness sometimes creates difficulties for the researchers. Experimental subjects try to work out the purpose of the experiment and "help" the experiment by gearing their responses to what they think the researchers want. Survey researchers have to be extremely careful that their questions do not imply a "correct answer" or carry undertones of approval or disapproval of certain responses. Field researchers have to blend into the background if they are nonparticipant observers, or become accepted as part of the group if they participate. In either case, field researchers have to avoid changing the social activities they wish to observe. However, several research methods avoid responses from the people being studied by using indirect or nonreactive means of gathering data.

The various indirect methods of research may be used in relation to either quantitative data—data expressed in numbers and organized and interpreted statistically—or qualitative data—data expressed in words or symbols and interpreted the same way. In addition, secondary data analysis—the reanalysis of data that have been collected and organized for another purpose—can be useful to your own research.

UNOBTRUSIVE MEASURES

Why Do We Need Unobtrusive Measures?

Unobtrusive measurement uses the physical evidence or traces left behind by human activities. Archaeologists and physical anthropologists depend entirely on physical traces, which are the only evidence available from long-dead civilizations and ways of life. Human bones and possessions in old graves, flint chips, arrowheads, burnt bones from leftover meals, and broken pottery, as well as the remains of castles, pyramids, and palaces, are all important in helping us build a picture of human groups that have left no or few documents for us to interpret.

Social researchers studying contemporary societies can speak to people or observe them directly in action. Consequently, unobtrusive measurement has not been widely used. After all, direct observation and questioning are easier and more reliable. Observations can be repeated and checked by others; questions can be pretested before being used in case study interviews or in a survey. Guessing or inferring human actions and beliefs only from the traces left by human actions is extremely difficult and fairly unreliable. You do not see people in action; you do not discover their attitudes or beliefs directly through questioning or speaking to them. Why use a more difficult and unreliable method when direct methods are available?

As we have seen in our student eating habits research example, a direct method of research might not be reliable in some circumstances. When this is the case, other methods may be substituted or used as checks on the less reliable aspects of the direct research technique. As in our example of eating habits research, sometimes people are "paying lip service" or trying to live up to a social norm that has wide appeal. Of course, some students may honestly believe that they are eating well. Some students may be right but others are not, perhaps because they are unaware of their habit of snacking between meals. Habitual activities that we are unaware of can be investigated using nonparticipant observational methods, as discussed in Chapter 7. When habits also leave physical traces, such as pop cans, chip bags, and candy wrappers, it is possible also to use unobtrusive measures as a way to demonstrate that these habits exist and perhaps to estimate their importance or to infer connections with other aspects of social life. In one such case, urban researchers have looked at paths worn into the areas outside apartment complexes to map networks of interaction and sociability among apartment dwellers and between them and the surrounding neighbourhood.

Another area in which unobtrusive measures might be useful is in the study of inarticulate groups. For example, child psychologists might visit daycares and examine toys to look for patterns of wear as a way of discovering patterns of play among children of different ages. These researchers might also adopt the technique of urban researchers and look for paths worn in playground areas and parks to see how children use the spaces provided for them.

Unobtrusive measures might also be useful where activities are socially disapproved of or direct research would arouse hostility and deception. Studies of urban youth gangs might be examples here. Such studies are difficult because of the gangs' distrust of outsiders, and researchers are too old to become participant observers. Police and juvenile authorities or social workers may provide some useful information, but their major concerns are the control or deterrence of gang activities, and these goals may bias their reports. Consequently, some researchers have used indirect methods, such as the analysis of graffiti, to keep track of gang activity. Each gang defines its turf with graffiti that has a specific style and content. The graffiti operates as boundary markers and as expressions of gang identity and solidarity, and the imagery reflects values and concerns of the time.

Graffiti on public toilet walls has been studied by social scientists in an attempt to trace beliefs and attitudes that are socially disapproved of or that are viewed as private and not for public expression or display. As you might expect, researchers have found that sexual fantasies and obsessions, infantile preoccupations with excrement and body parts, extreme political views, racism, and bigotry are recurring themes.

Finally, perhaps the most extensive application of unobtrusive measures to the study of modern social life is in the field of industrial archaeology—a discipline that has grown rapidly since the 1960s. Unobtrusive measures are appropriate for studying human technological activities where "traces" (the wastes created by mining or manufacturing, for example), "erosions" (physical changes in the landscape created by uses such as building roads, railways, or canals, or quarrying and mining, logging, farming, etc.), or "accretions" (leftover buildings, disused machinery and tools, etc.) are left behind as a result of using the technology. Since the industrial revolution in the 18th century, technology has become an increasingly dominant part of our lives and has changed at an increasingly rapid pace. Keeping track of these changes has become an important part of social research, and more and more social research has become focused on technological change.

In industrial archaeology, looking for evidence of different technologies in such forms as disused mines, railways, or factory buildings and fragments of machinery and materials involves triangulation, the use of different methods to observe or produce data. Direct fieldwork and careful observation of the physical evidence (unobtrusive measures) are impossible without preliminary work with documents, such as local historical records and old maps. These documents are needed in order to identify areas where the technology is likely to be found and to pinpoint the possible locations of surviving physical evidence. Once discovered, physical remains need to be analyzed and interpreted, and again documentary and even interview-based research is necessary. Government and business records, catalogues of bankruptcy sales, postcards, and other documents are used to identify the physical items discovered. Surviving workers and their relatives (if the research is on the fairly recent past) may be interviewed to get details of day-to-day work routines and a more precise understanding of how things worked.

Advantages and Disadvantages of Unobtrusive Measures

Unobtrusive measures developed primarily as techniques for studying human behaviour without arousing reactions in those being studied. The approach was developed by social scientists who "rediscovered" the approach used by archaeologists, physical anthropologists, and historians studying extinct societies.

Unobtrusive measures may extend the range of social phenomena to be researched, or they may provide additional data in a study that also uses direct methods of research. Such measures have the advantage of being nonreactive. Consequently, where problems of interviewer effect, demand characteristics, or access to a group are likely to occur, using unobtrusive measures has many advantages.

However, there are several disadvantages or limitations associated with unobtrusive measures:

1. Sifting through physical traces in garbage cans, touring neighbourhoods in search of wall graffiti, and visiting such sites as parks or apartment complexes are very time-consuming. This is especially true if you attempt to develop a sampling procedure in order to use statistical techniques.
2. Unobtrusive measures are similar to covert observation in the sense that you are "going behind people's backs" as a deliberate research strategy. Therefore, this kind of research raises important ethical issues concerning rights of consent and privacy.
3. Unobtrusive measures are perhaps best used in conjunction with other methods. As archaeologists know, physical evidence by itself is very limited. The meanings and motivations, thoughts and beliefs that make up the social and cultural context of the physical traces cannot be easily or reliably inferred without additional data.

CONTENT ANALYSIS

In introducing this chapter we suggested that responses to surveys might reflect conformity to popular ideas rather than genuinely personal responses. Looking for the sources of these ideas and understanding why these ideas are so widespread would require a different research technique. Where do these ideas come from? How are they transmitted on a large scale? How are the ideas presented? What form of communication is involved? These kinds of questions lead to research into the social organization of communication, mass media, and analysis of communicated messages, including their senders and receivers or targets.

Content analysis is the systematic examination of communications or messages. **Quantitative content analysis** involves counting certain types of words in order to arrive at conclusions about the **manifest content** or intended meaning, and the **latent content** or the implicit, hidden, or unintended meaning of messages. Much content analysis is done of the mass media, such as newspapers and radio and television programs. Because of the sheer amount of such material available, content analysis has often been identified as a task for quantitative techniques that break down communications or messages into standardized units and organize and interpret the resulting data statistically.

Content analysis can also be applied to documents other than those produced by the mass media. This kind of analysis has a long history; in the 17th and 18th centuries, the Swedish Lutheran Church counted certain words in hymns and sermons in order to search out heretical ideas. Content is not always quantitative; **qualitative content analysis** is an approach that instead of counting words, identifies repeated and consistent themes, images, metaphors, and other meaningful traits within documents and other communication media. Either type of content analysis can also be applied to works of art, film, photographs, music, and even body language. Both quantitative and qualitative content analyses are concerned with discovering repeated or consistent themes and meanings in the messages that are analyzed. (See Research Info 8.1, 8.2, and 8.3, as well as the Weblinks at the end of the chapter.)

 Research Info 8.1   *The Decline in SAT Scores*

ARE DECLINING EDUCATIONAL TEST SCORES DUE TO SCHOOL EXAMINATIONS?

The debate in recent decades about the long-term declines in student test scores has mainly centred on the Scholastic Aptitude Test (SAT). But not all test results have declined—the results of some problem-solving tests have in fact improved. This pattern has led researcher J. E. C. Genovese (2002) to ask whether schooling itself has contributed to the decline in SAT scores and the like because schools have shifted their emphasis on the cognitive skills expected of students.

Genovese, therefore, tested the hypothesis that there has been a major shift in the skills promoted by public education in the United States during the 20th century. He undertook a content analysis of Ohio high-school entrance exams between 1902 and 1913, and the Ohio Grade 9 proficiency exams for the years 1997 through 1999. To support his hypothesis he had to establish that the schools' expectations of students' cognitive skills had clearly shifted from one era to another. The school exams, then, became the "texts" that he subjected to quantitative content analysis.

He created computerized text copies of the original exams and analyzed them using software content analysis programs. They were analyzed by counting and establishing the percentage of certain basic *prepositions* in the exams. The level of these prepositions in a text has been found to be a consistent and valid indicator of the complexity of ideas or meaning in a text. The text copies were also analyzed to determine the percentage of *knowledge content,* such as the names of countries of the world. And a third analysis searched and counted *object words,* which refer to basic physical objects such as tools, and *interrelation words,* which deal with mutual relationships or interrelationships.

Comparison of the 1902–1913 tests with the 1997–1999 tests showed that there had been a substantial decline in sentences using content words, but a substantial increase in the object words and interrelation words. Genovese also noted that the more object and interrelation words were used, the higher the proportion of prepositions found in the text. These prepositions were key indicators of the level of abstraction or complexity of thought expressed by, and required by, the exam questions.

In addition to analyzing the text of the exams, Genovese investigated the history of the growth of American public education and educational policy to better interpret his findings. He found that the goals of public education had shifted over time. Thus, Genovese combined a quantitative approach to the analysis of texts with an historical approach to the interpretation of the larger meaning or social context of these texts.

Source: J.E.C. Genovese. "Cognitive skills valued by educators: Historical content analysis of testing in Ohio." *Journal of Educational Research.* Issue 96 (2) (2002), Pg. 101–116.

 Research Info 8.2 *Media Images of Nurses during a Crisis*

WHAT WERE THE MEDIA IMAGES OF NURSES DURING THE SARS CRISIS?

In 2003 the world faced a serious outbreak of SARS (severe acute respiratory syndrome), caused by a previously unknown virus. It first appeared in late 2002, and while scientists suspected it originated in an area of China, they knew very little about what caused it or how it spread. This added to a general fear that SARS could spread around the world, but although cases were reported in different parts of the world, most were in Asia. Outside of Asia, only Toronto had to deal with the serious consequences of the SARS crisis. The SARS outbreak in Toronto was mainly contained within health care facilities, and those who were infected primarily contracted SARS from another person in a health care setting.

A group of researchers wanted to explore whether this extraordinary event generated media images of nurses that were different from those that had been well documented in the past. Although much content analysis of mass media messages or text has used a quantitative approach, their research on media portrayals of nurses takes a qualitative approach.

The researchers analyzed reports of the SARS crisis by 23 local and 12 national news media sources over a seven-week period during the height of the crisis in 2003. The "qualitative content" from these sources was subjected to an "iterative and reflexive" content analysis. That is, the content analysis procedure involved repeated readings of the texts to yield coherent, meaningful, repetitive, and consistent themes about nurses' expectations and concerns.

On the basis of this qualitative analysis procedure, several clear and consistent themes emerged. Examination of these themes in the light of previous media reports on nursing yielded both continuities and discontinuities between the SARS crisis reports and earlier ones. An example of continuity was the frequent moral characterization of nursing activities during the SARS crisis—nurses' selflessness, bravery, and so on. An example of discontinuity was an emerging portrayal of nurses as crucial parts of the health care team, as indispensable, and as technically advanced as doctors and other specialists.

The authors place these findings in the broader context of the current political and economic debates about health care, shifts toward an emphasis on patients' rights, and other institutional changes and pressures in the Canadian health care system.

Source: L.M. Hall, J. Angus, E. Peter, L. O'Brien-Pallas, F. Wynn, & G. Donner, "Media portrayal of nurses' perspectives and concerns in the SARS crisis in Toronto," *Journal of Nursing Scholarships.* Issue 35 (5) (2003), Pg. 211–217.

 Research Info 8.3   *Wrongly Imprisoned*

WHAT IS EXPERIENCED BY PEOPLE WHO HAVE BEEN WRONGFULLY IMPRISONED?

In recent years there have been various reported cases of people who had been wrongly imprisoned. This has sparked a growing concern about wrongful imprisonment, reflected in an increasing amount of social research and policy writing. But Campbell and Denov (2004) point out there is very little input from those suffering from the process. They therefore explore the experiences of those who have suffered from wrongful imprisonment and in doing so "give voice" to this overlooked marginal group.

For Campbell and Denov, qualitative content analysis is "suitable for research that attempts to uncover people's experiences and the meaning they make of these experiences." They undertook in-depth interviews with five wrongful-imprisonment victims, all Caucasian males with an average of five years' imprisonment. Four had been entirely exonerated; one was still fighting a lesser charge. The "texts" for their content analysis were the transcripts of the interviews. The statements were not treated as necessarily true or objective, but as revealing the experience of the individuals and the meanings they attributed to these experiences.

Each researcher read and annotated the transcripts to identify repeated themes. These annotations were then compared to develop a consistent framework for a second reading, focused on fully mapping out the themes. Each reader's experience of the second reading was compared to validate the framework. A third reading looked for individual variations in the way the themes were manifested, so that both common and variable features of the group's experiences could be identified and explored.

Campbell and Denov found that, as might be expected, the arrest, trial, and imprisonment for crimes that individuals did not commit produced a profoundly unhappy and disturbing set of experiences. Indeed, several supposedly ameliorating features of the prison system—parole and other review arrangements—essentially required the prisoners to first state they were guilty before they could be considered for moderation of their sentences. Even after exoneration and release, these individuals experienced considerable stigmatization and barriers to their full rehabilitation into normal life. The researchers argue that the experiences of wrongfully imprisoned individuals focus on several problems that should be addressed by a variety of criminal justice professionals.

Source: K. Cambell & M. Denov, "The burden of innocence: Coping with a wrongful imprisonment," *Canadian Journal of Criminology and Criminal Justice.* Issue 46 (2) (2004), Pg.139–164

In searching for themes in messages, content analysts may be looking for clues to the intentions of the broadcaster, writer, or promoter of the communication; the kind of audience the message is being aimed at; the unconscious or semiconscious influence of culture; the kinds of issues or concerns that people are focused on; or social changes and trends. To explain this process more clearly, let us return to our example of research on students and their eating habits.

An Example of Content Analysis

You have dug out your old lectures and class handouts on dietary matters and fitness. Perhaps you find references to books and pamphlets, and even to a video or two. What could you do with this material? You could read the written material and obtain and watch the videos and get an overall impression of the ideas they contain. Perhaps you could summarize these ideas to your own satisfaction and even provide some "representative" quotations in a paragraph or two in your research essay. But this summary could be easily challenged. What is the evidence? Are the quotations really representative, or are they atypical? Are the quotations accurate, or is their meaning distorted by being taken out of context?

In response, you could tell your teacher to look at the material herself, but it is doubtful that this would be accepted as a legitimate defence of your summary! What is required is some systematic, reliable, and objective procedure for establishing and summarizing the main ideas contained in the pamphlets, handouts, and videos. This is where content analysis comes in. Content analysis requires repeated, careful reading or viewing so that you can break down the content into message components or units of meaning, themes, images, keywords, and so on, and if possible provide some measures of the weight or importance of these components.

Units of Meaning

To break down the material into units of meaning, researchers look for items of text and images that occur repeatedly. In our example, we may find that certain foods are likely to be positively endorsed and others referred to negatively. This message should be easy to identify. A second simple message might be the way eating habits are connected to a healthy lifestyle or health in general. In other words, what are the specific benefits that are presented as the results of healthy eating? A third fairly direct aspect of the communication could be the way health is explained. Is there a lot of jargon? Are scientific reasons or studies referred to? Are authorities on the subject—doctors, medical and biological researchers, dieticians—quoted or referred to or featured in some way?

So far the content analysis is quite straightforward, dealing with very explicit ways in which the messages are presented. A somewhat more complex analysis might look at the way messages are presented, rather than at their direct content. Are there, for example, different ways of directing the information to male and female students? There may be differences in the language of the message, such as an emphasis on slimness and attractiveness for females, as opposed to an emphasis on strength and athletic abilities where males and eating habits are discussed. This difference might appear in the videos as images showing males and females in clearly different activities.

Other aspects of the way messages are presented might be the use of humour and of "horror stories." What rhetorical or cinematic techniques are used? How does the text or video presentation try to be "cool" and non-adult?

Coding

Systematically breaking down documents, programs, and such, into meaningful units requires a **coding system**. A coding system is a set of rules that tells you how to distinguish the content from the medium—the meaning from the text and images. These rules are clearly laid out so that anyone following them will get identical or nearly identical results.

Once you have defined what messages or units of meaning you are looking for, you need to organize a coding sheet listing these units. As you read through the literature or look at videos, you check off the messages specified on the coding sheet as they appear. You will use an identical coding sheet for every document or video, which will give you a consistent framework for identifying repeated ideas, themes, and images.

Examples of coding sheets—one for literature and one for videos—are provided in Table 8.1. Both sheets record what benefits of healthy eating are mentioned. The literature coding sheet records the number of times these benefits are mentioned. The video coding sheet also directs you to look for the way males and females in the audience are targeted and adds a new message unit to be recorded: the way males and females are presented.

Table 8.1 Content Analysis Coding Sheet

I. Health Literature

Theme	Number of Mentions	Style: Humorous/Serious	Target Reader: Male/Female
Importance of a good diet			
Specific health benefits			
Athletic benefits			
Physical/social attractiveness payoffs			

II. Health Videos and Gender Images

Theme	Number of Minutes	Style: Humorous/Serious
Females		
Social attractiveness for females		
Athleticism for females		
Other benefits for females		
Males		
Social attractiveness for males		
Athleticism for males		
Other benefits for males		

Many rules in a coding system may be quite simple, such as those for counting the frequency of a single word, a phrase, an image, or the space taken up by an idea or theme. Although the rules for counting may be simple, there are problems with this simple type of content analysis. Do very brief appearances or mentions get counted? Is the absence of a theme or an image important as a sign of a bias?

Measures

Identifying the various message components is often seen as just the first step in content analysis. You have shown that a particular message or way of presenting a message exists in a document, broadcast, or video, for example, but you have not really established *how important* the message or way of presenting it is. Most content analysis attempts to show the importance of a message by quantifying the analysis. This may be done in several different ways. Counting the number of times a message, theme, or idea is stated within a medium is the simplest. The amount of time or space taken up by a particular idea or image within the total document, broadcast, etc., is another simple measure. Identifying keywords and counting the number of times they appear is yet another simple measure of importance or salience. In mass media items, the importance of a message may be reflected by positioning, such as a front-page location in a newspaper or prime-time broadcasting on radio or television.

The two coding sheets in Table 8.1 are organized in terms of units of meaning (theme) and quantitative measures such as number of mentions and the number of minutes. As well, qualitative measures such as humour and gender are coded for. In text documents, the relative importance of an idea might be measured by the amount of text devoted to the idea as a proportion of the entire text; in videos, the measure might be the time devoted to specific images as a proportion of the length of the whole video.

Sampling

Content analysis is most often used to analyze large amounts of written, broadcast, or visual material, such as newspaper reports, advertisements, films, records, and radio and television broadcasts. The only practical way of analyzing so much material is to analyze samples. In the case we have discussed, sampling may not be practical or even necessary. Only a few standard educational pamphlets and videos about healthy eating may be used in a region, and obtaining out-of-province materials may be difficult. Getting access to mass media publications or broadcasts is another matter entirely. The sheer volume of the material often forces sampling on you.

Sampling for content analysis may be a tricky matter, because the units you sample and the units of observation (what you analyze) are not necessarily the same. For example, you may be interested in the way conservative versus liberal newspapers report large-scale strikes. To study this subject, you have to select some newspapers of each political leaning, identify some major industrial strikes in a given period, and read the editorials and major articles dealing with the strikes. You have selected a sample of issues of two types of newspapers, and perhaps you have selected two or more newspapers as representing political types. Here, you are sampling newspapers. What you look at and analyze with the aid of coding procedures will be editorials and articles.

Similarly, in analyzing ethnic or gender stereotyping in magazine advertising, you will sample magazines but analyze only advertisements. In television studies, researchers often sample types of programs but study the types of commercials in, or the sponsors of, those programs.

The method of content analysis we have explored so far is a quantitative approach. In order to analyze your findings you will need to become familiar with the techniques of quantitative data analysis, which are outlined in Chapter 9. There is also much scope for qualitative analysis of communications images, and this is what we now turn to.

Qualitative Content Analysis: Historical Research

In the course of your research on students' health awareness and eating habits, perhaps you noticed certain images in the school videos on health. All the students appearing in these videos, male and female, were very healthy, attractive, tanned individuals who obviously worked out regularly. While very pleasing to look at, they were definitely an unrepresentative sample of the students you know. It occurs to you that something is going on here that is not adequately presented in your measurement of units of meaning. The videos were presenting cultural ideals—desirable norms that are rarely achieved in real life. You begin to wonder about these ideals of physical appearance. Where do they come from? Have they changed over time?

It also occurs to you that visual images are very powerful and that in earlier times when few people read, such images would be very important. This idea leads you to search your textbooks on sociology, history, and anthropology for references to body images in different societies and at different times in history. These preliminary literature searches might lead you to discover historical works that use medieval paintings as sources of information on childhood and the family, to explore the ways paintings distort the human body in conformity with popular stereotypes, and to examine the persistence of images and stereotypes found in classical mythology, in premodern art and modern advertising.

Inspired by these social interpretations of art, you decide to visit the nearest museum of fine art and try your hand at qualitative analysis of visual images. But what do you look for? You are aware that the kinds of people represented in European paintings until fairly recently were mainly religious figures or the rich and powerful. You suspect that, like contemporary fashion photographers who spend time making up their models prior to photographing them, artists were careful in portraying their religious subjects and were flattering to the rich and powerful. But this should mean that the paintings really did show ideal images rather than realistic ones, and it is the ideal that you are searching for.

So what do these paintings (and perhaps sculptures as well) tell you about ideals of physical beauty? What are the overall body types presented (thin or fat, short or tall, well muscled or soft)? What complexions and skin tones seem to be favoured? What are the textures of flesh like? Are these qualities the same for both males and females? Do they change over time? What kinds of poses, activities, and settings predominate in these paintings, and how do these change over time? Do you see continuities as well as changes over time? Are there continuities and parallels between images in art and contemporary advertising?

And how would you write out your findings? Would you use verbal, descriptive techniques, or would you try to quantify the information? How would you measure flesh textures, poses, or settings? If you look back to the references you have found, you will realize that historical writing is largely qualitative, even though most historians analyze textual communications or records, rather than visual images such as paintings.

Perhaps the most famous example of such historical analysis is Max Weber's *The Protestant Ethic and the Spirit of Capitalism* (1958), which examined the sermons, writings, and teachings of notable Protestant leaders to make the argument that Protestantism was a source of cultural values allowing and promoting the rise of capitalism. Weber looked at moral values, especially the things that were identified as "evil" or "sinful" in Protestant teachings, to build up a picture of the cultural world of the different Protestant groups emerging during the Reformation. This analysis was not based on word counting or on using coding sheets. Instead, Weber used a method he called *verstehen*, or interpretive understanding. That is, he immersed himself in the writings of the Protestants in order to empathize with them, to experience the world as they saw it, and to explain their actions in terms of their experiences and their worldview.

A similar piece of historical research is E. P. Thompson's *The Making of the English Working Class* (1963). In contrast to Weber's study, which had the advantage of looking at a literate elite (Protestant reformers, churchmen, and other leaders), Thompson was studying an underclass whose members were largely illiterate and who left little documentary evidence of their own making for the historian to analyze.

Thompson wanted to discover the roots of English working-class culture and political traditions. He argued that these roots were found both in their initial reactions to the new types of work being forced on them by the rise of industrial factory employment and in criticisms of industrialism that circulated in the wider society. Thompson looked at the elite newspapers, pamphlets, letters, and other documents that complained about the new factory workers' lack of cooperation, their lack of discipline, and their unwillingness to abandon habits of pre-factory life. For Thompson, these complaints were valuable as clues to the ways working people reacted and felt about the rise of industrial work.

At the same time, radical ideas had spread to England from the French Revolution and found a sympathetic audience among craftsmen and small-business owners, who were negatively affected by the rise of the new industrial system. Many of these people were literate, and their pamphlets, petitions, and other documents provided Thompson with clues to an emerging political culture of opposition to and criticism of the new industrial economy.

Both Weber and Thompson used documentary sources and analyzed the content of these documents, but the techniques they used were qualitative, not quantitative. While there is some overlap between the two approaches (Weber looked for repeated themes or images expressing certain values or identifying certain actions as evil or sinful; Thompson looked for consistent references to factory workers' behaviour or criticisms of industrialism in middle-class pamphlets), the aims of the analyses are rather different.

Like Thompson, other historians have gained a better understanding of the less powerful and prominent social groups by examining various documents of the past. Barbara Hanawalt's (1993) work on childhood in medieval London illustrates this historical approach. Hanawalt was dissatisfied by the widespread acceptance of Aries' (1962) famous argument that childhood was not recognized as a distinct stage of life in European societies before the 1500s. As a specialist in medieval English social history, Hanawalt decided to test this idea using the abundant archival data available in London. London, as the capital city and a major economic centre in medieval times, is well documented by a great variety of documents from that era. Many court records survived from the royal government, the mayoral council governing the city, the church, and the many prominent artisanal and trade guilds. Between them, these courts dealt with disputes between citizens, taxation, inheritance and estates, crime, accidents, major social problems such as poverty, treatment of widows and orphans, and so on. Consequently, they provided a picture of many aspects of everyday life affecting all social classes. Taxation and inheritance cases provided a picture of the standard of living of different classes. Marriage and family disputes often dealt with in church courts provided much information on family life, relations between spouses, the norms surrounding treatment of children, and so on. Guild courts gave a picture of work involving both adults and children. The mayor's council discussed special problems such as crime and lawlessness often including the misbehaviour of young apprentices and children.

For Hanawalt the breadth and variety of court records gave a more representative picture of medieval society than dependence upon the artistic and literary sources that Aries favoured. The latter sources tend to reflect the ideals and way of life of the upper classes. Despite the breadth and variety of these documents, however, Hanawalt notes that she had to look hard for evidence of the lives of girls and young women, since these persons had much less independence and were treated essentially as the property of their fathers or husbands. Searching for the place and experiences of females she characterizes as a literary process of interpretation—"listening for silences and looking for absences" (Hanawalt, 1993, p. 15). The sheer volume of documentation led Hanawalt to employ three research assistants, including a specialist in statistical analysis. However, her research is primarily qualitative, sorting and sifting through the records in order to develop a picture of life in medieval London and the place of children in it.

Advantages and Disadvantages of Content Analysis

Quantitative analysis aims at showing the existence of certain ideas and themes in newspapers, broadcasts, and other media, and establishing how much of a presence or weight these ideas and themes have. It does this by showing how often words, phrases, or images occur; how much space or time is devoted to them; and so forth. Qualitative content analysis tries to show how ideas and images are interrelated into a complex, meaningful whole, which in turn organizes the experiences, perceptions, and actions of individuals or groups in a particular time and place. In quantitative content analysis, the emphasis is on establishing the existence and weight of units of meaning. In qualitative content analysis, breaking down and identifying units of meaning is one step in the process of interrelating these meanings into a larger, more complex whole.

Content analysis can be applied to all forms of human communication. It is a method with enormous scope, and it can focus on very important aspects of contemporary life because of the importance of mass media in today's society. The data are all around us and are, for the most part, easily accessible. The technique is nonreactive and consequently avoids the problems of reactions to interviewers and questions found in interview-based surveys and case studies, the attempts by participants to understand and cooperate with researchers and other problems arising in experiments, and the reactions of groups subjected to observation in fieldwork.

Content analysis is an inexpensive technique, requiring patience and care rather than complex equipment or lengthy training (see Box 8.1). It is also a flexible technique in that it can be undertaken by a single researcher or by a team. The technique is flexible in another sense, as well: it can be used as the sole or main technique, or it can be used in conjunction with other research techniques. For example, content analysis is often used to code the responses to open-ended questions in surveys.

BOX 8.1   HOW TO DO CONTENT ANALYSIS

1. Develop a research question that requires content analysis as a major data-gathering technique. For example: What ideas or messages circulate about…? What is the target audience? How has this message changed over time? Are there both explicit and implicit levels of meaning in these ideas?
2. Identify a range of media sources generating or carrying the messages—newspapers, television, radio, magazines, movies, or popular song lyrics and videos, etc.
3. Identify and construct the appropriate units of meaning—phrases, themes, pictorial images, sound bites, etc. Distinguish explicit and implicit levels of meaning associated with units of meaning. Assess the validity of your units of meaning.
4. Operationalize the units of meaning and develop a coding system. Test for the reliability of coding.
5. Develop measures of the strengths, consistency, and salience of the messages.
6. Establish a sampling procedure for the selection of a manageable and, if possible, representative sample of messages.
7. Gather the data—the messages—and break them down into units of meaning by applying the coding system.
8. Summarize the results of the coding, and organize and interpret the results in light of your research question(s).

As with all research methods, content analysis requires careful preparation. The themes, ideas, or units of meaning have to be clearly identified and operationalized ahead of time. The sampling process has to be carefully thought through. What books, newspapers, magazines, and films will be selected for analysis, and why? What is going to be focused on in these materials? How is the coding sheet set up? Is it clear, logical, and unambiguous? The reliability of coding is a major issue in content analysis. Does the coding sheet give consistent results? Can two different coders come up with nearly identical results using the same coding procedure?

Finally, while content analysis has developed mainly as a quantitative technique, questions of interpretation are often a major problem. Human communication is very complex. Messages may contain several layers of meaning, meanings that are ambiguous or contradictory, or meanings that are deliberately or unconsciously hidden. Counting keywords or phrases or visual images can be done fairly *reliably*, but it may be difficult to *validly* reflect the meaning of the messages (see Box 8.2).

BOX 8.2   QUESTIONS TO ASK ABOUT CONTENT ANALYSIS

1. What is the aim of this study? What argument is being made about the message or ideas involved? What media of communication are being focused on? What audience is implicated or studied?
2. How are the ideas or messages broken down into units of meaning?
3. How are units of meaning operationalized so that coding becomes possible? How is reliability of coding ensured?
4. How are the results of coding interpreted? How is the validity of interpretation ensured?
5. What sampling of media and/or messages is undertaken? Are there any problems with sampling?
6. Do the results support the research question or hypothesis that is being explored by this study?

AVAILABLE DATA AND SECONDARY DATA ANALYSIS

The indirect methods we have discussed so far work with **primary data**. That is, they either gather the data (e.g., by sorting and categorizing cafeteria garbage), or they use social products, such as documents, videos, and television programs, as data for analysis. Another very important area of nonreactive research is the use of **available data**—data that have already been gathered by governments, corporations, and other organizations, such as censuses or financial statements. In social research the reorganization, analysis, and reinterpretation of such data are known as **secondary data analysis** (see Box 8.3). Secondary data analysis is also widely used to check on and often to reinterpret the published findings of other social scientists.

For example, a study shows that there is a rise in divorce rates among couples who were under 25 years of age at the time of their divorce. The study considers whether the couples had children and concludes that childless couples are more likely to divorce. But you note that although the researchers provide information on how many years the couples had been married, they fail to consider this factor in their analysis. So you analyze the data these other researchers collected and take into account how long the marriage lasted. You find that most divorces occurred in the first years of marriage. Consequently, you rearrange and reanalyze the data that other researchers collected and arrive at another explanation: most divorces occurred in the early years of marriage, when couples were more likely to be childless.

Another example might be that the media report that last month 44 000 new jobs were created. No mention is made of the type of jobs. You wonder whether many were part-time jobs. The source of the information is the Statistics Canada Labour Force Survey. So you decide to look up the data in the monthly publication on the labour force available online at the Statistics Canada website. As you look up the information, other questions come to mind. Has there been in the last decade a rise in the proportion of the labour force that works part-time? Has there been a rise in the proportion of part-time workers who would prefer full-time work? Do youths account for an increasing proportion of part-time workers? To answer your questions, you realize that you would have to analyze data from the last decade. You need not only to look up the information compiled by Statistics Canada but also to carry out further analysis of the existing data.

BOX 8.3   HOW TO DO SECONDARY DATA ANALYSIS

1. Select and develop a research topic for which data on key variables are available.
2. Familiarize yourself with the available data set or sets. Examine the purpose(s) for which the data were gathered, the techniques and instruments used in the study, the range of variables involved, the assumptions underlying the research, and any limitations and problems with the data.
3. Select the appropriate ranges of data for your research purposes and reorganize the data to fit the needs of your study.
4. Summarize, analyze, and interpret the reorganized data.
5. Write up the results in the appropriate research paper format.

Carrying out an analysis on available data involves much more than simply reporting the data. You need to arrange and analyze the data others have gathered and arrive at your own interpretations. For example, say your study is on crime in major cities of Canada. To carry out your analysis you will probably have to rely on crime statistics compiled by government and nongovernment agencies. Such data would already have been examined by researchers of these agencies. In other words, you would not be the first to analyze the data, but you would use the same information to carry out some other analysis.

Studies analyzing data that already exist are quite common in the social sciences. Consider, for example, the economist who relies on published government data for a study on the rise in foreign direct investments in Canada; the political scientist who reanalyzes various Quebec referendum polls to contrast the voting behaviour of francophone and nonfrancophone youths; the sociologist who uses existing hospital data to determine changes in the length of time newborn babies are kept in the hospital; the psychologist who relies on official data on suicide to analyze trends among different age groups; and the historian who uses 1881 and 1891 census data to establish the settlement pattern of ethnic groups in the Prairies.

Why Use Available Data?

There are at least three reasons why social scientists analyze data gathered by others. A major reason is that the data are available, and it may be costly and possibly require enormous resources and time to re-collect the data. This is especially true in carrying out national sample surveys. Many such surveys are undertaken by or for government agencies, such as Statistics Canada. An example is the monthly Labour Force Survey, which uses a complicated sampling procedure. (For more on the Labour Force Survey, see Box 5.2.) The survey's data are organized and published regularly, are readily available, and could be used to carry out various analyses, such as examining changing trends in women's employment or youth unemployment.

Another reason for using existing data is to contest the original researchers' interpretation of their data. (See Research Info 8.4.) Say researchers did a study on the time students spend on schoolwork. The researchers conclude that students spend very little time on homework, but you question this interpretation, since the study was carried out in the third week of the semester. Students have fewer assignments, and this might affect the results. You reanalyze the data and note that most students had no written assignments. Moreover, students who did have written assignments spent more time on their homework than students who did not have such assignments. Your reanalysis of the data others gathered suggests a different interpretation of the findings.

Research Info 8.4   *Teenage Employment and Delinquency*

DO EMPLOYED TEENAGE STUDENTS HAVE A HEIGHTENED RISK OF COMMITTING DELINQUENT ACTS?

There is general consensus in the literature that teenage employment, mainly "intensive" employment, has a harmful effect. School grades are lowered, academic aspirations diminish, and problem behaviours like smoking, drinking, and using illegal drugs, as well as committing delinquent acts, are more likely. But according to Paternoster, Bushway, Brame, and Apel (2003) these views have been based on thin empirical support. Their main criticism is that previous research did not deal in general with a selection effect. Teenagers who work intensively could have already been less successful in school and showed other antisocial and problem behaviour. If so, then school-time employment is not the cause of their behaviour, but rather can be attributed to preexisting differences between the youth who do intensive work and those who do not.

In their research, the authors focused on the relationship between adolescents with intensive employment and delinquency/problem behaviour. They used data from the National Longitudinal Survey of Youth (NLSY) in the United States, carried out in 1997 and again in 1998 and 1999 on a representative sample of approximately 9000 youths born between 1980 and 1984. The initial purpose of the survey was to find out more about the school-to-work transition.

They carried out a lengthy analysis of the data and concluded that the relationship between intensive school-year work and delinquency/problem behaviours reflects preemployment differences that need to be further investigated. Hence, their research puts into question results from prior research that stressed employment has a harmful effect. But Paternoster et al. recognize certain weaknesses in their research. For example, they had little or no information about the youths' jobs, including the kind of jobs and the skills associated with the jobs, and they did not consider gender or ethnicity differences. They conclude that further research should take these factors into account.

Source: R. Paternoster, S. Bushway, R. Crame, & R. Apel, "The effect of teenage employment on delinquency and problem behaviors," *Social Forces.* Issue 82 (1) (2003), Pg. 297–336

A further reason for relying on available data is to analyze a different issue from that studied by the original researchers. Suppose that for the past decade, local high schools gathered data on the destinations of their graduates. You can use this existing data to analyze and compare, say, the trends of female and male students who pursue postsecondary education.

Searching for Available Data

Where you search for existing sources of data will depend on what kind of data you are looking for. One useful place to find data is increasingly at the websites of national statistics agencies and international organizations. (See the Weblinks at the end of the chapter.)

Perhaps one of the most used sources of existing data in Canada is Statistics Canada, the nation's central statistical agency. It collects data on population, health, education, culture, employment, income, economic issues, trade, tourism, energy, and much more. Some of its widely known programs are the Census of Population, the Labour Force Survey, the Consumer Price Index, the Gross Domestic Product, and the International Balance of Payments. No other organization, and certainly no individual, has the resources to obtain and collect much of the data compiled by Statistics Canada.

To learn more about the array of information available from Statistics Canada see the Statistics Canada website listing at the end of this chapter. You will also find various publications based on research carried out by Statistics Canada. These include *Canadian Social Trends, Perspectives on Labour and Income, Focus on Culture, Health Reports,* and *Canadian Economic Observer.*

Newly released data and a list of Statistics Canada's most recent publications are found in *The Daily,* published from Monday to Friday each week. *The Daily* has been published since 1932, but it is now available only online at the Statistics Canada website. It usually provides a brief overview of the latest data being released and is accompanied by tables and charts, as well as sources that offer more data and analyses. You can also use the "Search" feature to look up topics in back issues. For example, if you want to find out if data exist on homicides in Canada, you can use "homicide" as a keyword. You will then get a list of back issues of *The Daily* in which the word appears, as well as links to other related documents.

Various databases are available that provide an enormous amount of information about Canada. A key database is E-STAT, an interactive learning tool of Statistics Canada available to the educational community. It contains significant information, both current and historical. The data can be rearranged and displayed in tabular or graphic form, or you can download and export the data into your own spreadsheet program. Also available on E-STAT are data from the Canadian Socioeconomic Information Management System (CANSIM), which allows users to track trends on a wide range of subjects, such as labour, trade, education, and much more.

Evaluating and Understanding Available Data

Any source of information, whether it is a source of ideas, or of qualitative or quantitative data, has to be assessed for its credibility and validity. All data, both qualitative and quantitative, are the product of research decisions and choices about the aims of the research, the data-gathering tools, sampling techniques, and so on. As a user of the resulting data, you have to live with the researchers' choices and decisions. The data would have been different if the choices were different, hence it is essential that you know how and why the data were generated. In other words, you have to take note of the research methods used to gather or create the data.

Credible research data sources provide information about the methods used in the research. In **peer-reviewed research**, the norm is that qualified consumers of the research (peers) should be able to replicate it so as to build up a body of strong, consistent, verified studies that become a firm foundation for further research. Government and international research bodies, such as national census departments, the International Labour Organization, and the World Health Organization, publish information on their research methods in order to establish standardized data-gathering approaches to support governmental policymaking, which requires high-quality data that are consistent over decades. Certain research foundations and survey organizations also routinely present both research findings and the methodology of their studies. Many other data sources, such as lobby groups, advocacy groups, and private think tanks, are less consistent. The less information that you have about the research methods underlying your available data, the more cautious you should be, and the more you should try to cross-check the validity of at least some of the data with other more reliable sources.

Data available online from government and nongovernment agencies, as well as international organizations, are most likely presented in tabular form together with visual, graphical presentations. However, data presented in tables and graphs should be "read" critically—that is, you should understand the material and assess its credibility, especially when you intend to reinterpret the data. The tables and graphs first need a "questioning" reading in order to interpret and assess them. To assist you in understanding the basic descriptive presentation of data in tables and graphs, which you are likely to find in publications such as those of Statistics Canada and in research reports, consider the following questions. (Further details on constructing tables and graphs from data are discussed in Chapter 9.)

- What is the subject of the table or graph? What does the title tell you?
- What are the sources of the data? Are these primary data gathered by the researcher, or are they data from other sources—for example, from the census—reorganized or extracted by the researcher?
- What variable or variables are the focus of the table or graph? This should be clear from examining the title and also the headings identifying the columns and rows of a table, or the axes of a graph.

- How are the variables measured, and what are the units of measurement? There may be notes identifying any complexities or difficulties associated with such measurement.
- What patterns emerge from the tabular or graphical display? Can you see or infer relationships from the numbers in a table, or trends in a line graph, or other patterns such as concentration or dispersion?
- Are the patterns that appear in the table or graph consistent with the statements made in the body of the report?

Problems with Available Data

Data you gather become meaningful only after you arrange, analyze, and interpret them. This is also true of data gathered by others. However, available data were gathered on the basis of the assumptions and concepts of the original researchers, and while you may prefer some other information, you can do nothing about it. You have to accept the data as given and, obviously, cannot change how the original researchers carried out the study. Therefore, you need to put much effort into finding the shortcomings and weaknesses of available data and original research. A good deal of your effort in analyzing the data is likely to rely on statistical procedures. (We cover some basic statistical techniques in Chapter 9.)

In using available data, therefore, it is essential to understand the aim and purpose of the original research (see Box 8.4). How were the data collected? What was the target population? What was the sample, and how was it selected? What were the operational definitions of key variables? What were the limitations of the original research and the data collected? For example, if you were to use the published unemployment data from the Labour Force Survey, it would be crucial for you to know how the survey was carried out and the survey's operational definitions of "labour force" and "unemployment" (sce Chapter 5, Box 5.2). Although you might prefer another definition, or to ask other questions, you have no choice but to accept their operational definition of unemployment if you want to use the data collected by the original survey.

BOX 8.4   QUESTIONS TO ASK ABOUT AVAILABLE DATA

1. What is the original source or sources of the data? What was the original purpose of the data? What assumptions and methods were involved in gathering or creating these data?
2. What are the differences between the purpose, assumptions, and methods of the original study and those of the current study?
3. When were the data collected? What is the time frame of the original data? What is the time frame considered in the current study?
4. Are there gaps or discontinuities in the data? How are these dealt with? Are any new data provided to update or fill in the gaps? If yes, have the new data been gathered using similar assumptions and methods as the original data?
5. How have the data been reanalyzed?

Archival Research

The recent growth in popularity of genealogical searches—the search for family ancestors—involves the use of information sources known as **archives**. Archival research is basic to historical research throughout the social sciences. The term *archive* originally referred to a storage place for public records or **government documents**. The term now may refer either to the location or to a set of documents stored together, and the location or storage is increasingly on the Internet. Library and Archives Canada, physically located in Ottawa, provides the most extensive collection of records about Canada's history, and is the usual starting point for both private individuals and professional social researchers concerned with Canadian issues. Except for certain records governed by privacy and access-to-information laws, most files are open to the public. Among the major documentary records available are:

- Census records up to 1901, which provide information on immigration, occupation, demographic indicators such as birth and death rates, and so on
- Land petitions and land ownership maps. The former are especially important for research into the United Empire Loyalists, for example.
- Passenger lists of ships bringing immigrants to major Canadian ports
- Early immigration files of both the Canadian government and other governments, such as the Russian and Chinese governments, which kept lists of emigrants from their lands
- Records of sick and dying immigrants admitted to the Grosse Île hospital
- Archives of music, art, photography, media broadcasts, and so on

There are many other archival sources located in university libraries, provincial and local government centres, religious and other community organizations, local historical societies, businesses, and other private organizations. In all these cases the material stored may take various forms: it might be in original form or stored on microfilm or other storage media, or in full-text or abbreviated or summarized form. Computerization has had several positive effects on archival storage. It has led to much more systematic and organized storage and recording of archival material in order to generate databases that are accessible through standard search procedures. It has opened up access to records because of the ease and cost effectiveness of electronic data storage. It has made archives more accessible through Internet research in place of travelling to the actual physical location of the archives. Of course, there are many private and community-based archives that may not have the resources to digitize their records and artifacts.

Social science researchers interested in demography, family organization, urbanization, race and ethnic relations, shifts in fashions, and mass culture routinely use various archival resources. Their research may be qualitative or quantitative or a combination of both. While there is an overlap between these social scientists and the ordinary citizen-genealogist in terms of their interest in archives, some significant differences exist in the way each assesses and uses the archival data. The lay genealogist's concern is a purely private one of discovering evidence for family connections in the past. Much of the information may be taken on trust with little attempt to systematically evaluate its credibility or to look at alternative supporting sources. The social science researcher has a professional obligation to be both rigorous and skeptical about the data. As well, the social researcher will have a broader analytical research interest than the lay genealogical researcher. Even when the research is for a historical biography focusing on the life of an individual, all sources have to be assessed for their credibility, and the individual's life is placed in a larger social, cultural, and historical context, which is itself carefully documented. Furthermore, the entire research process follows the same standards as set by the broader professional community.

The Historical Research Process

The process of historical research, as with social science research in general, begins with asking research questions. But whereas in other social science research the questions eventually lead us to select a research method to compile the data, in historical research the questions often direct us to search for specific documents. Hence, a significant characteristic of historical research is that it mainly rests upon searching, interpreting, and analyzing documents. A document in historical research refers to fragments from the past and can be almost anything, including letters, court records, newspaper articles, maps, and more.

In carrying out historical research the more focused your questions, the clearer the boundaries of your research and, at least initially, the documents you need to consider. Say you want to research the early development of an ethnic group at the turn of the 19th century in a certain city. You should begin by asking what you want to find out, and this in turn will guide you to consider the documents to search for. If you want to understand the ethnic group's settlement pattern, you may consider searching for census data and other documents of the period that illustrate where members of the ethnic group resided in the city. There are no constraints on the documents to look for—they can, for example, include church and business records, government reports, photographs, maps, autobiographies, and so on. But when using the documents you must always understand them as products of their time and place and recognize that the documents are not equally useful. So a key challenge is to take care as to which ones to use and how you use, interpret, and analyze them. As well, be aware that they will in turn likely generate further questions that will require the search for more documents to interpret and analyze.

Consequently, there are significant differences between much historical research and other social science research that tends to deal with contemporary life. The historical focus on past life means that archival documentary data and its interpretation and analysis is at the core of the research process, while interviews, field research, experiments, and surveys play a secondary, supportive role, if any. The search for appropriate documents and establishing their place in the time of their creation is a fundamental part of historical research and will likely require careful and systematic interpretation and cross-referencing with other data sources from the same time period. Documents cannot be used without establishing their authenticity: showing that they were produced by specific people in the time and place of the historian's concern. The other social science fields look at available data in terms of the reliability and validity of the methods that have been used. This is generally a simpler process than the historian's research into why a document or artifact was created, by whom, with what purpose, and in what context before accepting the validity of the data in order to proceed with the research question.

Other significant issues are marked in historical research. Going back in time, major shifts in language, experience, and culture have to be taken into account, and there may be gaps in the records and other documents. The researcher has to be constantly aware of limits to the documents and the extent to which inferences can be drawn despite the gaps. Further, although certain historical documents such as church and government records, land registers, and the like can be quantified with relative ease, many historical documents, such as paintings, diaries, and religious writings, are qualitative in nature. This qualitative character, in combination with the need for historical understanding of the documents' origins, the need for interpreting the documents in terms of the life and culture of the time, and the need for making reasonable inferences from fragmentary documents, requires the historical researcher to combine rigorous empirical and logical analysis with imagination and intuition more consistently than is the case perhaps for the other social sciences. (To try your hand at putting together pieces of historical material and acquire skills in learning how to do historical research, see the websites under Historical Research at the end of the chapter.)

Table 8.2 Advantages and Disadvantages of Indirect Measures

Unobtrusive Measures

Using physical evidence and traces of human activities

Advantages	Disadvantages
Useful where surveys cannot be used because of resistance to revealing the truth.	Physical evidence alone may be difficult to interpret without more context.
Reveal habitual, taken-for-granted, and unconscious elements in social activities.	Physical evidence is only a small part of the social action or social phenomenon studied.
Useful for studying inarticulate groups, such as children.	Often time-consuming.
Useful for studying deviant and disapproved-of activities.	May be physically hazardous to the researcher.
Provide additional, objective data to cross-check interviews and other data.	May involve serious ethical questions similar to those involved in covert observation.

Content Analysis

The systematic investigation of human communications and messages

Advantages	Disadvantages
Can be applied to all forms of communication: print, audiovisual, physical gestures, speech, etc.	Sampling issues need to be carefully considered.
Can be applied to explicit and implicit levels of communications.	Meaning in human communications is very complex, so coding procedures have to be developed that do not distort or confuse meanings.
Can use either quantitative and qualitative techniques, or both.	

(continued)

Table 8.2 *(Continued)*

Available Data (Secondary Data Analysis)

The exploration, reorganization, and analysis of data already gathered by others, generally for a purpose different from that of the social researcher now using it

Advantages	Disadvantages
Allows analysis of data of higher quality and scope than most individuals could obtain by gathering the data themselves.	Requires full understanding of the aims, assumptions, and techniques involved in the original research.
The data can involve long time periods or large samples.	The researcher may be unable to determine or may be unaware of large gaps in the data.
The time and other resources saved in data gathering can be used to analyze and interpret data more extensively.	Long-term time series data often involve shifts in operational definitions and of key variables.

What You Have Learned

- Special techniques have been developed to help overcome the problem of human reactions to being observed, surveyed, or experimented with.
- Unobtrusive measures are useful for studying habitual actions, inarticulate or hostile groups, attitudes and emotions that society disapproves of, and technological, architectural, and material traces of human life.
- Content analysis focuses on human communications and can be used in combination with both quantitative and qualitative data analysis.
- Available data use takes advantage of existing data generated for other purposes or for previous research and can be used to settle research disputes, to present alternative interpretations of other research data, or as the basis for an entirely different approach to the data.
- Computer and Internet technology is increasing the range of data available and making it easier to search for data.
- Archives are accumulations of records, documents, and data that are stored in an accessible way and valuable for many different social scientists.
- Much historical research relies on archival and documentary data, and interpretation and analysis of such data is at the core of the research process.

Focus Questions

1. Why have indirect and nonreactive methods of studying social phenomena been developed?
2. When are unobtrusive research methods useful? What is the major difficulty with such techniques?
3. Why have unobtrusive measures become increasingly used in the study of modern industrial societies?

4. What kinds of research topics may be explored through the technique of content analysis?

5. What are some important difficulties associated with content analysis?

6. What ethical issues might arise in using different types of unobtrusive or indirect methods research?

7. Why do social scientists use data gathered by others?

8. What are archives? Why are they useful to social scientists?

9. What are significant differences between historical research and other social science research?

Review Exercises

1. Select one of the articles in Research Info 8.1 to 8.3 and answer the following questions. Note that it is possible that not all of these questions are answerable, because of lack of information in the article, lack of clarity, or in some cases, technical complexity, as in statistical exploration of the data.

   a. What is the aim of this study? In other words, is it attempting to test a hypothesis, to explore an issue, or to describe some social process, event, or situation?

   b. Is the research based on a quantitative or a qualitative approach to content analysis? Do the researchers explain or justify their choice of approach?

   c. What is the "text" or the medium that is being subjected to content analysis? That is, is it textual content such as newspaper articles, website text, letters, etc.? Or is it audio, visual, or audiovisual content, such as radio or television programs, photographs, tattoos, graffiti tags, etc.?

   d. What sampling procedure and sample are involved in the research?

   e. What procedures are used to break down the "text" and to establish a standardized and valid framework for coding or interpretation?

   f. What are the results of the content analysis? That is, what patterns and relationships are found in or inferred from the resulting data?

   g. What conclusions do the researchers draw from the results? Do the results support a hypothesis if there is one? Do the results fill in or expose gaps in an area of research? Do the results raise questions for further research?

   h. Do the results have implications for an area of social policy?

   i. Do the researchers identify any limitations or problems with their research and its procedures?

   j. Do the researchers identify or address any ethical issues or problems in their research?

2. Study the article in Research Info 8.4 and answer the following questions.

   a. What is the aim of this study? Is it attempting to test a hypothesis, to explore an issue, or to describe some social process, event, or situation?

   b. Do the researchers explain or justify their research approach?

c. What data set is used in the research? What was the original purpose of the data set?

d. What sampling procedure and sample are involved in the research?

e. What conclusions do the researchers draw from the results? Do the results support a hypothesis if there is one? Do the results fill in or expose gaps in an area of research? Do the results raise questions for further research?

Weblinks

You can find these and other links at the companion website for *First Steps: A Guide to Social Research*, Fifth Edition: **http://www.firststeps5e.nelson.com**.

Content and Text Analysis

The first site is an introduction to content analysis and its different uses. It provides examples, including a content analysis of the depiction of robots in science fiction texts written between 1818 and 1988. The second site offers a directory of resources for content analysis and text analysis. It also provides information on numerous software programs that can assist you in doing the analysis, as well as links to research and other related information.

Writing Guide: Content Analysis
   http://writing.colostate.edu/guides/research/content/index.cfm

Resources Related to Content Analysis and Text Analysis
   http://www.content-analysis.de/

Historical Research

These sites offer information useful to anyone who wants to gain an understanding of how to do historical research. They illustrate how to research for primary sources and how to interpret and analyze them.

The first site below offers primary documents using the case study of Martha Ballard, a midwife who lived during the American Revolution. The site also offers excerpts from a film on Martha Ballard and provides a document on the research process that goes into making a historical film. It includes, as well, excerpts from a book written by historian Laurel Thatcher Ulrich to show how she used the diary of Martha Ballard as a primary source.

The second website is an introduction to historical research using examples from environmental history. It takes the reader through the various steps of the historical research process and shows how documents offer insights about the past.

The last website discusses the difference between primary and secondary sources, provides questions to ask about your sources, and offers advice on writing the historical research paper.

DoHistory (historical research)
   http://dohistory.org/home.html

Learning to Do Historical Research

        http://www.williamcronon.net/researching/

Writing about History

        http://www.writing.utoronto.ca/advice/specific-types-of-writing/history

Available Data

Below are links to websites of national and international agencies. These agencies are increasingly relying on the Internet to make their publications and data more easily accessible. Much of the data provided can be used to carry out secondary data analyses and is often in a format that can be easily downloaded.

    The first website is of Statistics Canada, the country's national statistics agency, which is responsible for collecting a wide array of socioeconomic data and offers numerous publications. The second site provides an impressive list of international statistical agencies. The third site offers an array of demographic and socioeconomic statistics for numerous countries and areas of the world. The last site is a long list of links to international organizations that offer data and publications, including the United Nations, World Bank, and World Health Organization.

Statistics Canada

        http://www.statcan.gc.ca/

International Statistical Agencies

        http://www.census.gov/aboutus/stat_int.html

International Data Base

        http://www.census.gov/ipc/www/idb/

International Documents Collection

        http://www.library.northwestern.edu/libraries-collections/evanston-campus/
        government-information/international-documents

Archives and Documents

Below is the website of Library and Archives Canada, which is a depository of numerous documents, records, and audiovisual artifacts organized to celebrate and preserve all major facets of Canada's history. This site is followed by various websites that offer links to archives, documents, and material of archival interest.

Library and Archives Canada

        http://www.collectionscanada.gc.ca/

UNESCO Archives Portal

        http://www.unesco-ci.org/cgi-bin/portals/archives/page.cgi?d=1

Archives Made Easy
>   http://archivesmadeeasy.pbworks.com

CBC Digital Archives
>   http://archives.cbc.ca/

Historical Text Archive
>   http://historicaltextarchive.com/index.php

EuroDocs: Online Sources for European History
>   http://eudocs.lib.byu.edu

Worldwide Guide to Museums Online
>   http://www.museumstuff.com/

Virtual Museum of Canada
>   http://www.museevirtuel-virtualmuseum.ca

Web Gallery of Art
>   http://www.wga.hu/

Architectural Archives Sites
>   http://www.andrew.cmu.edu/user/ma1f/ArchArch/netsites.html

List of Online Newspaper Archives
>   http://en.wikipedia.org/wiki/Wikipedia:List_of_online_newspaper_archives

What Are the Results?

What You Will Learn

- how to code primary data
- how to organize and analyze quantitative data
- how to organize and analyze qualitative data

INTRODUCTION

Once you have finished collecting the data, you are faced with the challenge of making sense of them and of answering your research question or determining if there is support for your hypothesis. This chapter explores how to summarize and present your data and briefly covers some basic statistical concepts. Because not all research is quantitative, the chapter is divided into two main sections: the first is concerned with the analysis of quantitative data, and the second with qualitative data.

ANALYZING QUANTITATIVE DATA

The preceding chapters have introduced you to several important research methods used to discover or produce factual information of good quality—information that is accurate, objective, reliable, and open to others for inspection, critical evaluation, and replication. However, facts do not speak for themselves; they are clear only when they are summarized and analyzed.

Research designs are so efficient at producing information that the sheer quantity of data can overwhelm researchers unless care is taken with organization and analysis. A small survey of 100 people using a 25-question questionnaire will have at least 2500 "bits" of information. This is too much to look at casually and find patterns and relationships among the bits at a glance. Laboratory experiments normally involve measuring the responses of dozens of subjects, which produces hundreds of pieces of data. Using available data, such as the census or the Labour Force Survey, brings you face to face with an immense accumulation of facts and figures, often extending back to the 19th century. These **raw data** (figures and information that have not yet been summarized or interpreted) need to be organized, condensed, and presented so that overall patterns in the data (along with any striking exceptions and puzzling deviations) are brought out clearly from the mass of details.

Coding

The first step in data analysis is to code or translate the primary data from their raw forms into forms that can be more conveniently analyzed and managed. Nowadays this usually involves reorganizing the raw data into a form that is easy to analyze using computers. Coding involves using a set of rules to regroup the data into categories identified by numbers. The codes lead to consistency and a standardized way of allocating each bit of raw data into the appropriate numbered category.

In some instances the coding process is relatively straightforward. For example, survey researchers often **precode** many of the questions in their questionnaires. The code categories are included in the questionnaire before researchers collect the data. The boxes for checkmarks to indicate the answers would be numbered, as in the following question:

What gender are you?

Male ❏ 1

Female ❏ 2

Males are coded as 1 and females as 2. Thus, precoding can save time when it comes to summarizing the data. It is not always possible to precode, and in such instances researchers will have to **postcode**, or create a coding system after the data are collected.

Regardless of when researchers develop the coding system, coding involves classifying responses into categories and assigning each a code. Let us say that you have information on the marital status of those who took part in your study. You first need to decide which categories to use. Will you include separate categories or a single category for divorced and separated? Will you include a separate category for cohabiting? And so on. Once you have decided on the categories, you then code them. For example, you might decide to use the following categories and codes:

1 = single, never married

2 = cohabiting

3 = married, living with spouse

4 = separated

5 = divorced

6 = widowed

9 = no answer

Thus, an individual who is divorced, for instance, is coded as 5 with regard to marital status. It is possible that you have no information on the individual's marital status. In such a case, you should use a number or combination of numbers that cannot be confused with other codes. In the above categories, 9 is used to indicate no answer.

Deciding on categories and assigning codes can be a simple process. But for some social science research, coding can be more difficult. It may not be possible to precode "verbiage" or textual information, such as answers to open-ended questions on a questionnaire. Laboratory notes describing reactions of individuals in solving tasks under competitive versus cooperative conditions may involve detailed descriptions of behaviour and interactions that require a lot of interpretation. In such cases, coding may involve identifying common patterns across sentences, phrases, and paragraphs. Coding responses to open-ended questions usually involves classifying a wide variety of individual statements into a narrow range of categories. This kind of coding is identical to content analysis, discussed in Chapter 8. As well, such coding has similarities with qualitative analysis, dealt with later in this chapter.

Consider an open-ended survey question such as "What type of music do you most enjoy listening to?" Respondents might identify their favourite group, artist, or recording, or they might describe the type of music they like. There could be almost as many different answers as respondents. Let us say that you want to identify a pattern between the music that respondents enjoy listening to and the amount of money they spend on CDs. You therefore need to develop a reliable way of combining responses into categories of music. You decide what these categories will be. Let us say you are specifically interested in comparing the spending habits of those who enjoy listening to rock, jazz, classical, and country. You may decide to use the following categories and codes:

1 = rock

2 = jazz

3 = classical

4 = country

5 = other

9 = no answer

You now face the task of reading the answers to the open-ended question and deciding in which category to code the music the respondent most enjoys listening to. If the respondent lists a group, an artist, or a recording, it is your task to determine which category the response belongs in. Suppose a respondent enjoys listening to worldbeat or tango music—here, the category would be "other," number 5. But what if the respondent gives as an answer, "CD compilation of different singers," which includes other artists whom you would categorize under "classical," "rock," and "jazz"? You may have to read the answers to the question first, and then reconsider your categories. For example, you may decide to include another category, called "mixed."

Coding, then, involves processing raw data using a standardized set of codes to classify the data into numerically labelled categories. The coding scheme should be organized in terms of the specific focus of the research. There is also one other basic rule of coding: you should always leave the data in a more detailed state than you need for your analysis. The data can always be simplified at a later stage. For example, if you find few differences between separated and divorced respondents in a survey of family life as you analyze your data, you may decide to combine them into a "separated or divorced" category. But had you coded this group together at the outset and then found that there seemed to be differences between them, you would have to scrap your analysis and return to the initial coding stage.

Data Summary

Now that you have determined the categories and coded the data, you are ready to summarize the data. There are various ways to do this. One possibility is to use a computer spreadsheet program (see Box 9.1). For example, suppose you were doing a data summary of the three questions above. Assume that question one establishes the gender of the respondent; question two, the marital status; and question three, the type of music the respondent enjoys listening to. Your data summary sheet might look like this:

Respondent No.	Gender	Marital Status	Music
01	2	3	1
02	1	5	4
03	2	1	1

BOX 9.1 THE SPREADSHEET

The computer has become a basic tool in organizing, summarizing, and analyzing data. Various powerful statistical software programs exist for data analysis, including the Statistical Package for Social Sciences (SPSS). However, for less sophisticated statistical analysis, a **spreadsheet** program may be sufficient, such as Excel. The spreadsheet allows you to place the data in rows and columns. As illustrated below, the rows are numbered from one on, and the columns are labelled with letters from A on. The number of rows and columns available will depend on the program, but most have hundreds of rows and columns, far more than you are likely to need.

	A	B	C	D	E	F	G
1							
2							
3							
4							
5							

Cell D2

The place in which you store information is known as a **cell**. It is identified by the position of the column and row. For example, D2 is column D, row 2.

The information included in the cell can be numbers, labels, or formulas. The numbers can be whole numbers or decimals. The labels can be symbols or words. For example, in a particular cell you may want to include the heading of a series of numbers in either a column or a row. One of the main advantages of the spreadsheet is its ability to provide a formula that will display the calculation of the values contained in the cells. For example, you can total the numbers in a column by simply providing the formula in a cell and indicating which cells to add up. The program rapidly adds the values in the other cells and displays the sum in the cell in which you stated the formula.

Spreadsheets have powerful capabilities that allow for quick organization and calculation of the numeric data. They have a variety of summary statistics that, for example, enable you to easily find the mean, the standard deviation, and more. In addition, spreadsheets allow you to create line graphs, bar charts, and pie charts.

Thus, the spreadsheet can simplify the task of organizing, calculating, and presenting your data. But you still need to be clear about how you are organizing the data, why you are carrying out certain calculations, and whether the data are clearly presented.

Thus, respondent number 01 (row 01) is coded 2 for gender and is therefore female, is coded 3 for marital status and is therefore married and living with spouse, and is coded 1 for music and therefore enjoys listening to rock. What could you say about the second and third respondents? (As you may have noted, the summary sheet we are using is similar in design to a computer spreadsheet. See Box 9.1.)

The next challenge is to organize the data. Assume you want to summarize the information on one variable, say the music that respondents enjoy listening to. When summarizing and analyzing data derived from one variable, we are carrying out a **univariate analysis**. You can simply add up the number of respondents for each category and present what is called a **frequency distribution**. (The symbol for frequency is "f.") However, rather than present the raw data, you may prefer to also summarize the information as a percentage distribution. For example, assume we have 83 respondents in our sample and 31 enjoy listening to rock music. You could turn the raw data (31) into a percentage (i.e., $31/83 \times 100 = 37$). The distributions are illustrated in Table 9.1.

Table 9.1 Frequency of Type of Music Enjoyed

Type of Music	f	%
Rock	31	37.3
Jazz	10	12.0
Classical	9	10.8
Country	8	9.6
Other	20	24.1
No answer	5	6.0
Total	**83**	**100.00**

Note that the table is numbered (Table 9.1) and titled, and that the units of measurements are stated (f = frequency and % = percent). Consider how you were able to organize the data in such a way that it facilitates what you will say about the results. At a glance, for instance, you can say that 12 percent of respondents enjoy listening to jazz.

With such data as type of music, which are simply labelled, it may not matter in which order they are listed in the table. However, some categories have to be presented from lowest to highest, or from highest to lowest, such as with grades, income, or levels of education. There may be many scores, and these may be widely spread out. In such instances, a simple frequency distribution would be quite difficult to read. You would want to present a cumulative frequency distribution. For example, suppose you had the grade distribution of students in a class. (We will assume they all scored over 50 percent.) If you wanted to provide a simple frequency distribution, you would make the following table:

Table A Frequency Distribution of Grades (35 students)

Grades	f	Grades	f	Grades	f	Grades	f	Grades	f
100	0	90	0	80	1	70	0	60	0
99	0	89	0	79	0	69	0	59	0
98	0	88	0	78	2	68	2	58	0
97	1	87	1	77	1	67	2	57	0
96	0	86	2	76	1	66	1	56	1
95	1	85	1	75	2	65	1	55	1

(*continued*)

Table A (Continued)

Grades	f	Grades	f	Grades	f	Grades	f	Grades	f
94	0	84	0	74	1	64	0	54	1
93	1	83	2	73	1	63	3	53	0
92	0	82	0	72	2	62	1	52	0
91	0	81	0	71	0	61	1	51	1

As you can see, this table is quite lengthy, and many scores have no frequency. If it is not necessary to show the exact score of each student in your table, you could make it more presentable by condensing the scores into small groups. For example, you could group the data into smaller equal groups, such as the number of scores that fall between 91 and 100, 81 and 90, and so on. In this way, you could construct a grouped frequency distribution, as in Table B below. (What would be the grouped frequency distribution if the groups were 96 to 100, 91 to 95, 86 to 90, and so on?)

Table B   Grouped Frequency Distribution of Grades (35 students)

Grade	f	%
91–100	3	8.6
81–90	6	17.1
71–80	11	31.4
61–70	11	31.4
51–60	4	11.4
Total	35	100.0

The frequency is indicated for each group. For example, three students had scores that were between 91 and 100. As with the simple frequency distribution, it might be more appropriate to also present the percentage distribution, especially if you were making a comparison between two or more sets of scores. The percentage distribution is shown in column 3. (Note that the total percentage of the numbers displayed in the table is 99.9 percent. However, we write 100 percent, since the difference of 0.1 percent was lost in the rounding.)

Often it is more useful to present the frequencies in a cumulative way. For example, you could present the distributions to show how many students scored above or below a certain score. A **cumulative frequency distribution** (cf) allows you to locate the proportion of observations above or below a given point very easily (see Table C below). As you move up from the lowest score to the next-lowest, you add the number of students who obtained each score. To illustrate, we will continue to use the scores in the above grouped frequency distribution in Table B. The cumulative frequency for a group is obtained by adding up the scores in that group with the scores of all the other groups below it. For example, four students scored from 51 to 60, and, since no other group of scores exists below this group, 4 is its frequency distribution. The number of students who scored from 61 to 70 is 11. The cumulative frequency for this group is therefore 15—that is, 11 for the group plus the 4 of the group below it. Consequently, we can tell that 15 students scored 70 percent or lower. (How many students scored 80 percent or lower?)

Table C Cumulative Frequency Distribution of Grades (35 students)

Grade	f	%	cf	c%
91–100	3	8.6	35	100.0
81–90	6	17.1	32	91.4
71–80	11	31.4	26	74.3
61–70	11	31.4	15	42.9
51–60	4	11.4	4	11.4
Total	**35**	**100.0**		

We can also construct the cumulative percentage (c%) of the distribution. Instead of showing the number of students, we can show the percentage of students who had a certain score or lower. To determine the c% we calculate the percentage of the cf. For example, the grouped frequency distribution of the scores 51 to 60 is 4. Therefore, we determine what is the percentage of 4 over the total 35 ($100 \times 4/35 = 11.4$). The group frequency distribution of the scores from 61 to 70 is 15 and we calculate the percentage of 15 over the total 35, which gives us 42.9 ($100 \times 15/35 = 42.9$). Consequently, we can say that about 43 percent of the students scored 70 percent or lower. (What can we say about the other groups of scores?)

Levels of Measurement

The data collected for our research are not all measured in the same way. In addition to the already noted variables of gender, marital status, and type of music, you may have information on age, how much is spent on CDs, the respondents' preferred radio stations, their occupations, incomes, and so on. Clearly, an individual is either male or female, but one can be older or younger by so many years than another individual, or earn thousands of dollars more or less than someone else. These data (gender, age, income, etc.) differ in terms of their levels of measurement. **Measurement** refers to a standardized system for describing changes in the value of a variable. Levels of measurement differ, depending on the amount and kind of information they provide (see Figure 9.1).

With a variable such as type of music preferred, respondents indicate their preferences in words: "rock," "jazz," and so on. The musical preference then is measured by sorting and labelling it with these words—a process of classification. This process is measuring data at a **nominal level of measurement**; respondents are identified by their main musical preference, and each is put into the appropriate musical preference category. Similarly, when you ask for the gender of respondents in a questionnaire, you divide the responses into male and female. Nominal measurement organizes the data into groups or categories that are mutually exclusive (rock, jazz, classical, or country). It allows the researcher to count the numbers in each group, as was done in Table 9.1, for example. Hence, you can apply simple counting and percentages to variables at the nominal level of measurement.

Now suppose you have information on whether individuals strongly agree, agree, disagree, or strongly disagree with banning a CD that has violent lyrics. You are able to count how many respondents are in each category, but you are also able to rank the categories in terms of a range of agreement/disagreement. Some strongly agree, while others agree, and so on. This categorization allows you not only to divide people into groups, as with nominal

Figure 9.1 Levels of Measurement

Nominal Scale Example: Type of Music Listened To

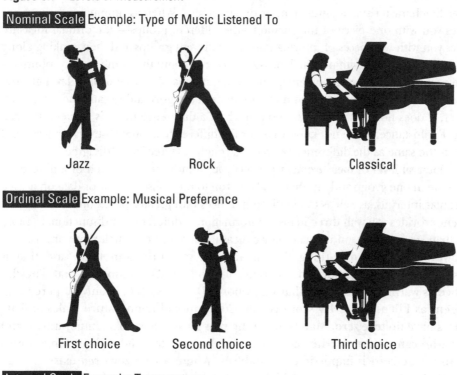

Jazz   Rock   Classical

Ordinal Scale Example: Musical Preference

First choice   Second choice   Third choice

Interval Scale Example: Temperature

–5°C   5°C   15°C   25°C

Ratio Scale Example: Income

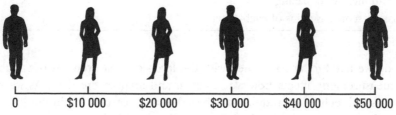

0     $10 000 $20 000 $30 000 $40 000 $50 000

measurement, but also to make a "more or less" comparison between the groups in terms of the specific characteristic of agreement or disagreement. Whereas nominal measurement provides you with one piece of information—the different groups— the ordinal measure provides you with two pieces of information—the different groups and their ranking along a specific dimension. As with nominal measures, you can count the numbers of respondents in each group and calculate a percentage of these numbers. This is an **ordinal level of measurement**. It allows you to count how many are in each category and to rank the categories. However, it does not allow you to say how much of a difference there is between the categories. For instance, we do not know whether the difference between "strongly agree" and "agree" is the same as the difference between "strongly disagree" and "disagree."

The **interval level of measurement** permits you to indicate how much of a difference exists between one group and another. It allows you to rank observations or data on a scale with regular intervals, as well as to identify differences with greater precision. Interval measurement provides you with three items of information: a difference or distinction, a "more or less than" comparison, and a quantitative measurement of how much more or less.

However, there is one thing that an interval measure cannot do, and that is express one score as a proportion or ratio of another. We cannot say that Tuesday was twice as warm as Monday or that Ian, whose IQ score is 100, is only 80 percent as intelligent as Ellen, whose IQ score is 125. Conventional temperature scales and IQ measures have no **true zero**. Many days in the year have below-zero temperatures, and anyone who can read and write can take an intelligence test and will get some points on it, so a true zero is impossible to establish. Where a true zero *can* exist, we can use a **ratio level of measurement**, which specifies both the difference in rank and the ratio numerically. For example, if John spent $25 on CDs and Mary spent $50, we can say that Mary spent $25 more than John and also that she spent twice as much. Or, again, if Ellen has had 14 years of formal education while Ian has had only 7, we can say both that Ellen has 7 more years of education than Ian and that Ian has had half the education that Ellen has. The characteristics of different levels of measurement are summarized in Table 9.2.

Table 9.2 Characteristics of Different Levels of Measurement

Characteristics	Nominal	Ordinal	Interval	Ratio
Classifying into mutually exclusive categories	X	X	X	X
Rank ordering		X	X	X
Equal and measured spacing			X	X
Comparing scores as ratios of each other				X

Why do we need to be concerned with the levels of measurement of variables? The level of measurement affects how we organize and analyze the data. As we shall see, certain statistical techniques require variables that meet certain minimum levels of measurement. For example, we noted earlier that the variable "type of music respondents preferred listening to" is at the nominal level of measurement. Look at the frequency distribution in Table 9.1. Can you say what type of music the respondents enjoy listening to on average? Or is it more appropriate to ask which music *most* respondents enjoy listening to? What if, instead, you asked the same question about the amount of money they spent on CDs? To explore these questions further, let us consider measures of central tendency.

Measures of Central Tendency

At times researchers may want to have a single number that summarizes certain information. One such number is known as a measure of **central tendency**, since it is usually located toward the centre, around a value where most of the data are concentrated. An example would be the class average on an exam—a single number that summarizes all the grades in the class. In everyday conversation, the term *average* is used loosely to compare things. We get a sense of how our friends performed on a class test in order to judge whether our own performance was above or below average. We may compare a new pizza service with our old favourite and say the new one is "about average." Therefore, the term *average* may mean "like" or "similar to," or it may mean "not very exciting" or "disappointing." Researchers require more precise terms and use several different kinds of measures of central tendency, or average, which can differ in value for the same set of data. The three best-known measures of central tendency are the mode, the median, and the mean.

THE MODE   The **mode** is the most frequent value in a distribution. For example, in the frequency distribution in Table 9.1, more respondents listened to rock music than to any of the other types of music. For another illustration, suppose you had the following information on the daily number of hours that people who enjoy country music listen to the radio: 1, 1, 5, 6, 1, 5, 2. The mode is 1, since it occurs more often than any other score.

The mode can be used with nominal, ordinal, interval, and ratio levels of measurement. It is the only measure of central tendency available for nominal level variables such as the "type of music respondents enjoy listening to," gender of respondents (male, female), or marital status (single, widowed, divorced, etc.).

However, the mode is only a rough measure of central tendency. In some instances there may be no mode, or there may be more than one mode; in such a case all of them need to be noted.

THE MEDIAN   The **median** is the middle point in a frequency distribution. It is the point at which half the cases or scores are above and half are below. Let us take the following distribution of the daily number of hours that a set of people listens to the radio: 1, 1, 5, 6, 1, 5, 2. To determine the median, you would first rank the scores from the lowest to the highest, including all the scores: 1, 1, 1, 2, 5, 5, 6. Next, you would locate the score that is in the middle, which splits the distribution into two equal parts. In the total seven scores, 2 is the median, since there are three scores below it and three above.

But what if there is an even number of cases, say eight scores? Again, the median separates the scores down the middle; you need to find the two middle values and calculate their average. To illustrate, say we have the following eight scores: 1, 1, 1, 2, 5, 5, 6, 7. The two middle values are 2 and 5. The median is their average—that is, $(2 + 5)/2 = 3.5$.

The median can be used to measure ordinal, interval, or ratio data, such as age, how much respondents spend on CDs, or how many hours they listen to the radio. It is especially useful when you need to know a distribution's representative middle number.

THE MEAN   A popular measure of central tendency is the arithmetic **mean**, commonly known as the *average*. The mean is calculated by adding up the values of the distribution and then dividing by the total number of values. The mean of the earlier distribution of daily hours that people listen to the radio is:

$$\frac{1 + 1 + 5 + 6 + 1 + 5 + 2}{7} = \frac{21}{7} = 3$$

The calculation of the mean can be summarized by the following formula:

$$\bar{X} = \frac{\Sigma X}{N}$$

$\bar{X}$ = The symbol for the mean, pronounced "X bar"
Σ = the Greek capital letter "sigma" or the "sum of," which indicates that we must add up the values that follow
X = the raw score in the set of scores
N = the total number of scores

In other words, the arithmetic mean ($\bar{X}$, or 3) is equal to the sum of (Σ) raw scores (X_1, X_2, X_3, and so on, or 1, 1, 5, and so on) divided by the total number of scores (N, or 7).

Although the mean is widely used, in certain instances it may not be particularly helpful. For example, suppose you wanted to calculate the mean amount of money respondents who enjoy country music spend on CDs in a month. Your eight respondents each spend the following (expressed in dollars): 10, 15, 20, 20, 10, 10, 15, 300. What is the mean? The answer is $50, but does this appropriately reflect the distribution? One respondent spent $300, far more than the other seven added together. If we omitted the extreme score of $300, the mean for the remaining seven would be $14.29, a figure that seems much more representative. In other words, the mean can be affected by extremely large or small scores.

COMPARING THE MODE, MEDIAN, AND MEAN   Which measure social scientists use depends on the research objective and especially on which measure is appropriate. The mode can be used with nominal, ordinal, interval, and ratio levels of measurement. For example, you would want to use the mode to determine the preferred radio station of the largest number of respondents. The median requires that we rank the categories from lowest to highest and therefore cannot be used for nominal data, but it can be used for ordinal, interval, and ratio data. For example, you can determine the median income of respondents, but it would be impossible to give a median of nominal data such as gender, preferred radio station, or type of music. The mean is used with interval and ratio data. For example, it makes sense to compute the mean age of the respondents, but it would be meaningless to compute the mean gender of the respondents.

When the mode, median, and mean are equal, the scores have a **normal distribution**—that is, they follow a bell-shaped curve (see Figure 9.2). Many physical and psychological characteristics are normally distributed. Few of us are very small (for our age) or extremely large, or very obtuse or very smart. Most of us are "around the average" for common physical and psychological characteristics.

Many social characteristics, however, are not symmetrical in their distribution patterns. More people live in poverty than are very rich; the proportions of the population in different age groups or with different levels of education are not symmetrical. In such cases, the mode, median, and mean are quite different. If such features of social life were to be represented graphically, the results would be asymmetrical, or skewed (see Figure 9.2). Some scores pile up in one direction, giving the curve a pronounced "bulge" and an equally marked tail.

Figure 9.2 Normal and Skewed Distributions

Normal distribution curve

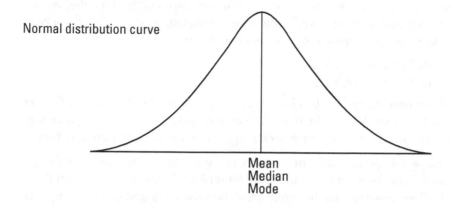

Mean
Median
Mode

Negatively skewed distribution curve

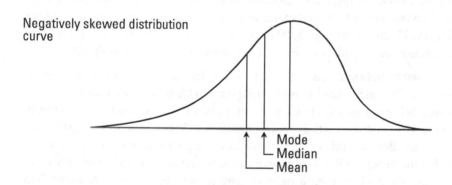

↑ ↑ Mode
└ Median
└ Mean

Positively skewed distribution curve

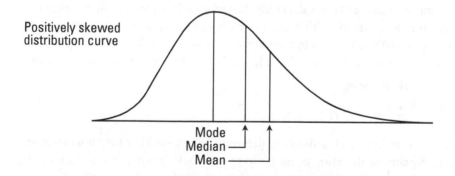

Mode ↑
Median ─┘ ↑
Mean ──────┘

A **negatively skewed distribution** pattern has a much larger tail on the left. This distribution might represent our aging population, with a very small proportion of infants and children, and the aging baby boomers prominently showing up off-centre to the right. A **positively skewed distribution** is the reverse, with a clear tail on the right and the bulk of scores on the left. This would probably be the shape of income distribution in our society, showing tiny numbers for the very rich, and with the bulk of the population in the middle- and lower-income groups.

Measures of Dispersion

Although the mode, median, and mean can be useful summaries of the data, they do not tell us how widely spread the values are. To illustrate the usefulness of knowing the spread of values, let us calculate the means of these two sets of scores:

Set A: 10. 10. 14. 16. 17. 20. 25

Set B: 12. 13. 14. 16. 18. 19. 20

The two have the same mean ($\overline{X} = 16$). However, as you can see, the scores in the first set are more spread out than those in the second. Thus, it helps to know what the spread was, and for that there are at least three measures: range, variance, and standard deviation.

RANGE   The **range** is simply the difference between the lowest and highest scores. Although the two sets above have the same mean, the range for set A is 15 ($25 - 10$), and for set B it is 8 ($20 - 12$). If all we knew was that the two sets had the same mean and that the range was smaller in set B, we might be inclined to see the mean of set B as more reflective of the scores in that set than is the mean of the scores in set A. But the range gives only a rough sense of the spread of the values. Although the range is simple to calculate, its main disadvantage is that it takes into account only the two extreme scores. For the earlier example of money spent on CDs (10, 15, 20, 20, 10, 10, 15, 300) the range of $290 gives us an impression of enormous differences in expenditure. Yet in all but one case the range is only $10.

VARIANCE AND STANDARD DEVIATION   As we noted earlier, the mean is commonly used in everyday situations. One such situation is when we want to know the class mean (average) on an exam, usually because we want to know how our own mark varies from the class mean. If the class mean is 70 and our mark is 78, we proudly say that we scored 8 points higher than the class mean. But does this imply we have scored higher than most students in the class? Recall that the mean is influenced by extreme scores. A few very extreme scores may account for a class mean of 70, even if most students scored either higher or lower than 70. Is there a way of taking into account the difference from the mean of every score in the distribution and coming up with a measure that reflects the variability of the scores?

We can obtain such a measure by calculating the **variance**. It is quite simple to calculate: you would take the deviation (the difference) of each score from the mean, square each one, add them up, and divide the sum by the number of scores. Let's illustrate. Suppose you had the following distribution of scores: 8, 9, 11, 5, 2. Here is how to calculate the variance:

1. Calculate the overall mean.

$$\frac{8 + 9 + 11 + 5 + 2}{5} = 7 \quad \text{or} \quad \overline{X} = \frac{\Sigma X}{N} = \frac{35}{5} = 7$$

2. Subtract the mean from each score (i.e., determine the deviation of each score from the mean). Square the deviation by multiplying it by itself (2 squared is written as 2^2, which is $2 \times 2 = 4$). Note that squared numbers are always positive (e.g., -2^2 is $-2 \times -2 = 4$, the same as 2^2).

$X - \overline{X}$	$(X - \overline{X})^2$
$8 - 7 = +1$	1
$9 - 7 = +2$	4
$11 - 7 = +4$	16
$5 - 7 = -2$	4
$2 - 7 = -5$	25

3. Then add the squared deviations and calculate their mean.

$$\frac{1 + 4 + 16 + 4 + 25}{5} = \frac{50}{5} = 10$$

What you have calculated is the variance, symbolized as s^2.

In our example above, the variance (s^2) is 10. But there is a problem with this number: it is not expressed in the original unit of measurement. If our scores represented the monthly amount of money spent on CDs, the variance we have calculated is expressed in squared amount of money spent. This figure is quite difficult to interpret, so we do another calculation to return the measure of variability into its original unit of measurement. We simply take the square root of the variance and have what is called the **standard deviation**.

1. Calculate the square root ($\sqrt{}$) of the variance to determine the standard deviation.

$$\sqrt{10} = 3.2$$

The standard deviation is symbolized by s or SD, and the formula is

$$SD = \sqrt{\frac{\Sigma(X - \overline{X})^2}{N}}$$

In other words, the standard deviation (SD, or 3.2) is equal to the square root of the sum of squared deviations from the mean, $(X_1 - \overline{X})^2 + (X_2 - \overline{X})^2$ and so on, or $(8 - 7)^2 + (9 - 7)^2$, and so on, divided by the total number of scores (N, or 5). Computer spreadsheet programs, such as Excel, and statistical computing programs, such as SPSS, will compute the mean and standard deviation. Certain calculators also have keys that can calculate the mean and standard deviation.

THE MEANING OF THE STANDARD DEVIATION   Both variance and standard deviation are measures of the distribution of scores (or values of variables such as IQ, income, or amount spent on CDs) around a mean. An advantage that the standard deviation has over the variance is that it is in the appropriate unit of measurement. But what is its usefulness? How does it help us to know the standard deviation of a set of scores? Recall that we wanted to measure the spread of scores—for example, the various grades on an exam, the number of hours a person listens to the radio, or the amount spent on CDs. Thus, we are essentially asking, "How spread out are the scores?" To better understand the usefulness of the standard deviation, let us suppose that we have two sets of marks on a similar exam given to two sections of a course.

	Section A	Section B
	10	11
	16	12
	16	13
	16	15
	16	19
	16	21
	22	21
Mean	16	16
Range	12	10
Standard deviation	3.2	4.0

Now, assume that all you knew about the two sets of marks were their means. Can you say anything about the spread of the marks knowing only the mean? The mean summarizes the data, but it does not tell us about the spread of the scores. Assume you were also told the ranges. The ranges suggest a smaller spread in section B, and we might be tempted to conclude that more scores are clustered around the mean in section B than in section A. But now consider the standard deviation, which is lower for section A than section B. This tells us that there is less of a spread in section A than in section B, as is evident from a quick look at the raw scores. In other words, the scores in section A are closer to the mean than are the scores in section B. Clearly, the standard deviation is a more useful measure of the distribution of the scores. This example suggests that it is insufficient to judge the overall performance of a class on the basis of the mean; it is best to also know the standard deviation.

The standard deviation is especially useful when the sample is representative of a population and the sample data are normally distributed. In a normal distribution, the mode, median, and mean coincide with 50 percent of the scores above and 50 percent below (see Figure 9.2). Consequently, if the class grades are normally distributed and you know the mean is 65 percent, then you know that half the scores are (or half the class scored) below 65 percent and the other half above 65 percent. The standard deviation is a measure of the distribution of scores around the mean. If the standard deviation is large, then the scores are widely distributed and most scores are away from the mean; if it is small, many scores are close to the mean (see Figure 9.3(A)). Mathematicians have established that by knowing the mean and the standard deviation of normally distributed scores, we are able to tell the proportion of the scores that are within certain limits.

Let us assume that you administered an IQ test to a large sample of college students and the scores were normally distributed with a mean of 100 and a standard deviation of 15. Here we would say that *one* standard deviation equals 15. A characteristic of normally distributed scores is that about 34 percent of the scores will fall between the mean and one standard deviation above the mean (see Figure 9.3(B)). Thus, in our example, the mean is 100 and one standard deviation is 15; therefore, about 34 percent of the scores would be between 100 and 115 (i.e., 100 + 15 = 115). Since the normal distribution is symmetrical, it holds that about 34 percent of the scores will fall one standard deviation below the mean, or rather between 100 and 85 (i.e., 100 − 15 = 85). Hence, we can also say that about 68 percent of the IQ scores fall between 85 and 115.

Stated differently, about 68 percent of the scores lie between one standard deviation above the mean and one standard deviation below the mean, or rather between +1 and −1 standard deviation from the mean. As the normal distribution in Figure 9.3(B) shows, about 95 percent of the scores would lie between +2 and −2 standard deviation from the mean. Therefore, about 95 percent of the IQ scores in your study would fall between 70 (i.e., 100 − 30 = 70) and 130 (i.e., 100 + 30 = 130). You could also state that only about 5 percent have IQ scores that are either below 70 or above 130. Again, this holds only if the scores are normally distributed. To fully grasp the usefulness of the standard deviation and all its implications requires more knowledge of statistics than we can cover here.

Tabular Analysis

Although measures of central tendency and dispersion are useful in understanding particular variables, researchers are usually interested in understanding the relationship between at least two variables. Let us assume your research asserts that males enjoy different types of music from females. You would then want to present the data in such a way that the data show the relationship between the independent variable (gender of

Figure 9.3 Standard Deviation and Normal Distributions

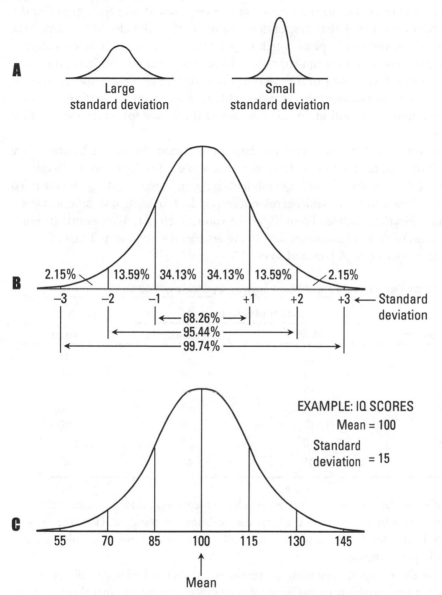

respondent) and the dependent variable (type of music enjoyed). Thus, you want to split the data according to gender. The skeleton of the table would look like this:

| | Gender of Respondent | |
| | Male | Female |
Type of Music		
Rock		
Jazz		
Classical		
Country		
Other		
No answer		

For each respondent on your data summary sheet, you would put a tick in the blank cell that corresponds to the gender of the respondent and the type of music enjoyed. Earlier we noted that respondent number 01 is female and enjoys rock music. Therefore, you would place a tick in the top row of the right-hand column. You would continue to tick the appropriate cells for the other respondents, then add up the ticks in each cell and put the total in the cell. The result is a special type of table known as a **cross-tabulation** (or **cross-tab**). In effect, you are setting up more than one frequency distribution, in this case two of them: one for males and another for females.

Suppose you added up the ticks and have constructed the cross-tabulation (see Table 9.3). Note that the total in the last column of Table 9.3 is similar to the frequency distribution of raw data on the music respondents enjoy in Table 9.1. We have expanded the table to include another variable: gender of respondent. In using two variables we are carrying out a **bivariate analysis**. From Table 9.3 you can tell that 31 respondents enjoy rock music, and of these 12 are males and 19 are females. Or you can tell that of the 41 males, 12 enjoy rock music; 6 jazz; and so on.

Table 9.3   Cross-Tabulation of Type of Music Enjoyed, by Gender of Respondent

| Type of Music | Gender of Respondent | | Total |
	Male	Female	
Rock	12	19	31
Jazz	6	4	10
Classical	5	4	9
Country	5	3	8
Other	11	9	20
No answer	2	3	5
Total	**41**	**42**	**83**

The logic of data summarization is to select categories, code the data, and then present the results in a manner that allows you to interpret the data. However, as with one-variable frequency distributions, it is generally better to express the results of cross-tabulations in percentages.

There are three ways of presenting percentages; the method selected will depend on what is being interpreted. For example, the three tabular percentages in Table 9.4 are all based on the data in Table 9.3.

Take a while to examine each of the tables. Carefully read the titles and note the direction in which the percentages are calculated. In Table 9.4(A) the percentages are obtained by dividing the frequency in each cell by the total number of respondents, as in the case of males who enjoy rock music (12 ÷ 83 × 100 = 14.5 percent). In Table 9.4(B) the percentages are obtained by dividing each frequency by the total number of respondents in that *column*. For example, the frequency in each cell under males was divided by the total number of male respondents, as in the case of males who enjoy rock music (12 ÷ 41 × 100 = 29.3 percent). In Table 9.4(C) the percentages are obtained by dividing each frequency by the total number of respondents in that *row*. For example, the frequency in each cell of the row for rock music was divided by the total number of respondents in that row, as in the case of males who enjoy rock music ($12 \div 31 \times 100 = 38.7$ percent). Attempt to interpret the figures for males who enjoy rock music in each table.

Table 9.4   Cross-Tabulation Percentages

A. Cross-Tabulation of Type of Music Enjoyed and Gender of Respondent, with Total Percentages

| Type of Music | Gender of Respondent | | Total % |
	Male %	Female %	
Rock	14.5	22.9	37.3
Jazz	7.2	4.8	12.0
Classical	6.0	4.8	10.8
Country	6.0	3.6	9.6
Other	13.3	10.8	24.1
No answer	2.4	3.6	6.0
Total	**49.4**	**50.6**	**100.0**
Number of respondents			(83)

B. Cross-Tabulation of Type of Music Enjoyed and Gender of Respondent, with Total Column Percentages

| Type of Music | Gender of Respondent | | Total % |
	Male %	Female %	
Rock	29.3	45.2	37.3
Jazz	14.6	9.5	12.0
Classical	12.2	9.5	10.8
Country	12.2	7.1	9.6
Other	26.8	21.4	24.1
No answer	4.9	7.1	6.0
Total	**100.0**	**100.0**	**100.0**
Number of respondents	(41)	(42)	(83)

C. Cross-Tabulation of Type of Music Enjoyed and Gender of Respondent, with Total Row Percentages

| Type of Music | Gender of Respondent | | Total % | Number of Respondents |
	Male %	Female %		
Rock	38.7	61.3	100	(31)
Jazz	60.0	40.0	100	(10)
Classical	55.6	44.4	100	(9)
Country	62.5	37.5	100	(8)
Other	55.0	45.0	100	(20)
No answer	40.0	60.0	100	(5)
Total	**49.4**	**50.6**	**100**	**(83)**

Each table has a slightly different story to tell. Which one is appropriate? Recall that you expected that the type of music enjoyed (the dependent variable) depended on the gender of the respondent (the independent variable). You want to know, for example, what percentage of females enjoy country music as compared to males. Therefore, Table 9.4(B), the cross-tabulation with total column percentages, would serve your purpose. The table shows that 12.2 percent of males enjoy country music, as against 7.1 percent of females. Take note that since the comparison was between males and females (the independent variable) and these were in columns, you calculated column percentages. If the independent variable had been placed in the rows instead, you would have carried out row percentages. In addition, a general rule in reading and interpreting tables is to compare the percentages in the direction opposite to that in which they were calculated. For example, you wanted to compare males and females (independent variable) and placed the variable in columns to determine the percentage of total males and percentage of total females who like the different types of music. Therefore, you calculated column percentages, but you compared across rows (percentage of males who like a particular type of music compared to percentage of females).

Correlation

Organizing data in a table on the basis of two variables at the same time, as in Table 9.4, is common practice in social analysis. Cross-tabulation can become very elaborate, as you will discover if you pursue social research further. At this point we shall take up one key idea in cross-tabulation: the statistical relationship between variables.

Look at Table 9.4(B). How could you describe what you see? You might answer: "A greater proportion of females than males preferred rock as opposed to each of the other types of music." In other words, there is a relationship between gender of respondent and musical preference. This is a *statistical* relationship simply because it shows up as a result of our grouping the data as percentages. Careful examination of percentage tables allows us to discover any statistical relationships that may be present, but other techniques can also be used, such as visual presentations and statistical measures.

For a visual impression of the relationship between two variables, we can produce a **scattergram** (also known as a *scattergraph* or *scatterplot*)—a diagram with each point representing a score on two variables. It gives us an indication of the kind of relationship that exists between the two variables.

For example, let us assume your research asserts that there is a relationship between the number of hours a student spends studying for an exam and his grade on the exam. Suppose we plotted the results, with each point representing a student's time spent studying and his grade. Let us say we did this for three classes and got the following scattergrams.

What can you say about how close the relationship is between time spent studying and grades received for each of the three classes? In other words, what is the **correlation** between study time and grades? The scattergram for class 1 shows a strong **positive correlation** (or **association**) the grade increases with an increase in the time spent studying. A strong relationship also exists in class 2, but the grade increases with a decrease in the time spent studying! This is a **negative correlation**. Looking at the scattergram for class 3, it is quite clear that no relationship exists between the time spent studying and the grade; there is a **zero correlation**.

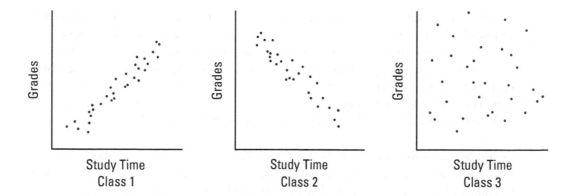

For classes 1 and 2, or situations in which the points are roughly on a straight line, researchers may calculate a numerical value of the correlation. The measure is sometimes called the *Pearson's product moment correlation coefficient*, or simply *correlation coefficient*, and the symbol is usually r.

Again, note that the correlation coefficient is a measure of a linear relationship. The closer the points are to a straight line, the higher the correlation (whether positive or negative). The more scattered the points, and the more difficult it is to imagine a line around which the points cluster, the smaller the correlation. A correlation coefficient ranges from -1.0 to $+1.0$, with -1.0 indicating a perfect linear negative correlation, and $+1.0$ a perfect linear positive correlation. Suppose a study tests the association between respondents' years of schooling and their fathers' levels of schooling. The relationship can be summarized with a correlation coefficient. If the result was $r = 0.75$, we would say that there was a strong positive correlation. In other words, people whose fathers were more educated were themselves more likely to be more educated. Be careful how you interpret this, though: correlation is not the same as causation; rather, it is a relationship that has to be explained (see the section below). Many factors will influence the level of someone's schooling other than the father's level of schooling.

Pearson's correlation coefficient is only one of several **measures of association**—a single number that summarizes the strength and sometimes the direction (positive or negative) of a relationship between variables. Look again at Table 9.4. Knowing the gender of the respondents, would you have some chance of correctly guessing their musical tastes? Which taste(s) are you more likely to be able to guess? Which musical taste would most likely lead to a lot of wrong guesses? The ability to guess or predict some variable based on your knowledge of another variable is the basis of most of the commonly used measures of association.

A measure of association will tell us the *amount* of relationship between the variables. That is, it will tell us how strong or weak the relationship is. If we are dealing with nominal variables (such as the relationship between gender and musical taste, or gender and party vote), the strength or weakness of the relationship is all that the measure can tell us. Analysis of ordinal, interval, and ratio level variables (IQ, age, income, or years of schooling, for example) can tell us both the strength and the direction of the relationship. For example, students who get As in one course are very likely to get As in another course—a positive relationship between ordinally measured grades. Or the more years of formal schooling they have, the higher graduates' average incomes are—a positive relationship between ratio measured variables. A negative or inverse relationship might be found when looking at educational background and the experience of unemployment (i.e., the more highly educated a person is, the shorter the time spent unemployed).

It is beyond the scope of this book to cover the method of calculating correlation coefficients and to describe all the different types of correlation coefficients that exist. However, statistical computing programs will easily calculate the correlation coefficient, and some calculators provide keys that will calculate them for small sets of data.

CORRELATION DOES NOT IMPLY CAUSATION   A common everyday error is to quickly conclude that a correlation reflects a causal relationship, meaning one factor causes another factor to happen. For example, assume that in his research on public transportation, a student wanted to better understand the drop in the number of bus passengers in recent years. He collected and examined various data over a 10-year period, including the overall cost of gas, the cost of bus fares, and the number of bus passengers. He found that the increases in the cost of gas were accompanied by increases in bus fares and declines in the number of bus passengers. He therefore concluded that the rise in the cost of gas caused the increase in bus fares, and the increase in bus fares caused a decline in the number of bus passengers. But was he correct to conclude that what he found was a causal relationship?

As noted in an earlier section, *correlation* refers to the extent that a relationship exists between two or more variables. If a relationship exists, then it could be a *positive* correlation. In other words, as the value of *x* increases, the value of *y* increases, or they decrease together. For example, over the previous 10 years, with every major increase in the price of gas, there was an increase in the price of bus fares. The relationship seems to suggest that the rise in one variable is associated with the rise in another. Assume that the same study also found a *negative* correlation. In other words, as the value of *x* increases, the value of *y* decreases. For example, with every major increase in bus fares, the number of passengers decreased. The relationship seems to suggest that the rise in one variable is associated with a decline in another.

Plotting the data on a graph will often provide a good indication of whether the correlation is positive or negative. Nowadays one can quickly calculate a correlation using a statistical program or even a spreadsheet program like Excel. However, whatever the results, it is essential to recognize that the calculated correlation indicates only what *might* be occurring.

Regardless of whether there is a strong positive or negative correlation, the results might not be reflecting the actual relationship. Indeed, other factors might be influencing the correlation. In our example, it is possible that the increase in bus fares was also due to aging buses requiring more bus repairs, causing a rise in the cost of maintaining the buses and therefore an increase in bus fares. The need for more bus repairs contributed to delays, fewer buses in service, and therefore longer waits for passengers, which in turn discouraged people from using public transportation.

Hence, even when the calculated correlation seems to support a causal relationship between two variables, the issue must be further explored. You should ask: Why are the two variables related? Is it a temporary or permanent relationship? Is there another factor or other factors that connect the two variables and affect them both? In other words, even after having calculated a correlation we need to recognize that correlation does not imply causation.

Visual Summaries

If you had the choice between reading a table and looking at a visual presentation of the same data, which would you prefer? A quick perusal of magazines and newspapers indicates that it is far more popular to summarize data using graphic forms than tabular forms.

Computer spreadsheet programs have made it relatively easy to construct graphs. However, you need to be careful to select the appropriate type of graph. The information in the graph is based on information that could be provided in tabular form, but it is often simplified to make the data more readable or more "interesting." You should make sure that the graph you select still conveys the information efficiently. Graphs are useful in summarizing the data and helping others grasp the significance of the information, but only if they present the data accurately.

For example, take the frequency distribution of types of music in Table 9.1. An appropriate graph for such data is the **pie chart**, whose pieces add up to 100 percent (see Figure 9.4). However, the pie chart should contain a limited number of sections; otherwise, its usefulness is lost. (Software packages generally allow for only a few sections.) In certain instances, if you want to stress a particular category, such as rock music, you could highlight that section by moving it outward, or "exploding" the section (see Figure 9.5).

The same data can also be presented in a **bar graph** (see Figure 9.6). But unlike the pie chart, the bar graph can be used for many different types of data. As with the pie chart, we have an understanding of the relative preferences for the different types of music. For

Figure 9.4   Pie Chart of Preferred Type of Music

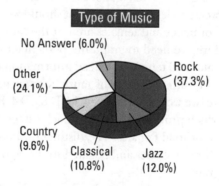

Figure 9.5   Pie Chart with Exploded Section

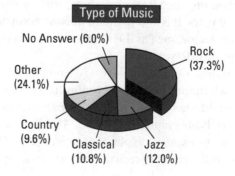

Figure 9.6 Bar Graph of Type of Music

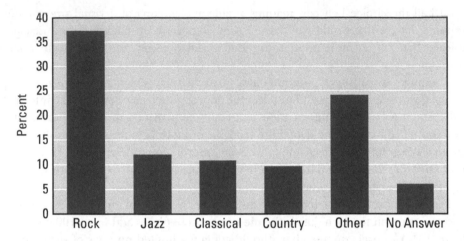

example, we can tell that the most popular music among the respondents is rock music. Note that the widths of the different bars are the same in order to make an appropriate comparison. Furthermore, there are fixed gaps between each bar to demonstrate that we are referring to different types of music, which are listed below the horizontal axis (or *x*-axis). The vertical axis (or *y*-axis) on the left side indicates that we are referring to the percentage of respondents. (In this example, the bars run vertically. Horizontal bar graphs are generally used when there are many different categories.)

An advantage of the bar graph is that it allows for comparisons, such as the difference in music preferences between males and females. But should we use frequency or percentages? Since the numbers of males and females are not the same, frequency would give a distorted picture. Therefore, we need to graph the percentages. Furthermore, since the comparison is between males and females, we need information on the percentage breakdown of preference for the different types of music among males and the breakdown among females. Therefore, we would use the data in Table 9.4(B). These data give us the bar graph in Figure 9.7, which provides two bars for each category: the percentage of total males and the percentage of total females. The distinction between male and female is clearly shown by using different shades and is stated in the *legend*, which in our example is placed inside the bar graph.

Another widely used visual summary is the **line graph**, which displays values that change across time. It allows us to look for trends and to make comparisons. A simple line graph has information on some factor over time, such as the number of rock music records sold in the last 10 years. It is also useful in making comparisons by plotting information on more than one factor, such as the number of rock records and the number of jazz records sold over the past 10 years.

Suppose we have information on the average amount of money that males and females spent on CDs each month in the last 12 months. We could plot the information on a bar graph, or we could select a line chart. Compare the two graphs in Figure 9.8. Which would you prefer? Both graphs are correct. In many respects, your choice will depend on what you want to express visually about your data. The bar graph gives us a clearer perspective of the difference between the average spent by males and females each

Figure 9.7 Bar Graph of Type of Music by Gender

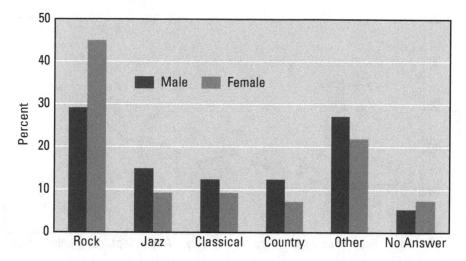

month. The line graph provides a better perspective on the trend in the average spent on CDs over the 12 months.

What does the line graph tell us about the trend in the amount of money that males spent on CDs over the last 12 months? What does it tell us about the trend in the amount of money that females spent on CDs? The line graph helps us to visualize the trends in spending for males and females, and it allows us to compare their spending trends.

Thus, when using a graph to summarize data, it is important to have a clear understanding of the purpose of the graph. For example, the information plotted in the pie charts (Figures 9.4 and 9.5) could also be presented as a bar graph, but not as a line graph. The information plotted as bar and line graphs in Figure 9.8 would lose its meaning if plotted in a pie chart.

Figure 9.8 Bar Graph and Line Graph of Average Spent on CDs by Gender

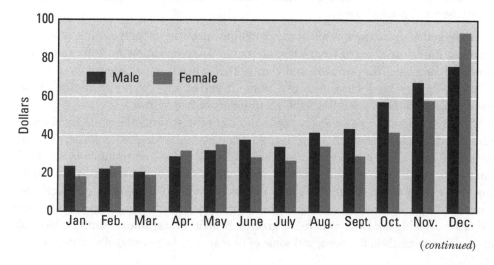

(*continued*)

Figure 9.8 (*Continued*)

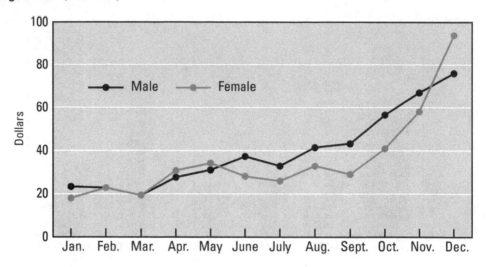

What Is Statistics?

Statistics is a branch of applied mathematics that is concerned with two areas of application. **Descriptive statistics** deals with the collection and classification of information as numbers. **Inferential statistics** applies mathematical ideas to the organization of numerical data in order to draw conclusions or to draw inferences from it.

More simply, statistics can be seen as a language or a way of thinking about information. As a way of thinking and expression, it allows us to say complicated things briefly and with *precision*. Can you express the following statements without using numbers and still be as brief and precise?

> My car's gas consumption seems to be 20 percent less per fill-up since its last tuneup. That's saving me over $15 per week.

> My grade average is 10 percent lower than last term.

> Government revenues from alcohol and tobacco taxes are 8 percent higher this year, while income tax revenues are 13 percent lower.

Purely verbal restatements will lose something in terms of pinpointing *how much* something differs or changes from something else. There is simply no substitute for a "numerical language" for precision and clarity. This is not to say that everything has to be statistically presented, but simply that where it makes sense to do so, statistics have considerable advantages. (See Box 9.2 for a summary of basic terms.)

The three statements above are examples of descriptive statistics. They use numbers to describe and compare situations. As the first two statements showed, whether we are aware of it or not, we often think in descriptive statistical terms. We are also surrounded by statistical information that is often very important for us as citizens and as consumers, as well as in almost every other aspect of our lives. Public opinion surveys are often used by political parties competing for our votes. Economic statistics may be useful in helping us to decide which areas of job training are likely to have the greatest benefit for our future earnings. Medical statistics are constantly in the news, and some of us may even be persuaded to abandon our

BOX 9.2 BASIC TERMS IN QUANTITATIVE DATA ANALYSIS

Term	Symbol/Formula	Description
Frequency	f	The number of cases, scores, or observations in a group or of the same type
Cumulative frequency	cf	The total number of cases having any given score or a lower score
Arithmetic mean	$\bar{X} = \dfrac{\Sigma(x_1 + x_2 + \ldots + x_n)}{N}$	The sum of a set of scores divided by the total number of scores in the group
Median	mdn	The midpoint of a set of scores set up from lowest to highest scores
Mode	mo	The most frequent, most typical, or most common score in a distribution
Range	r	The difference between the highest and lowest scores in a distribution
Variance	$s^2 = \dfrac{\Sigma(X - \bar{X})^2}{N}$	The mean of the squared deviations from the mean of a distribution
Standard deviation	$s = \sqrt{\dfrac{\Sigma(X - \bar{X})^2}{N}}$	The square root of the mean of the squared deviations from the mean of a distribution

bad habits as a result of seeing or hearing such reports. Many of us actually enjoy reading sports statistics!

We often think in the more complicated inferential statistical manner, as well. What are the chances of my passing the next test if I postpone cramming until the night before? Are an outline and the first three lectures of a course enough to conclude that this course is not for you? Inferential statistics involves interpreting data to see what they mean in relation to a hypothesis. Do the data show what we would like them to show—that is, do they support our hypothesis? How do we know our results are not due to pure chance or coincidence?

For social scientists, statistics is like a toolbox from which they select the right tool to help them in their data analysis. Therefore, if you have not already done so, and need to be familiar with more statistical tools, you may want to acquire knowledge of more concepts and applications of statistics than are covered in this chapter. (See the Weblinks at the end of this chapter.)

QUALITATIVE DATA ANALYSIS

In Chapter 7 we pointed out that in field research, data collection and analysis are less separated than in other research methods. This is just one of several characteristics of qualitative data analysis.

> *Qualitative data analysis ... is a breaking up, separating, or disassembling of research materials into pieces, parts, elements, or units. With facts broken down into manageable pieces, the researcher sorts and sifts them, searching for types, classes, sequences, processes, patterns, or wholes. The aim of this process is to assemble or reconstruct the data in a meaningful or comprehensible fashion. (Jorgenson, 1989, p. 107)*

Two things should be noted about this definition. First, the idea of causality—what factor or factors are responsible for the existence, appearance, or change of other factors—is underemphasized. Instead, stress is placed on the meaning, comprehension, or interpretation of facts and their interconnections.

Second, qualitative data analysis is presented as a two-stage process: first, facts are broken down in the search for "types," "classes," and so on; and second, the resulting patterns are reorganized into larger units of meaning. In reality, this process is often more like a repeated cycle than two distinct stages. Analysis of qualitative data might be compared to a dialogue or conversation between the researchers and their data—it is a process of continual questioning of the data through which patterns gradually emerge.

Research based on qualitative data tends to proceed in a more exploratory fashion than quantitative data analysis. For social science researchers who use fieldwork or documentary interpretation, the main task is often to *describe* what they are studying. These researchers also argue that describing historical eras, tribal or peasant communities, or contemporary subcultures should be done in terms of the people's own language and outlook—how the participants view their own way of life. Imposing an outsider's view of their lives will distort our understanding rather than aid it.

In addition, qualitative researchers often argue that social causation properly refers to the intentions and understandings that motivate human actions. Consequently, even exploratory research must first create a reasonably complete picture of a way of life so that the context of actions and relationships can be understood and used in explaining social action and social organization.

This concern with understanding and interpreting the **subjective meaning** of social situations, groups, and ways of life involves a complex process of reading and rereading qualitative data. In the early stages of exploring historical archives or working in the field, attention is focused on two things: first, the researcher has to locate sources of data or gain access to and acceptance by a group; second, the researcher has to develop a clear, basic understanding of the sources or social group.

As the researchers become more at ease with their subjects or sources, they can focus on specific issues of study and on a new and deeper understanding of complexities and depths of meaning. As materials are collected in greater number and become familiar and understood, the collection of more documentation or additional observation becomes less important. There is an increasing need to integrate the material into a coherent pattern. Bringing the observations and interpretations together into a coherent picture that combines the "basic sense" of the era or group with its depths and complexities is a tricky process of moving back and forth between the original documents or field notes and the various interpretations that emerge from successive analyses of the notes or documents.

Filing and Coding

Consider for a moment how you would conduct field research on a classroom. Think of the many things you would have to observe and carefully take notes on: the different types of classes (large/small, lectures, discussions, presentations, guests, video/films), the different types of teachers (male/female, young/old, strict/informal, humorous/serious), the different types of students (male/female, groups/isolates, nerds/goof-offs), various types of behaviour, important or different incidents, and so on. All of these items would have to be carefully observed and documented in detailed field notes.

As you can see, the primary data of qualitative research involve an enormous mass of notes. The analysis of such data creates more notes, notes on notes, and further notes! In addition to the primary data, you would also note your interpretations, queries, hunches, and maybe even associations. All this note taking generates a great amount of material, which has to be organized, filed, and cross-referenced. The success of qualitative research, then, depends on the quality of the primary data and on the researcher's ability to analyze them and to organize them using an appropriate, regularly reviewed, flexible filing system. There are three types of files.

Data files contain primary data in the form of field notes or copies of archival or documentary material, along with notes identifying research details such as location, sources, dates, and any other identifying matter.

Analytic files contain the data in broken-down and reassembled form—that is, coded and interpreted. For example, analytic files of classroom fieldwork might be broken down into subsections on large versus small classes, male versus female teachers, and types of student roles. Such files might also contain discussions and ideas about the coding (see the next section) and interpretation principles involved, the ideas on which they are based, and so on.

Special-purpose files are needed for materials such as bibliographies and references, literature review material, correspondence with other researchers in the field, and notes on the history and development of the research project as it shifts perspectives.

Managing these files as they expand requires regularly reviewing them, identifying each file's purpose, labelling emerging new issues, and so on. As you work through your data, trying alternative interpretations and finding new patterns, sequences, or classes, you need to modify your filing system. But you should not simply throw out the discarded interpretations! You may want to return to them later on. You might register the changes of your interpretations in a special-purpose file and date them so that you can keep track of these changes and clarify the grounds for your views as your interpretations shift and develop.

Sorting and Sifting

It is no simple task to interpret qualitative data so that they emerge as a comprehensible whole, combining overall patterns of meaning with underlying complexities. It involves repeatedly coding the data, then breaking the data down, recoding them, and comparing alternative results.

The coding of data is a process of reorganizing raw data into groups or classes according to some research question, issue, or theme. For example, if you are interested in how teachers cope with "class clowns," you would look through and copy all your notes on clowning incidents into a special file or section of a file. Another concern may be differences in the behaviour of male and female students, so you sift through your field notes in search of observations on this aspect of the classroom. You then copy these observations and place them in a separate, appropriately labelled file.

Sometimes the process of sorting and reorganizing involves insights or hunches; sometimes it is suggested by other research that you have read. At other times a theme is clearly suggested by the data. Usually, however, the principles of coding emerge from sheer hard work, trying first one scheme and then another until patterns, sequences, repeated processes, cases, or relationships begin to emerge.

The first stage of data analysis involves **open coding**. Here, the researcher slowly reads through historical documents or field notes looking for critical terms, key themes, or important events. These are noted and become a basic framework for initial analysis.

A second reading uses this framework to explore connections, classifications, sequences, and cases. Are the critical terms, themes, or events systematically connected with others, or found in association with other terms, events, and themes? Are there variations or subtypes of the initial terms, themes, or events or their connections and associations? Are there exceptions, deviations, or situations where some parts exist without others?

In reading the data and asking these questions, the researcher moves toward a more selective coding approach. The researcher begins to focus, tracking down cases that illustrate patterns and themes, and then making comparisons and contrasts to refine and elaborate these patterns.

For example, as you review your material on how teachers cope with class clowns, you may discover different patterns of responses by teachers. These patterns may relate to other aspects of their teaching styles, so you begin to shift your focus onto the latter, pulling in other observations unrelated to class clowns. On the basis of these observations you may begin to develop a classification of types of teachers, or teaching strategies. You may then see that some teachers are very clear examples of these teaching types; you write profiles of these individuals and add them to your analysis. Such profiles are perhaps no longer just summaries of your observations, but instead are "ideal types" —that is, descriptions that are exaggerated for the sake of clarity.

The coding process, then, moves from a situation in which the codes are determined largely by the primary data to one in which concepts increasingly determine the coding categories. These concepts may be partly built from the data and partly developed from ideas and theories drawn from other research. Coding becomes more elaborate as the issues in the study are defined and focused. Isolated concepts become part of a general framework, subthemes are developed, and commentaries on the data move away from pure description toward analysis and explanation. Note that this process is frequently checked against the primary data files to ensure that the analysis never becomes entirely separated from the perspectives of the individuals or the groups being investigated.

The enormous amount of textual material that is created by qualitative research and analysis of qualitative data has led to the development of software designed specifically for qualitative data analysis. These programs can help researchers search for and retrieve sections of text by using keyword searches similar to those involved in library and Internet searches. This makes some of the multiple coding and filing tasks somewhat easier than purely manual sorting. Other qualitative data analysis software programs have been developed to help the researcher organize and code data, annotate the data, and more (see the qualitative research websites listed at the end of Chapter 7). However, many qualitative researchers emphasize the limits of computer-assisted qualitative data analysis. They argue that, while computers can be helpful in organizing data so that the researcher can recognize patterns in the data, machine coding is less helpful in exploring the potential meanings in the data.

Analytical Strategies

The movement away from description toward analysis and explanation is helped by using certain **analytical strategies**.

LOOK FOR ESSENTIAL COMPONENTS When you identify an idea or pattern in a group of historical documents or in your field notes, try to break it down into a cluster of essential subcomponents. Mentally experiment with these subcomponents. If you were to take away any of them, would this change the overall character significantly?

If changes are likely, what changes might there be? For example, how would a class operate without a class clown? Without a lecture from the teacher? Would things be different if lectures were preprinted and distributed for reading, and classes turned into question-and-answer sessions or discussion sessions? If the teacher brought in guest lecturers, would there be a different dynamic?

LOOK FOR RELATIONSHIPS Are there any detectable relationships among particular pieces of information or observations? Is there some larger sequence of events or processes at work? Are there repeated connections or combinations associated with specific actions or situations? For example, do some types of teachers seem to provoke some types of students? Do some subjects open up opportunities for talking back, being rowdy, or clowning around? Do things happen in classes more at the beginning of term than at the end, or vice versa?

COMPARE AND CONTRAST Is what you see in one case the same or similar to others? Is it different? If it is different, how is it different? Identifying similarities and differences among phenomena enables you to arrange facts into classes or types. For example, how do college classes differ from high-school classes? How do older teachers differ from younger ones? Male teachers from female? Large classes from small classes?

CONSIDER NEGATIVE EVIDENCE Are there exceptions or deviant cases—events, actions, or ideas that do not seem to fit the emerging patterns? What do they mean? Why are they there? How do they challenge the positive evidence or its interpretation? What is hidden or overlooked in the documents or field notes? What is taken for granted or assumed in any explanations by participants?

ASK ALTERNATIVE QUESTIONS Questions can be asked negatively or positively. Why did something happen? Why this event and not something else?

By constantly taking apart and reanalyzing data through the use of these conceptual strategies, the researcher gradually moves toward a **synthesis**—a coherent picture of a past period or event in history or of a way of life of a particular social group or subculture. The synthesis simultaneously shows life as it is lived from the perspective of the historical society or the particular social group and as it is interpreted and classified by the researcher.

BOX 9.3 SUMMARY OF QUANTITATIVE AND QUALITATIVE DATA ANALYSIS

Quantitative Analysis	**Qualitative Analysis**
Research normally guided by hypothesis to be tested	Research normally guided by a series of questions and issues to be explored
Research strategy aim: quantitative data	Research strategy aim: rich, descriptive, qualitative data
Conceptual framework organized in terms of independent and dependent variables	Conceptual framework often deliberately open or unclear about causation and variables
Variables defined operationally at the outset of the research	Variables left open and modifiable by information and insights gained as data gathering proceeds
Data analysis coded once so that raw data are transformed into numerically labelled groups as a basis for quantitative description and interpretation	Data analysis continuously coded and recoded, usually at a nominal level of measurement, according to shifting interpretations and concerns
Coded data organized into tables and graphs or numerical summaries	Alternative coding frameworks applied in connection with different analytical strategies that view data from different angles and ask different questions
Tabular, graphical, and summary measures are used in turn to guide selection of appropriate inferential statistical tests	Repeated data breaking down and recoding, and use of different analytical strategies leads to a synthesis or interpretation showing the depth of complexity alongside some overall patterns

---

What You Have Learned

- Data need to be organized, summarized, and interpreted in order for patterns and relationships to be clearly seen or demonstrated.
- The first step is to change the raw data into a more manageable form by coding them; this takes different forms that depend on whether the study is quantitative or qualitative.
- In the case of quantitative data, coding allows the data to be organized as a frequency distribution and to also summarize the information as a percentage distribution.
- The resulting data will vary in terms of the levels of measurement, and this will affect the kinds of statistical organization and analysis you may be able to do.

- Different kinds of tables, graphs, and measures have specific advantages and limitations, which have to be taken into account when searching for the best ways to organize and interpret data.
- The coding process for qualitative data is complex and involves a cycle of progressively more focused filing of the data in relation to particular analytical concerns.
- This interpretative process involves several analytical strategies, which aid the researcher in moving toward a synthesis of the data.

Focus Questions

1. What is coding? What are the basic rules of coding?

2. What are the different levels of measurement? Find two examples of social science variables for each level.

3. What is a measure of central tendency? Why is it useful?

4. What are the differences between the mode, the median, and the mean?

5. What is a measure of dispersion? Why is it useful?

6. Why is it necessary to calculate the standard deviation from the variance?

7. What is the purpose of cross-tabulation?

8. What kinds of data are produced by qualitative research?

9. What is the "two-stage process" of qualitative data analysis?

10. What are the different kinds of files used in qualitative data analysis?

11. What different kinds of coding are used in qualitative data analysis?

12. What are the analytical strategies used in qualitative data analysis? At what stage in the analysis are they used?

Review Exercises

Describe what the data in the following table show.

Smokers (Reporting Smoking Daily) and Age Groups

	1989	1995
15–19	421 636	458 228
20–44	3 885 343	3 478 230
45–64	1 660 569	1 485 478
65 and over	508 493	394 965

Source: Statistics Canada, Health Indicators, Catalogue No. 82–221-XWE.

1. Calculate the percentage of each group for each year.

2. Produce a pie chart for each year.

3. Interpret the results of the reworked data. Compare the visual presentations with the raw data. Do you arrive at the same interpretation(s)? Elaborate.

4. Discuss the limitations of using only visual presentations to interpret data.

Weblinks

The various sites below provide online learning resources that familiarize you with basic statistical analysis, terms, and graphs and provide examples and explanations on when to use them. The sites also include links to online statistics books, tutorials, downloadable statistical software packages, and related resources. The last site below contains numerous videos on major topics covered in an introductory statistics course. For sites on qualitative research, see the Weblinks at the end of Chapter 7.

You can find these and other links at the companion website for *First Steps: A Guide to Social Research*, Fifth Edition: **http://www.firststeps5e.nelson.com**.

Statistics: Power from Data! (Statistics Canada)
> **www.statcan.ca/english/edu/power/toc/contents.htm**

Mathematical Thinking: Statistics
> **http://www.open2.net/sciencetechnologynature/maths/menu_statistics.html**

Gallery of Data Visualization
> **www.math.yorku.ca/SCS/Gallery**

Introduction to Statistics (videos)
> **http://www.khanacademy.org/**

The Research Report: The End and the Beginning

10

What You Will Learn

- how to review and draw conclusions from your research data
- the basic parts of the research report
- differences between qualitative and quantitative reports
- the APA and MLA styles of documentation
- planning an oral presentation

INTRODUCTION

The journey started some time ago, when you puzzled over some aspect of the social world. You began by collecting background information, which made you more aware of various facets of the issue and helped refine the focus of your investigation. Next you clarified the variables, chose a sample, and selected the research method(s). Then you began the task of collecting the data. Once completed, you had what seemed to be a mountain of evidence. Painstakingly, you summarized and organized the data. Throughout the research process you had a good feeling of accomplishment, as you became increasingly more knowledgeable about the issue. But probably the greatest excitement came at the stage when you summarized and organized the data. You finally had the information you had been looking for. You now had an answer to the research question or hypothesis that first launched you on this journey of discovery.

At this stage you will want to crown this achievement by writing a report to let others know what you found. You may be tempted to rush immediately into writing the research report, but before doing so, consider this: What are your interpretations of the data? What can you conclude? What else needs to be examined? What should others be told about the research? How should the research report be organized? Thus, before drafting the report it is well worth spending some time reviewing your work. It also helps to make notes as you answer these questions; these notes may be useful in writing the draft copy of the report. In this chapter we will suggest ways to clarify your ideas, bring the research to a close for now, discuss the preparation and organization of the research report, and consider what to do next.

REVIEWING AND CONCLUDING

What Did I Want to Find Out?

In the process of collecting and organizing the data, it is easy to get bogged down in detail and lose sight of the original research question or hypothesis. Thus, before we draw conclusions, we need to take note of the initial research objective, the findings, the interpretations of the findings, and the limitations of the study.

In Chapter 3 we suggested that you write down the research issue you were interested in and then ask a series of clarifying questions. The purpose was to arrive at a research question or hypothesis about the issue that would be the focus of your research. Here we suggest you keep the focus on the general issue and the specific research question or hypothesis. Suppose, for example, your initial interest was in the relationship between gender and academic aspirations. Your interest was sparked by a media report implying that males were more likely to drop out of school than females. This prompted you to ask various questions, which helped isolate a particular aspect of the issue for your research. After further consideration and a literature review, let's say you settled on a specific question to research:

> Among university students, are females more likely than males to express an interest in pursuing graduate studies?

Thus, the general issue was the relationship between gender and academic aspirations. But your research focused on a particular aspect of the issue, that of male and female university students and their future interest in doing graduate studies. Moreover, what was the population of your study? How did you select your sample? Suppose your study was a social survey of a randomly selected group of university students in their last year, and the results showed that more females than males planned on doing graduate studies. Could you claim on the basis of the results of your study that females are less likely than males to drop out of university? The answer is no. The question is quite broad and asks about the relationship between gender and academic aspirations. Your study did not concern the possibility of dropping out of university. Again, it is important to take into account the larger issue, but you will want to remain especially focused on the research question or hypothesis.

What Did I Find?

After you decided on a particular research question or hypothesis that related to the general issue, you clarified the variables and selected a research method. Why? You decided to compare the academic aspirations of female and male university students. Why? You decided to measure academic aspirations in terms of whether a respondent intended to do graduate studies. Why? In other words, you had reasons for concentrating on a particular aspect of the general issue, and your main findings should reflect that focus.

What do the data tell you? For example, say your survey found that females were more likely than males to continue their studies after completing their undergraduate degree. You could of course simply state what you found. But that is not enough.

While the information is interesting, you need to interpret what you found. How do the data relate to your research question or hypothesis? For example, in examining academic aspirations among university students you need to do more than simply state the proportion of females that intended to do graduate studies; you need to give information on the male students and then contrast the academic aspirations of the female and male students. Or suppose you wanted to compare the social behaviour of boys and girls in a daycare. To say that boys push and shout at other children does not fully relate to your research objective. You will want to describe the behaviour of girls and then compare the behaviour of boys and girls in the daycare. Likewise, in interpreting the results of an experiment using an experimental and a control group, the analysis would focus on both groups. The results of the experimental group only partly relate to your research objective—you also need to take into account the findings on the control group. You will also want to relate the findings to the research question or hypothesis. Thus, you need to interpret the data by comparing the results of the control and experimental groups.

Did you answer the research question or find support for the hypothesis? If your findings did not provide a definite answer or support, that in itself is important information. However, it is essential not to interpret vague and confusing data as providing a clear answer to the research question or support for the hypothesis. Recall that you focused on a specific aspect of a larger issue. Your research is a small part of many research journeys around that larger issue. Once you have considered what you found, it is time to take note of the limitations of your research.

What Were the Limitations of My Research?

Early in the research process you made several decisions about the operationalization of the variables, the population and sample, the research method, and so on. Possibly there were constraints that affected your decision, such as time and resources. You would have liked to carry out a social survey on a random sample of university students, but lacked the resources and therefore used a nonrandom sample of students at only one university. You wanted to compare how women were portrayed in advertisements in two magazines from 1939 to 1946, but had access only to the copies published from 1943 to 1946. You would have preferred to do an experiment, but were unable to get the necessary permission and so did an observational study of children in a daycare. Clearly, as with virtually all research, your research had certain **limitations**.

In addition, you are likely to note some limitations when the research is in progress or is completed. Again, this is true of most studies. For example, after completing the data collection you realized that you neglected to ask the respondents a particular question. Or you realized too late that the control group was in a room next to a noisy class, which may have distracted their attention. Or you were unable to control certain variables that you believe might influence the results. The limitations of your research do not necessarily make your study less worthy. On the contrary, recognizing the limitations of the study allows you to arrive at a more appropriate **conclusion** and to make suggestions for future research.

What Can I Conclude?

Recognizing the limitations of your research will help you to arrive at a proper conclusion. You did not carry out a social survey on a random sample of all university students, but only on a nonrandom sample of students in one university. You compared only the advertisements of two magazines from 1943 to 1946, and not the copies from 1939 to 1946. You observed children in a daycare; you did not do an experiment. You neglected to ask respondents for their ages. And so on. Whatever the limitations, they need to be taken into account as you consider the conclusion of the study. For example, if you selected a nonrandom sample for your survey then you cannot claim that the results are representative. In this case you have to argue that your findings are suggestive, indicating directions for further inquiry in subsequent research.

BOX 10.1 CHECKLIST: REVIEWING THE RESEARCH

The aim of the review is to bring the entire research process into focus. Write down notes for yourself; they will be useful when you write the draft copy of the report. Below are suggestions of questions to consider.

REFLECTION AND REVIEW

- What was the general issue?
- What was the specific issue (research question/hypothesis)?
- Was the research exploratory, descriptive, or explanatory?
- What was the research design and why?
- What was the population and sample?
- What were the findings?
- What do the findings say about the issue or hypothesis?
- What were the limitations of the research?
- What can be concluded?
- What are the implications for future research or analysis?

PLANNING THE REPORT

- Who will be the readers of the report?
- What language, style, and presentation are appropriate?
- Is there a specified length for the report?
- Is there a specified style of documentation (e.g., parenthetical, footnote, or endnote)?
- How much space should be given to the different sections of the report?

There are two main facets to a conclusion: the specific and the general. First, note the main results and relate them to your research question or hypothesis. What are the main results? What do they say about the research question? Do they clearly support or reject the hypothesis? Next, note how the main results relate to the larger issue you were interested in. Suppose the general issue was the relationship between gender and academic aspirations. You limited the study to finding out whether female university students are more likely than male university students to express an interest in doing graduate studies. What do the results say about the relationship between gender and academic aspirations? The principal purpose of the conclusion is to connect your main findings back to the research question or hypothesis and to the larger issue. (We have more to say about the conclusion later in this chapter.)

THE RESEARCH REPORT

You put a lot of effort and energy into carrying out the research; its accomplishment will have given you great satisfaction. In some small way you came to better understand our social world. But others need to be informed about your research if it is going to contribute to the further development of knowledge on the issue. The widely accepted way of informing others is through a **research report**. The research report tells others what you found, allows them to evaluate your work, and suggests directions for future research. Before discussing the writing and formatting of the research report, let us briefly consider some guidelines. (See also Research Info 10.1.)

Who Are Your Readers?

As with all writing, the research report has to clearly communicate the information to the relevant **audience**. A report directed to a general audience will differ in language, style, and presentation from one intended for an audience of experts on the topic. Suppose you carried out an experiment on cognitive development among grade-schoolers. If your report is intended for a nonspecialist audience, say the parents of grade-schoolers, you will want to tailor your report for them. However, you will want to adopt a different approach if your readers are child psychologists. If the report is for a course, you may want to discuss the intended audience with your professor. Thus, before drafting the report, consider who will read it and adopt the language and style that is appropriate for that audience.

Research Info 10.1 *Tips on Writing*

After having collected, organized, and analyzed the data, our next major challenge is to write the report. But writing the report may seem like a formidable task, partly because for many, writing is in itself a struggle. Consequently, we postpone the writing process or we rush to complete it and give little importance to revising and rewriting the report.

Writing a research report is unlike writing a personal journal. As McFarland and Rhoades note, "Scientific writing...requires factual accuracy, attention to detail, and repeated editing. There is no one right way to write, and there is no one formula for writing. However, in professional writing some rules do apply most of the time." The authors offer 25 useful suggestions on professional writing that they have compiled after years of experience in receiving criticisms of their own writing and critiquing the writing of others. Their tips on ways to begin and complete the writing process can perhaps save you time and even some aggravation.

Source: K. McFarland & D. Rhodes, "Tips for professional writing," *Annals of the American Psychotherapy Association.* Issue 6 (3) (2003), Pg. 31–34. InfoTrac record number: **A110807189**

What Is the Aim of the Report?

We noted in Chapter 2 that your research could be exploratory, descriptive, or explanatory. The aim of the research led you to select a particular method and to search for certain data. It will, therefore, have an impact on what you tell the reader about the research.

The principal aim of exploratory research is to find out more about an issue on which there is little firm information and to point the way to more research. Suppose there was a lack of research on what happens to high-school dropouts in small towns. Perhaps your research was to find out more about the topic, including whether it is a serious problem. Key aims of your research report would be to provide more information about the issue and to indicate the direction that more detailed research should take.

A descriptive study would usually follow an exploratory study. Assuming research has shown that there is indeed a problem with high-school dropout rates in small towns, the aim of your descriptive research could have been to describe the variations of the problem. You may have described the difference between males and females who are high-school dropouts, or compared the high-school dropout rates of small towns and metropolitan areas, for example. Thus, be especially careful to tell the reader what applies to a sample and what applies to the population.

An explanatory study goes beyond description and instead attempts to find out why the variations exist. For example, the aim of your research could have been to explain why the high-school dropout rate was higher among males than females in certain types of small towns. Explanatory research tends to be quantitative in approach and therefore the results are often expressed in the form of statistics. Consequently, your report is likely to be organized around tables and figures that will clarify your conclusions. You will want to be especially concerned with the level of knowledge your readers have about the statistics used in the report. A nonspecialist audience, particularly one not comfortable with statistical analysis, may need more verbal explanation of the results than a specialist audience.

How Long Should the Report Be?

The length of reports varies and usually depends on their purpose and audience. Some reports are brief research notes of about one to five pages. In such a report you mainly highlight the findings and only briefly state the method. However, the most popular research report is the one whose intent is to tell readers the research question or hypothesis, discuss earlier studies that formed the basis of the research, explain how the data were collected, present and analyze the data, answer the research question or hypothesis, and suggest future research. Although the length of such reports varies, your report will have to be long enough to provide the necessary information. Such a report is basically modelled on the articles found in academic journals, as discussed in Chapter 3. They are generally equivalent to between 20 and 30 double-spaced typed pages. We suggest you see articles in leading academic journals in your discipline for an indication of the length and other expectations, such as whether to include an abstract. If the report is for a course, you may want to discuss the length and other requirements of your report with your instructor.

PRESENTING THE RESEARCH REPORT

Now that you have spent some time reviewing why you did the research, what you found, what you concluded, and for whom you want to (or have to) write the report, let us turn to the organization and writing of the report. In most research reports, the information is provided in the order of the basic steps of the research, although how much importance is given to the various sections may vary. The following parts are usually included in all research reports:

1. Title
2. Introduction and Literature Review
3. Research Design

4. Data Analysis
5. Discussion and Conclusion
6. References or Works Cited

Other parts may also be included, such as acknowledgements, the abstract (which is usually found in research reports of academic journals; see Chapter 3 and Box 10.2), notes if necessary, and an appendix. We will discuss the research report in the order of the parts presented above. (If the report is for a course, you may want to discuss with your instructor what additional elements are required.) The parts can also help you to think through the information you need for the report. However, you will not necessarily write the report in this order. For example, you may decide on a title only after you have finished the draft of the entire report. But whatever order you find easiest, the final report should include the various parts in their logical sequence.

Title

Deciding on a title for the report might seem relatively easy. But a title must not be simply catchy; it must provide enough information to give the reader a sense of the general and specific issues of the research. All of this should be expressed in as few words as possible. What would you understand was the subject of the study on the basis of the following title?

Academic Aspirations

This title gives no sense of what the study was about, except that it had something to do with academic aspirations. Now consider the following title:

Academic Aspirations of University Students

This version is an improvement; we now know the study concerns academic aspirations among university students. But if it focused on students in their final year of university, you might want to convey that information in the title:

The Academic Aspirations of Students in the Final Year of University

Can we further improve the title? Possibly. What is important to remember in deciding on a title is that you want to convey enough information to give the reader a good sense of what the research is about.

BOX 10.2 THE SECTIONS OF A RESEARCH REPORT

While much effort goes into the content of the report, its appearance is also important. Consider how willing you would be to read a report that looked disorganized. Your audience will respond to your report based partly on how it looks. The key to an acceptable report is clear language and text that is easy for the readers to follow. Here are some suggestions.

TITLE

In addition to the title of your report, the title page should include your name and other relevant information. If it is written for a course, include the name and number of the course, your professor's name, and the date. Do not number the title page.

continued

Box 10.2, *continued*

ABSTRACT

An **abstract** is a brief summary statement describing the essential points of a research report. Abstracts are required for research published in academic journals, especially those using APA style of documentation, and for research to be presented in professional meetings. When searching databases for our own research purposes, the titles and abstracts are usually key in determining whether or not we decide to read the entire article. Hence professional researchers take great care in writing the abstract and titles for their work. Such careful preparation is reinforced by the fact that journals impose a length restriction on the abstract of usually less than 150 words.

Doing justice to the contents of a research paper within the word limit is, then, the central challenge of writing an abstract. However, since it is the last section to be written you have something to work with: the trick is to extract the core statements from a few key sections of the paper. The key sections are the introduction, which identifies the question, issue, or hypothesis to be addressed by the research; the research design, which reviews the methods used to investigate the question; and the results or findings of the research. If you look for sentences and phrases that reasonably summarize these sections, you will have begun to extract the "essence" of the paper. By using what you have already written, even if you have to further shorten the statements, you ensure that your abstract stays close to the language and ideas of the main body of your work. It is likely that at least a couple of rewrites will be necessary in order to obtain a brief but adequate abstract.

INTRODUCTION AND LITERATURE REVIEW

This follows the title page and abstract if included. From this page on, a number is allotted to each page of the report. There is no need for a major heading for this part, but you may want to use subheadings that reflect its organization. Suppose your literature review first concentrated on Canadian works and then on U.S. works. Subheadings might be used to distinguish the sections. You will have to be the judge of whether subheadings are needed. But make sure they are appropriate and enhance the appearance of the report.

How many pages you dedicate to the introduction will depend on various factors; there is no standard length. Even though there are no restrictions on the length of the introduction, keep in mind the overall length of the report. However, consider the expertise of the readers and how much information they need. If the readers are nonspecialists, for example, more clarification is usually needed.

RESEARCH DESIGN

This part has a major heading. Again, there is no standard length for this part of the report, but take note that much effort went into the research design and gathering of the data. While you want to tell the readers only what is necessary,

continued

it is important that they clearly understand the various steps you went through. Subheadings may be useful to emphasize the importance of each topic. For examples, glance through one of the leading journals in your discipline.

DATA ANALYSIS

A major heading clearly identifies this section, which is the heart of the paper. Although there is no standard length, this section deserves much importance. If you have no restrictions on the length of this section, we suggest at least one-third of the report be on the data analysis. Subheadings are especially helpful in this part of the report. Again, for examples, we suggest you refer to one of the leading journals in your discipline.

DISCUSSION AND CONCLUSION

This part is clearly identified by a heading. There are usually no subheadings. Although this is the last section in the report, it may be the first that some readers will turn to. It should give them a quick sense of what the research report is about and its main conclusions, so they can decide whether to read the rest. Thus, its appearance can be critical.

REFERENCES OR WORKS CITED

The **references** or **works cited** section provides the complete bibliographic citation of the sources mentioned in your report. The information on the sources begins on a separate page after the text of the report and before any appendixes that you may have included. The section is clearly titled "References" if you use the APA style, and "Works Cited" if you use the MLA style. (See the sections in text on APA Basic Guidelines and MLA Basic Guidelines.)

Introduction and Literature Review

In some reports the introduction and the literature review are treated as separate sections. Here we treat them as one section.

The introduction leads the reader into the paper. The opening paragraphs should give the reader a general understanding of what the research is about, including an understanding of the research question or hypothesis.

Your report is a small contribution to the larger body of knowledge about an area, and you will want to tell the reader where your research fits in. Thus, much of the introduction is a review of the literature, whereby you assess previous research that is especially relevant to your research question or hypothesis. In Chapter 3 we provided some advice on preparing the literature review. However, here the literature review needs to be incorporated into the research report, and there will probably be limited space available. In deciding which literature review items to concentrate on, keep in mind that you want to tell the reader what is already known, and that review items have to be relevant to your research question or hypothesis.

The introduction closes by building on the literature review. State your research question or hypothesis clearly and connect it back to what you determined to be an important gap in the previous research you reviewed. After reading the introduction the reader should have a clear sense of why you decided to focus on a certain research question or hypothesis.

It is your responsibility to make certain that readers clearly understand what you wanted to find out. Suppose you specifically focused on the relationship between the types of magazines people read and their voting behaviour. Your readers need to know more about this relationship; that is, they need to be able to distinguish the dependent and independent variables. Are you expecting that people's voting behaviour is dependent on the magazines they read? Or do the magazines people read depend on the people's voting behaviour? According to your study, which comes first?

Research Design

Now that the readers know *what* the research question or hypothesis is and *why* you selected it, you need to tell them *how* you collected the data. Thus you need to describe the sample, the type of measurement instrument used to gather the data, any ethical issues, and any other information regarding the context of the research. Keep in mind that you should provide enough detailed information for readers to be able to replicate the study. To help you in writing this section, we suggest different questions to ask yourself about the research. They do not necessarily have to be presented in the order noted here. However, all these points should be covered.

HOW WERE THE DATA COLLECTED? It is important that readers be clearly told the type of instrument selected to gather the data. Do not simply state the research method used. What instruments did you use? How did you conduct the experiment, the nonparticipant observation research, the content analysis, and so on? If all you told your readers was that you conducted a social survey, they would be unable to replicate the study. You will want to briefly tell the readers how the questionnaire was prepared and pretested; how you administered the questionnaire; whether it was administered to individuals or groups; whether you sent it by mail or handed it to individuals and waited until they completed it; and where the questionnaires were administered and under what conditions. To be completely clear, you may also want to include in an appendix a copy of the questionnaire, or a more detailed description of an experimental arrangement, or the details of the fieldwork process.

When in doubt about including some information, we suggest including it, especially in your draft of the report. Suppose while you were carrying out your experiment an instrument broke and had to be replaced. This fact may be important to readers in properly appraising the procedure and evidence.

The readers also need to know how the key variables were measured. In most cases these involve the dependent and independent variables. Take the earlier example of the magazines people read and their voting behaviour. How did you measure what magazines people read? How did you measure their voting behaviour? Assume you measured voting behaviour by means of a questionnaire that included a question asking respondents which political party they would vote for if an election were held. Respondents were then categorized according to the political party they supported. Assume you measured magazines people read by asking respondents which magazines they subscribe to and which magazines they bought in the last month. All this information is of interest to your audience, because it gives them a clear understanding of what you mean by your variables and how you measured them. They can also better judge your research and appraise the validity and reliability of the research.

WHAT WAS THE SAMPLE AND HOW WAS IT SELECTED?   Recall that all research is based on some sort of sample. The readers need to know what was the sample of your study, and in turn what was the target population. Further, they need to be told how you selected the sample. Try not to leave out necessary information. For example, if you carried out an experiment in which you used control and experimental groups, state how you selected the participants and how they were assigned to the groups. Do not embellish your sample to make it seem more perfect than it really was. If you had difficulties with selecting the sample, inform your readers. Suppose you carried out a survey and initially planned to select a random sample but had difficulty obtaining the necessary list of the population from which to select the sample. Inform the reader why you were unable to select a random sample.

Nor is it enough simply to state the type of sample used. Suppose you selected a simple random sample of 100 students from the population of second-year students at your college or university. Telling your audience that it was a simple random sample is insufficient. How was the sample selected from the population? Did you have a list of all the second-year students? What kind of list was it and were there flaws with the list? Why did you choose the particular sampling procedure? How many people did you have to select from the list to arrive at the actual number of people who agreed to take part in the study?

WHAT WERE THE ETHICAL CONSIDERATIONS?   As we have noted throughout this book, social scientists have responsibilities to the people they research. You will now want to tell the readers what ethical issues your research faced. Did the people you studied volunteer to take part in your research? What did you tell them your research was about? Did you deceive them? Were people told their responses would be kept confidential? How did you ensure confidentiality? Did you infringe on people's privacy? Did your study have any psychological consequences on the participants? Did you seek proper professional advice before carrying out the study? And so on. Do not casually dismiss the ethical considerations; consider how you would feel if you were the respondent of the questionnaire, the participant in the experiment, or the subject of covert observation.

Data Analysis

This is the main section of your paper. Your readers know the research question or hypothesis, the reasons for carrying out the research, and how you gathered the data. Now tell them the findings and your interpretations of the findings. In the process you will also be building a response to your research question or hypothesis.

The method used in the research will largely determine how you present and interpret the data. As noted in other chapters, social science research can be broadly divided into quantitative studies and qualitative studies. The two approaches may greatly differ in their presentation and interpretation of data. For the sake of clarity, we discuss the two separately.

QUANTITATIVE DATA   Your aim in analyzing quantitative data is to inform the reader. As noted in Chapter 9, tables and figures are helpful in summarizing and organizing this type of information. Make sure they are properly labelled, have titles that clearly describe the content of each, and are numbered consecutively. For example:

Table 4: Earnings of Part-Time Working University Students

Since it is labelled "Table 4," it is the fourth table in the report. But is the title clear? Is there room for improvement? What "earnings"? Do you mean weekly wages, hourly wages, or what? And which university students? For example, if the data are restricted to university students registered as full-time, say so. Now consider the following title:

Table 4: Average Hourly Earnings of Full-Time University Students Who Work Part-Time

The improved title tells more about the data in the table. Remember that the title of the table has to make sense to your readers.

The data in the table should be arranged in an appropriate format. Keep in mind what interpretations you want to make of the data, and pass these on to the readers. Always ask yourself what information you are trying to convey.

Suppose you want to create a percentage table on gender and support for candidates running for president of the student government. Should you do row percentages or column percentages? The choice will depend on what information you want to get across. Tables need to be formatted in such a way that they convey the kind of information you want the readers to consider. For example, is the aim of your table to tell the readers the breakdown of support among males and females for each candidate? If so, your table might look like example Table 10.1. In this table, you are telling the readers, for example, that among the supporters of the candidate Lee, 25 percent are females and 75 percent are males. In addition, the total number of respondents that support Lee is 88.

Table 10.1 Breakdown of Candidates' Support by Gender

	Candidates		
	Lee	Santoni	Marshall
Gender	%	%	%
Females	25	50	70
Males	75	50	30
Total	100	100	100
(Number of respondents)	(88)	(64)	(80)

But what if you want to describe the support for the various candidates among females and males? Then your table would be quite different, as illustrated in example Table 10.2. This table shows the readers that among females, 20 percent support Lee; 29 percent, Santoni; and 51 percent, Marshall. In addition, there were 110 female respondents.

Table 10.2 Breakdown of Support for Candidates by Gender

	Candidates				
	Lee	Santoni	Marshall	Total	(Number of
Gender	%	%	%	%	Respondents)
Females	20	29	51	100	(110)
Males	54	26	20	100	(122)

Clearly, the decision on which tables you choose to include in the report depends on the interpretations you want to make of the data. The above examples also illustrate the importance of identifying the variables and the type of data in the table. If you use percentage tables, also remember to indicate the base ("Number of Respondents" in the above examples) on which you computed the percentages (see Chapter 9). This allows readers to recalculate the data. For example, in the second table, knowing that 20 percent of the 110 females supported Lee, the reader can reconstruct the raw numbers from the percentages and possibly arrive at a different interpretation.

Whatever tables or figures you include in the report, make certain you discuss them in the text. But try to avoid simply repeating the data in the table. For example, do not simply repeat what support each candidate has; instead, guide the readers in interpreting the data. Which candidate is more popular among females? Which candidate is more popular among males? Which candidate is more likely to win? Your tables serve as evidence to support your claims.

QUALITATIVE DATA   Unlike quantitative data analysis, which presents findings numerically, qualitative data analysis describes and interprets a social phenomenon. The basic findings are not numerical, but are instead details about an aspect of the social world that occurred in a real-life situation. This is not to say that there will be no tables or figures. But in large part, you describe and interpret what you observed and communicate the information to the reader by means of the written word. Be careful—do not turn the report into mere reportage of what you observed.

Suppose you carried out a one-month nonparticipant observational study of the social behaviour of a class of "gifted" high-school students. You would not only *describe* what you saw, but also *discuss* what you saw. Let's say you gradually noted a different pattern in the behaviour of the teachers that seemed to be connected to which students they were interacting with. What did you observe? What was the behaviour of the teachers when they interacted with male students and with female students? If there were definite differences, what were they? Based on what you observed, why do you think there were differences in the teachers' behaviour? The reader has to understand what you observed, and how you interpret your observations.

The organization of this section will partly depend on the particular type of qualitative study you are describing. For example, say your study was on how one becomes accepted into a group of gifted high-school students. You might choose to give a chronological account of the process of becoming accepted by the group. You may also want to break down the description of the process into stages and describe and discuss what occurred at each stage.

Qualitative research involves keeping extensive field notes, which may also include direct quotations. You may find that in some instances you want to include direct quotations, either from your field notes or from participants. In either case, make sure it is clear to the readers that these are separate from your discussion of them. One suggestion is to indent and single-space quotations that are lengthy.

In describing a place or people, make sure that you do not identify them. In the study on the gifted high-school students, you might choose simply to refer to it as a high school in a large metropolitan city, which in your study is called City High. Similarly, do not identify the people in the study by name.

Discussion and Conclusion

Now that the readers have been told all they need to know about your research, you need to tell them what your main conclusion is. And since there is undoubtedly more to learn about the issue, you may suggest directions for future research.

What have you concluded about the research question or hypothesis? Should the hypothesis be accepted or rejected, or is there no clear evidence? What is your answer to the research question? What do the results say about the larger issue of which your research was a small part?

Suppose the research question or objective was to find out whether there was enough support among students to justify setting up a new student club. The conclusion would state whether students were for or against setting up the club. But perhaps the results showed that social science students expressed more interest than pure and applied science students. You may want to also mention this in your conclusion. Finally, is there enough overall support to justify setting up the new club?

Suppose the study was to test a hypothesis derived from a theory on child development. Assume the hypothesis was that four-year-olds who have siblings are more cooperative than those who have no siblings. The conclusion would state whether or not you found support for the hypothesis. It is also possible that from the data collected, you are unable to arrive at a clear decision. That in itself is a conclusion. Next, ask yourself what your conclusion about the hypothesis says about the theory of child development. Does it lend support to the theory? Should the theory be modified?

After you have drawn conclusions about your findings, you will want to briefly consider suggestions for future research. The conclusion was restricted by the limitations of your research. How could future research overcome some of these limitations? Ask yourself how you would do the study all over again, if you had the time and resources. Would you focus on the same hypothesis? Would you operationalize the variables differently? Would you take into account other variables, such as age, ethnic background, or marital status? Would you select the same research method? Your suggestions are helpful, since no study is the final word on an issue. Your work contributes to the accumulation of knowledge about the issue. It becomes part of the existing research, which guides future research.

References or Works Cited

Although the research process follows the same basic steps, the various disciplines use different documentation styles. If the report is for a course, you may want to ask your instructor whether there is a preferred style for the course. You can also identify the more widely used approach in your discipline by looking at a leading journal in your discipline. Whatever style you use, be consistent throughout your report.

The two commonly used documentation styles are the **APA documentation style,** recommended by the American Psychological Association, and the **MLA documentation style,** recommended by the Modern Language Association.

CITATION MANAGEMENT GUIDES

As we emphasized in Chapter 3 when we discussed the writing of the literature review, it is essential to follow certain formal rules of social science writing. Many of these rules focus on the citation of sources that have been read or used in the process of undertaking research. Research is more credible to the extent that it takes into account the accumulated research by others in a particular field or area—that is, you show yourself to be knowledgeable in the field. Research is also credible in terms of the researchers' honesty in indicating their debts to other researchers and the acknowledgement of others where their ideas have been used or adapted. For these reasons social science researchers attempt to clearly indicate all the sources they have used in any write-up of their research.

The following guidelines show the forms of the APA and MLA styles. This is far from an exhaustive list of the various possibilities of sources that you will need to document and is not a replacement for the APA manual or MLA handbook. We suggest you always consult with your instructor as to which style of documentation to use. Further, ask your instructor about how the style should be applied in particular cases and whether to include more information than that indicated by the style. For more information, consult the latest edition of the APA manual or MLA handbook. See also the Weblinks at the end of this chapter, as well as those in Chapter 3.

APA BASIC GUIDELINES

The following guidelines are taken from the *Publication Manual of the American Psychological Association*, sixth edition.

Essay Paper Format
- Use letter-sized paper (8.5" x 11"). Type the paper, double-spaced, with 1-inch margins on all sides and use standard font, such as 12-point Times New Roman.
- On the top of every page insert as a "running head" the title of your paper set flush left, and the page numbers set flush right.
- There are four main sections to an APA paper: title page, abstract, main body, and references.

In-Text Citations

Use the author—date method—that is, the author's last name and the source's year of publication are separated by a comma and placed directly in the text in parentheses, as in the following example:

(Ariely, 2010)

More information about the source should appear in the reference list at the end of the paper. If you quote from a work, place a comma after the year and include the page number preceded by "p." For example:

"The importance of experiments as one of the best ways to learn what really works and what does not seems uncontroversial" (Ariely, 2010, p. 292).

If instead you introduce the quotation with a signal phrase that includes the author's last name, include the source's date of publication in parentheses and place the page number at the end of the quote in parentheses preceded by "p." For example:

According to Ariely (2010), "The importance of experiments as one of the best ways to learn what really works and what does not seems uncontroversial" (p. 292).

If the work is by two authors, name both authors at all times. Use "and" when referring to the authors within the text and use the ampersand (&) in the parentheses when you cite the text or the source. For example:

As Bibby and Posterski (1992) found…

…" (Bibby & Posterski, 1992, p. 116).

Or if you are only citing the source with no quote, use the following format:

(Bibby & Posterski, 1992)

If the work is by three to five authors, include all the authors' last names the first time you cite the source, and in subsequent citations use only the first author's family name followed by "et al." whether in the signal phrase or in the parentheses. For example:

(Krahn, Lowe, & Hughes, 2011)

(Krahn et al., 2011)

If the work is by six or more authors, use the first author's last name followed by et al. whether in the signal phrase or in the parentheses. For example:

(Jacob et al., 2012)

If the source you cite is taken from another source, name the original source in your signal phrase and indicate the source from which it is cited in the parentheses as well as in the reference list. For example:

Oakes states that "many believed she was correct" (as cited in Warren, 2012, p. 135).

Footnotes and Endnotes

The APA style does not generally use footnotes and endnotes unless to provide explanatory notes. We suggest you discuss with your instructor what is acceptable.

Reference List

The reference list is placed at the end of your paper and begins as a separate page with the heading "References" centred at the top of the page. It consists of the complete list of sources cited in your paper, and the text is double-spaced. The line that follows the first line of each source should be indented by one centimetre or five spaces (i.e., hanging indentation).

The sources are listed in alphabetical order according to the last name of the first author of the source, and if there is no author, by the first word of the title (other than "The," "An," or "A").

Capitalize the major words in journal titles, and the first letter of the word of a title and subtitle of books, chapters, articles, or web pages. Italicize titles of long works such as books and journals, but do not italicize, underline, or put in quotation marks journal articles and articles in edited collections.

Below are various rules about presenting your sources in the reference list.

GENERAL FORMAT

Author's last name, Initials). (Year). *Title of book.* Location: Publisher.

SINGLE AUTHOR

Kirton, J. J. (2007). *Canadian foreign policy in a changing world.* Toronto, ON: Thomson   Nelson.

Two Authors   Present the names of the authors in the order in which they appear on the title page. Use only initials for the first names; separate the names with a comma and ampersand.

> Oakes, L., & Warren, J. (2007). *Language, citizenship and identity in Quebec.* New York: Palgrave Macmillan.

Three to Seven Authors   Present the names of the authors in the order in which they appear on the title page. Use only initials for the first name and separate the names with commas, except the last two authors, which are also separated by an ampersand. Note that if the book, regardless of the number of authors, is other than the first edition, indicate the number of the edition followed by "ed." in parentheses after the title, as in the example below.

> Krahn, J. H., Lowe, G. S., & Hughes, D. K. (2011). *Work, industry and Canadian society* (6th ed.). Toronto, ON: Nelson Education.

Edited Book (Anthology or Compilation)   If you refer to an edited source (e.g., a book that contains articles by various authors), but not to a specific author in the source, it is listed according to the name of the editor followed by (Ed.) as in the following example.

> James, C. E. (Ed.). (2005). *Possibilities and limitations: Multicultural policies and programs in Canada.* Halifax, NS: Fernwood.

Article in an Edited Book   If you refer to an article that is in an edited book, list the article by its author's name. A new sentence begins after the article title with the word "In" followed by the editor's initials and family name and (Ed.). Note that only the title of the book is italicized and the pages of the article are indicated after the book's title.

> Author's last name, Initials. (Year). Title of article. In Editor's initials. Editor's last name (Ed.), *Title of book* (pp. xx–xx). Location: Publisher.

> Jansen, C. J. (2005). Canadian multiculturalism. In C. E. James (Ed.), *Possibilities and limitations: Multicultural policies and programs in Canada* (pp. 21–33). Halifax, NS: Fernwood.

Organization as Author   This can include government departments, professional associations, agencies, and other organizations whose publication does not indicate an author. If the author is also the publisher, use "Author" in place of the organization's name.

> Canada Mortgage and Housing Corporation. (2011). *Housing market outlook: Canada edition, first quarter.* Ottawa, ON: Author.

Article in a Journal   List an article in an academic journal by the author's last name. Only the volume number is indicated if the periodical uses continuous pagination throughout the volume. If each issue of the volume begins with page one, indicate the issue number in parentheses. Note that only the title of the journal and volume are italicized, as shown below.

> Author's last name, Initials. (Year). Title of article. *Title of Journal, Volume* (Issue number), page numbers.

> Crush, L. (2007). When mediation fails child protection: Lessons for the future. *Canadian Journal of Family Law, 23*(1), 55–91.

Article in a Popular Magazine   The article is listed by the author's last name. Note that the date of publication is also indicated and that only the title of the magazine and volume number are italicized. If the volume number and issue number are known, indicate either the volume or both, as shown below.

Author's last name, Initials. (Year, Month). Title of article. *Name of Magazine, Volume number*(Issue number, if available), page number(s).

Gwin, P. (2011, March). Battle for the soul of Kung Fu. *National Geographic, 219*(3), 94–113.

ARTICLE FROM AN ONLINE JOURNAL   Follow the same guidelines as if it were a printed journal article, followed by the retrieval information.

Author's last name, Initials. (Year). Title of article. *Title of Online Periodical, Volume number*(Issue number if available). Retrieved from http://www.xxx

Ghaziani, R. (2010). School design: Researching children's views. *Childhoods Today, 4*(1). Retrieved from http://www.childhoodstoday.org/download.php?id=48

JOURNAL ARTICLE WITH A DIGITAL OBJECT IDENTIFIER (DOI)   The DOI is an alphanumeric string that identifies journal articles and other documents. The APA recommends using the DOI, if available, instead of the URL or the database name. Indeed, the APA also recommends using the DOI for print sources if available. URLs are likely to change, while DOIs are unique links for electronic documents and do not change. Many publishers include the DOI on the first page of the document or with the journal's title and volume. Once you know the DOI of a document, you can search for it at the DOI site (**http://dx.doi.org**).

Author's last name, Initials. (Year). Title of article. *Title of Online Periodical, Volume number*(Issue number if available), *page numbers*. doi:xx.xxxxxxx

Cunningham, S., & Kendall, T. D. (2011). Prostitution 2.0: The changing face of sex work. Journal of Urban Economics, 69(3), 273–287. doi:10.1016/j.jue.2010.12.001

DATABASE NAME   The APA states that it is not necessary to include database information. However, you should check with your professor in case there is a need to include the database information.

NON-PERIODICAL DOCUMENTS ON THE INTERNET   Include the author, title of the document, and the date the material was updated, if available, or when it was posted online. If there is no date, replace the date with (n.d.). If there is no author, place the title in the author position. Include the URL of the document cited.

Library and Archives Canada. (2010). *The early Chinese Canadians 1858–1947*. Retrieved from http://www.collectionscanada.gc.ca/chinese-canadians

If the page is likely to be moved, provide the date you retrieved the document and indicate the URL, that is, Retrieved Month, Day, Year, from http://www.xxx.

If this is a particular source such as PowerPoint slides, video files, and other material that needs to be clarified, add the description in square brackets after the title: [PowerPoint slides] [Video file].

Kahneman, D. (2010, March). *The riddle of experience vs. memory* [Video file]. Retrieved from http://www.ted.com/talks/daniel_kahneman_the_riddle_of_experience_vs_memory.html

EXAMPLE OF THE REFERENCE LIST   Start the reference list on a new page with the title "References" centred at the top, neither underlined nor italicized. List, double-spaced, all the sources mentioned in the paper in alphabetical order by author, as shown below.

References

Ariely, D. (2010). *The upside of irrationality: The unexpected benefits of defying logic at work and at home*. New York, NY: HarperCollins.

Campbell, C. (2007, March). Fears of an "epidemic" of abortions. *Maclean's, 120*(9), 22.

Crush, L. (2007). When mediation fails child protection: Lessons for the future. *Canadian Journal of Family Law, 23*(1), 55–91.

Cunningham, S., & Kendall, T. D. (2011). Prostitution 2.0: The changing face of sex work. *Journal of Urban Economics, 69*(3), 273–287. doi:10.1016/j.jue.2010.12.001

Kahneman, D. (2010, March). *The riddle of experience vs. memory* [Video file]. Retrieved from http://www.ted.com/talks/daniel_kahneman_the_riddle_of_experience_vs_memory.html

Library and Archives Canada. (2010). *The early Chinese Canadians 1858–1947.* Retrieved from http://www.collectionscanada.gc.ca/chinese-canadians

MLA BASIC GUIDELINES

The following guidelines are taken from the *MLA Handbook for Writers of Research Papers*, seventh edition.

Essay Paper Format

- Use letter-sized paper (8.5" x 11"). Type the paper, double-spaced, with 1-inch margins on all sides and use standard font, such as 12-point Times New Roman.
- Do not include a title page, unless requested by your instructor. Instead, in the upper left-hand corner of the first page on separate, double-spaced lines, write your name, your instructor's name, the course name, and the date. Double-space again and centre the title of your paper, capitalized but not underlined, italicized, or placed in quotation marks. (You can, however, use quotation marks or italics for any part of the title that refers to another work.)
- On each page include a header in the upper right-hand corner with your last name followed by a space and the page number. Place the header about a centimetre from the top and flush with the right margin. (Your instructor may request that the header and page number do not appear on the first page.)
- When dividing the essay into various sections, number each section, adding a period and then a space followed by the title of the section.

In-Text Citations

Use the author-page method, that is, the author's last name and the page, if necessary, placed directly in the text in parentheses, as for example:

(Ariely)

If the page number is required, add it after the author's name, separated by a space. More information about the source should appear in the Works Cited list at the end of the paper. For example:

"The importance of experiments as one of the best ways to learn what really works and what does not seems uncontroversial" (Ariely 292).

If instead you introduce the quotation with a signal phrase that includes the author's last name, include the page number at the end of the quote in parentheses.

According to Ariely, "The importance of experiments as one of the best ways to learn what really works and what does not seems uncontroversial" (292).

Likewise, if the author's name is mentioned in the text and you feel it necessary to indicate the page reference, include it in parentheses at the end of the statement before the period.

If the work is by two or three authors, name the authors every time. For example:

(Bibby and Posterski 116)

(Krahn, Lowe, and Hughes 273)

If the work is by more than three authors, include the last name of the first author followed by "et al." and the page number if required.

(Martinez et al. 342)

If the source you cite is taken from another source, name the original source in your signal phrase and indicate the source from which it is cited in the parentheses. Use the abbreviation "qtd. in" to indicate that is was quoted in another source.

Oakes states that "many believed she was correct" (qtd. in Warren 135).

Footnotes and Endnotes

The MLA recommends limited use of footnotes and endnotes unless to provide explanatory notes. We suggest you discuss with your instructor what is acceptable.

Works Cited Page

The works cited list is placed at the end of your paper and begins as a separate page with the heading "Works Cited" centred at the top. The text is double-spaced. The line that follows the first line of each source should be indented by one centimetre or five spaces (i.e., hanging indentation).

The sources are listed in alphabetical order according to the last name of the first author of the source. Both the last name and complete first name of the author are indicated. If there are various sources by the same author, list the sources by the year of publication, beginning with the earliest year. If there is no author, list the source by the first word of the title (other than "The," "An," or "A").

Capitalize each word in the titles, except words that are articles, prepositions, or conjunctions, unless they are the first word of the title or subtitle.

Italicize titles of works such as books and journals. (Note that previous editions of the MLA handbook indicated to underline the titles.)

For each entry indicate the medium of publication. Most are likely to be print, but you could also have others such as web, film, DVD, or television.

Do not include URLs for web entries, unless your instructor requires them. If needed, include the URL in angle brackets after the entry, followed by a period, for example, ‹http://www....edu/›.

If you cite a publication that was originally issued in print but that you retrieved from an online database, indicate the online database name in italics.

Below are various other rules about presenting your sources in the works cited page.

General Format

Author's last name, First name. *Title of book.* Location: Publisher, Year. Medium of publication.

Single Author

Kirton, J. John. *Canadian Foreign Policy in a Changing World*. Toronto: Thomson Nelson, 2007. Print.

Two or Three Authors   Present the names of each author in the order in which they appear on the title page. The name of the first author appears with the family name followed by the first name, while the names of the other author(s) appear as first name followed by family name. If the book is of an edition other than the first, indicated the edition after the title, as shown below.

Krahn, Harvey J., Graham S. Lowe, and Karen D. Hughes. *Work, Industry and Canadian Society*. 6th ed. Toronto: Nelson Education, 2011. Print.

Four or More Authors   List only the first author followed by "et al.," or you list all the authors in the order in which their names appear on the title page.

Acharya, Viral V., et al. *Guaranteed to Fail: Fannie Mae, Freddie Mac, and the Debacle of Mortgage Finance*. Princeton: Princeton University Press, 2011. Print.

OR

Acharya, Viral V., Matthew Richardson, Stijn Van Nieuwerburgh, and Lawrence J. White. *Guaranteed to Fail: Fannie Mae, Freddie Mac, and the Debacle of Mortgage Finance*. Princeton: Princeton University Press, 2011. Print.

Two or More Books by the Same Author   If there are various sources by the same author, list them by year of publication, beginning with the earliest year. For subsequent sources, replace the author's name with three hyphens, as shown below:

MacMillan, Margaret. *Paris 1919: Six Months That Changed the World*. New York: Random House, 2003. Print.

---, *The Uses and Abuses of History*. Toronto: Penguin, 2008. Print.

Edited Book (Anthology or Compilation)   If you refer to an edited source (e.g., a book that contains articles by various authors), but not to a specific author in the source, list it according to the name of the editor, followed by "ed.," as in the following example:

James, Carl E., ed. *Possibilities and Limitations: Multicultural Policies and Programs in Canada*. Halifax: Fernwood, 2005. Print.

Article in an Edited Book   If you refer to an article that is in an edited book, list the article by its author's name. Enclose the title of the article in quotation marks. Set the title of the book in italics, followed by "Ed." and the editor's name. Note that the pages of the article are indicated after the year of the book's publication. The basic format is as follows:

Author's last name, First name. "Title of article." *Title of book*. Ed. Editor's name. Location: Publisher, Year. Page numbers. Medium of publication.

Jansen, Clifford J. "Canadian Multiculturalism." *Possibilities and Limitations: Multicultural Policies and Programs in Canada*. Ed. Carl E. James. Halifax: Fernwood, 2005. 21–33. Print.

Organization as Author   This can include government departments, professional associations, agencies, and other organizations whose publication does not indicate an author.

Canada Mortgage and Housing Corporation. *Housing Market Outlook: Canada Edition, First Quarter*. Ottawa: Canada Mortgage and Housing Corporation, 2011. Print.

ARTICLE IN A JOURNAL   List an article in an academic journal by the author's last name, followed by the title of the article in quotation marks. Italicize the title of the journal and follow it with the volume number as well as issue number, if available.

> Author's last name, First name. "Title of article." *Title of Journal* Volume.Issue number (Year): Page numbers. Medium of publication.

> Crush, Linda. "When Mediation Fails Child Protection: Lessons for the Future." *Canadian Journal of Family Law* 23.1 (2007): 55–91. Print.

ARTICLE IN A POPULAR MAGAZINE   List the article by the author's last name. Place the title of the article in quotation marks and italicize the title of the magazine. Note that the month of publication is abbreviated.

> Author's last name, First name. "Title of Article." *Name of Magazine* Day Month Year: page numbers. Medium of publication.

> Gwin, Peter. "Battle for the Soul of Kung Fu." *National Geographic* Mar. 2011: 94–113. Print.

ARTICLE FROM AN ONLINE PERIODICAL   Follow the same guidelines as with a printed journal article, followed by the medium of publication you used and the date of access. If the journal article has no page numbers, use "n. pag." (no pagination) before the medium of publication.

> Author's last name, First name. "Title of Article." Title of Online Periodical Volume.Issue number (Year): Page numbers. Medium of publication. Day Month Year of access.

> Ghaziani, Rokhshid. "School Design: Researching Children's Views." *Childhoods Today* 4.1: 1–27. Web. 15 May 2011.

If the article was retrieved from a database, follow the same guidelines as with a printed journal and add the name of the database in italics, the medium of publication (usually Web), and the access date.

> Paek, Hye-jin, Albert C. Gunthe, Douglas M. McLeod, and Thomas Hove. "How Adolescents' Perceived Media Influence on Peers Affects Smoking Decisions." *Journal of Consumer Affairs* 45.1 (2011): 123–147. *InfoTrac College Edition*. Web. 15 May 2011.

NONPERIODICAL DOCUMENTS ON THE INTERNET   Collect as much of the following information as possible.

> Author's last name, first name. "Article or Document Title if Available." *Title of the Overall Website.* Version or edition numbers, including posting dates, volumes, or issue numbers, if available. Publisher or n.p. if no publisher, publication date or n.d. if no date. Web. Date of access. <URL (if required by your instructor)>.

Include the information that is available.

> "Why the Chinese Came to Canada." *The Early Chinese Canadians 1858–1947*. Library and Archives Canada. 16 Jan. 2009. Web. 30 Mar. 2011.

EXAMPLE OF THE WORKS CITED LIST   Start the works cited list on a new page with the title "Works Cited" centred at the top, neither underlined nor italicized. Include double-spaced and in alphabetical order according to author all the sources mentioned in the paper, as shown below. Indicate the medium of publication (e.g., print, web, television, and so on) for each source.

Works Cited

Acharya, Viral V., Matthew Richardson, Stijn Van Nieuwerburgh, and Lawrence J. White. *Guaranteed to Fail: Fannie Mae, Freddie Mac, and the Debacle of Mortgage Finance*. Princeton: Princeton University Press, 2011. Print.

Canada Mortgage and Housing Corporation. *Housing Market Outlook: Canada Edition, First Quarter*. Ottawa: Canada Mortgage and Housing Corporation, 2011. Print.

Krahn, Harvey J., Graham S. Lowe, and Karen D. Hughes. *Work, Industry and Canadian Society*. 6th ed. Toronto: Nelson Education, 2011. Print.

MacMillan, Margaret. *Paris 1919: Six Months That Changed the World*. New York: Random House, 2003. Print.

---, *The Uses and Abuses of History*. Toronto: Penguin, 2008. Print.

"Why the Chinese Came to Canada." *The Early Chinese Canadians 1858–1947*. Library and Archives Canada. 16 Jan. 2009. Web. 30 Mar. 2011.

ORAL PRESENTATIONS

At college and university it is likely that you will be required to make oral presentations in some of your courses. This requirement often arouses considerable anxiety in many students. However, it is a useful skill to develop and, with appropriate preparation, can be a rewarding and even enjoyable experience. Normally an oral presentation will be about a particular piece of research or writing that you have undertaken for the course. Keep in mind that it is your research or topic, so you are the expert on the subject. Your knowledge will be greater than your fellow students' and may even be more detailed and up to date in some respects than your teacher's knowledge of the particular subject matter.

The most essential preparatory step is to obtain clear directives and guidelines from your instructor about the length of time permitted for presentation; the overall format of the approach to be taken (e.g., highly formal, discussion provoking, etc.); particular points or issues that are to be covered; whether or not audiovisual aids are permitted; the location and audience if these are different from the regular class; and so on. These guidelines will help to frame and develop your approach to the presentation. Time limits are very important to note and are often quite short (10 to 20 minutes), forcing you to set clear priorities because you will not be able to say everything you know about the topic within the allotted time.

It is likely that your instructor has indicated some aspects or issues that should be focused upon, and these should be prioritized. The course itself is organized around certain key ideas, issues, and topics, and these too should be used as points of reference for selecting and focusing the elements of your presentation. For example, a research methods course is likely to require emphasizing information and discussion of research design, whereas a substantive course on a particular field such as ethnicity and ethnic relations is likely to emphasize description of patterns and shifts in demographic and other social data. Other courses might require reviews of theories, debates, and their intellectual history.

A successful presentation not only follows the instructor's requirements and relates to the themes of the course, but is coherent in its own terms. That is, it hangs together in a meaningful way for both you and your audience. This is perhaps the trickiest part of an oral presentation since your subject matter will be academic—in other words, somewhat abstract and distant from everyday experiences, personal interests, and feelings. Yet successful presentations often depend on the personal link between presenter and audience, and are perceived as more meaningful and interesting as a result. Of course, the more you are interested in and enthusiastic about the topic, the more likely it will come across as interesting for your audience as well. Your own comfort with the presence of an audience, as signalled by eye contact, for example, also gives the sense that you are interested in and value their response. Above all, then, a personable way of presenting the material is invaluable.

Avoid boring your audience through speaking in a monotone; presenting a large proportion of your talk as lists, statistics, and definitions; or using jargon-laden vocabulary. Illustrate your points with striking examples; use clear, everyday language with a minimum of jargon; and present statistical information with appropriate tabular and graphical summaries. Where useful, connect your material to points raised by other student presenters, as well as to the major themes and issues of the course. If using audiovisual aids, ensure that they support rather than upstage your presentation (see below). Note that you'll need to take all this into account without losing sight of the major issues determined by your subject matter or topic!

Bringing together the academic material and presenting it in an involving and interesting manner is difficult. It is impossible without careful preparation—doing your research thoroughly; writing and rewriting the results to "boil it down" to the required time-limited presentation format; developing a narrative technique involving appropriate examples, audiovisual elements, and perhaps some commentary on personal aspects of the research; and so on. Finally, it is impossible to undertake an oral presentation without rehearsing and timing your planned talk, preferably with someone who can give you supportive feedback and advice. It need not be a colleague in the same class. If you can give a talk to a complete outsider who still finds it clear and interesting, you will have done very well and will likely succeed in the class itself. Bear in mind that an oral presentation, like most complex tasks, is an activity that improves with practice.

Audiovisual Aids and PowerPoint

Computer technology and the Internet have led to the "wired classroom" and the extensive use of audiovisual techniques and components as part of student presentations. Nowadays it is common to use audiovisual aids by incorporating them in a PowerPoint presentation. In their simplest form, PowerPoint and other presentation software substitute for the blackboard or overhead projector and should merely present the key points or outline of a presentation and permit the speaker to elaborate further. However, because of the capabilities of programs such as PowerPoint combined with other technological aids, we too often forget that they are tools to assist us and should not be a replacement for or central focus of the presentation.

In other words, your presentation should be more than reading the PowerPoint slides. A lot of care should go into the preparation of the slides and how they contribute to the presentation. Limit the use of slides to a number that will allow you to complete the presentation in the time permitted. The text in the slides should be simple phrases, key points, outlines, or definitions of basic terms. The slides are merely a part of the presentation, so lengthy texts should be used sparingly, if at all. Further, restrict what you read from the slides, especially text, for reading aloud is a sure way to bore your audience.

PowerPoint and other similar programs have become widely popular in part because they offer multimedia capabilities. But it is essential that they be used appropriately to enhance what you want to say. Don't get carried away with lots of splashy visuals and videos that will upstage the presentation and overwhelm or confuse the audience. Indeed, the purpose of the presentation is to engage the audience about the issues being presented, not to dazzle them with distracting visuals.

In sum, use audiovisual aids only if they help clarify and illustrate what you want to say; avoid making them the focus of the presentation. View your talk as the central and primary means of communication. Preparation and practice are of the essence in building your confidence in doing the oral presentation and feeling at ease when in front of the audience. Using a software program as PowerPoint makes practice even more necessary to ensure a successful presentation. Indeed, you should be prepared to do the presentation without PowerPoint not only because the technology may fail, but also because doing so will help you to focus on your verbal communication. Once you are well prepared, you can concentrate on how you communicate and engage the audience. Be enthusiastic and confident, for you have put much time into preparing your presentation. Never forget the audience, speak clearly, keep eye contact, smile, and lastly enjoy their applause!

WHAT NEXT?

Recall that when you thought of an interesting researchable issue, your first step was to explore what other research existed on the issue. You then clarified your initial issue and built on the research of others, thereby making a contribution to our understanding of the social world. But your quest to better understand the issue does not end there. As you noted in the conclusion of the report, there is more to know. Many other research journeys will be undertaken to help us better understand the social world.

We hope that this book has given you a sense of the serious effort and thrill that goes into social science research. The information you have gained has, we hope, made you more critical of what is acceptable as social science research. More important, we hope you will be able to begin to carry out your own research and experience for yourself the challenge and exhilaration of doing social research. Research is the very foundation of the social sciences; as one research project ends, another begins. Enjoy!

What You Have Learned

- Before we draw conclusions, we need to review the initial research objective, the findings, the interpretations of the findings, and the limitations of the study.
- Before drafting the report, we need to consider who will read it and adapt the language and style to that audience.
- All research reports include the following parts: title, introduction and literature review, research design, data analysis, discussion and conclusion, and references (or works cited).

- Although different types of research reports exist, the main difference is how much importance is given to the various sections.
- It is essential to use a proper form of documenting sources and to provide a complete bibliographic citation of the sources mentioned in the report.
- Oral presentations of successful research are often required and involve particular kinds of preparation, including planning the appropriate use of audiovisual aids.

Focus Questions

1. What is the aim of writing a research report?
2. Why should you take into account the audience of your report?
3. What are the main sections of a research report?
4. Why is it important to properly document the sources and provide a reference list at the end of the report?
5. Why should you take into account the limitations of your research?
6. What are some differences between writing a quantitative data analysis and a qualitative data analysis?
7. What should you include in the conclusion of the report?
8. What are some of the issues to be considered when planning an oral presentation?

Weblinks

The following websites offer practical advice on writing a research report, information on avoiding plagiarism, guides to using the APA and MLA documentation styles and sample papers using either style of documentation, and suggestions on how to do a presentation. (See also Chapter 3 for more on applying the APA and MLA styles of documentation, as well as on how to determine what is plagiarism and how to avoid it.)

You can find these and other links at the companion website for *First Steps: A Guide to Social Research*, Fifth Edition: **http://www.firststeps5e.nelson.com**.

Writing Resources

General Writing Resources
> **http://owl.english.purdue.edu/owl/section/1/**

The Elements of Style by William Strunk, Jr. (book)
> **http://www.bartleby.com/141**

Advice on Academic Writing
> **http://www.writing.utoronto.ca/advice**

Tips to Writing an APA Research Essay
> **http://www2.capilanou.ca/programs/psychology/students/apa/tips.html**

APA Style of Documentation

General APA Guidelines: Purdue Owl
> http://owl.english.purdue.edu/owl/resource/560/1/

A Guide for Writing Research Papers (APA style)
> http://webster.commnet.edu/apa/apa_index.htm

Frequently Asked Question about APA Style
> http://www.apastyle.org/learn/faqs

APA Format Citations—Sixth Edition (Word tutorial video)
> http://www.youtube.com/watch?v=9pbUoNa5tyY&feature=related

MLA Style of Documentation

MLA Formatting and Style Guide: Purdue Owl
> http://owl.english.purdue.edu/owl/resource/747/01/

A Guide for Writing Research Papers (MLA style)
> http://webster.commnet.edu/mla/index.shtml

Citation Wizard (creates citations)
> http://workscited.tripod.com/index.html

MLA Style Essay Format—Seventh Edition (Word tutorial video)
> http://www.youtube.com/watch?v=22CPQoLE4U0&feature=channel_video_title

Oral Presentations and PowerPoint

The websites below provide advice on how to do a presentation and how to use and avoid misusing PowerPoint.

Oral Presentations
> http://www.writing.utoronto.ca/advice/specific-types-of-writing/oral-presentations

Life After Death by PowerPoint 2010 (comedy sketch video)
> http://www.youtube.com/watch?v=KbSPPFYxx3o&feature=related

Glossary

abstract Brief summary statement describing the essential points of a research article. (55)

academic journals Regularly published journals featuring reports, debates, discussions, and reviews of professional social science research, written for other professionals and often published by professional associations or institutions such as universities. (55)

accidental sample Nonrandom or nonprobability sample made up of available data, such as historical documents and other sources surviving over time. (85)

analytic files In qualitative studies, files containing data that have been broken down or taken out of context for purposes of analysis and interpretation; the data are no longer in their primary, original form. (237)

analytical strategies Various techniques for interpreting and analyzing qualitative research data in order to develop an overall interpretation or synthesis. (239)

anonymity In survey research, the impossibility of identifying the respondents through the submitted questionnaires. In field research, anonymity requires that the researchers make it impossible to identify individuals from the published research. (126)

anthropology A social science discipline that studies the physical evolution and variety of human beings, as well as the nature and variety of human cultures. (9)

APA documentation style The system of formatting in-text citations, endnotes, and references, as well as the general style and organization for research papers, articles, and books; developed by the American Psychological Association. (256)

archive The location or collection of documents, audiovisual recordings, or other materials, stored so that they are accessible to those interested in these materials as potential data for their research. (202)

audience The readership of your research report; it may consist of a teacher, clients, the general public, or other researchers. Each type of audience requires a different style of communication. (247)

available data Information or data already gathered and used by other researchers or by governments and other institutions, which can be accessed, analyzed, and interpreted by researchers interested in different questions from those of the researchers who created the data. (196)

bar graph A visual presentation of quantitative data in which one axis represents classes, categories, or groups, and the other axis represents the quantity or size of the groups as bars. Bar graphs may be either horizontal or vertical. (231)

bibliography A list of books, articles, and other sources used in writing a research report or research proposal and appended to the end of the text. A bibliography with brief summaries of each item is called an annotated bibliography. (57)

bivariate analysis Analysis that looks at data in terms of the relationships between two sets of variables; for example, level of education and income, or gender and television-viewing tastes. (226)

case study method In-depth investigation of one or a few subjects or other phenomena, involving a combination of data-gathering techniques. (160, 177)

casual (or lay) research Unsystematic fact gathering, often focused on a problem of everyday life, such as buying a new car. (5)

cause and effect Understanding phenomena in terms of systematic connections, which can be discovered through the research process. (26)

cell (spreadsheet) The space for data entry at the intersection of a column and a row in a spreadsheet. (213)

central tendency Any measure of the typical, most numerous, or "average" value or group in a set of data. The mode, median, and mean are the most widely known measures of central tendency. (219)

closed-ended questions Questions in interviews and questionnaires that require the respondents to select from a list of alternative responses. (105)

cluster sample A sample obtained by a random or probability sampling technique, usually used when dealing with large populations that have no lists of their members but that do have lists of subgroups of the population. For example, there is no list of all Canadian schoolchildren, but there are class lists. In this case, researchers may randomly sample schools, randomly sample classes within schools, and then randomly select individual children from the selected classes. (83)

coding/coding system A system of changing raw or primary data to a set of standardized subgroups for either qualitative or quantitative analysis by applying classification techniques. Used especially in analyzing questionnaires and in content analysis of mass media. (190)

computer-assisted telephone interviewing (CATI) A technique of using the computer to randomly dial telephone numbers and of displaying survey questions on-screen while the interviewer uses the keyboard to enter responses to the questions as they are answered. (122)

conclusion The final section of the text of a research report, which highlights the main findings, evaluates the strengths and weaknesses of the research, and indicates possible directions for further research. (245)

confidence interval The range within which the true value of some population characteristic is expected to fall. (91)

confidence level The probability that an estimate of a confidence interval is correct. (91)

confidentiality The researcher's promise to keep the knowledge of research participants' identities restricted to the research team. (126)

content analysis The systematic coding of ideas, themes, images, etc., in mass media. The content is usually quantitatively analyzed, but content analysis may also refer to the qualitative coding or interpretation of documents and artifacts as encountered in historical research. (183)

contingency question A survey question that leads some, but not all, respondents to another question (or questions) that follows up on a particular response. (112)

control group In experimental research, those participants who will not be exposed to the experimental variable. However, the experimental and control groups resemble each other in every other way possible. A control group is used to control for possible effects, such as the passage of time, that may lead to changes but are not effects of the experimental variable. (137)

convenience sample A nonrandom or nonprobability sampling technique that consists of selecting for research the most convenient and accessible respondents or subjects. Often used by the media, this technique is sometimes necessary in social science when there are no alternative sources of information. (86)

correlation The idea of a relationship between variables such that changes in the value or strength of one or more independent variables are associated with changes in value or strength of one or more dependent variables. The strength and the direction of these changes may be assessed by various statistical measures. Correlation does not imply causation. (228)

covert observation Fieldwork that is carried out without the knowledge or consent of those being observed. (178)

cross-sectional survey/study A survey or study that takes place at a single point in time, giving a "snapshot" of a state of affairs that may later change. (100)

cross-tabulation (cross-tab) A table that presents the frequencies and percentages of two or more variables. (226)

cumulative frequency distribution A presentation of a data set so that any given score and scores lower are presented as a percentage of the total number of scores. (215)

data analysis Going beyond summary and organization of data to interpretation of patterns within data. In quantitative research, this tends to be a distinct stage following the gathering and organization of data; in qualitative research analysis, it tends to be intertwined with data gathering and organization. (37)

database Large collections of data organized for ease of computer search and quick retrieval, mainly available via the Internet. Such databases may be quantitative information such as census statistics, archival collections of documents or photographs, or the full text of articles in periodicals and journals. (51)

data collection The stage in the research process when information is gathered through surveys, experiments, fieldwork, or indirect methods to generate data through which to explore a research question or to test a hypothesis. (171)

data files In qualitative research, files containing the primary data of firsthand observation, field interviews, video footage, etc., along with notes identifying exactly when and where the information was gathered. (237)

debriefing Explaining the procedures of a research project to participants who have been deceived about any part of it. This procedure was developed in response to ethical criticisms of psychological experiments and covert fieldwork. (150)

deductive reasoning Drawing ideas from other ideas or theories; moving logically from general principles or assumptions to more specific ideas. (25)

dependent variable The variable that is assumed to depend on or be caused by another variable, the independent variable, in a hypothesis. (27)

descriptive statistics The techniques used to organize and summarize data quantitatively, including tables, graphs, and measures, such as those of central tendency and variation. (234)

descriptive study A study that systematically observes and carefully describes the social phenomena of interest. Description is one of the major aims of social science, along with explanation and prediction. (35)

dichotomous question A question that offers two options, generally yes/no, true/false, or agree/disagree; also known as a binary question. (103)

direct research methods Techniques of data gathering that involve interaction with the individuals or groups being studied, such as through surveys, experiments, and participant observation fieldwork. (183)

double-barrelled question A question that asks two or more questions simultaneously, resulting in an ambiguous response that is difficult to interpret. (107)

economics The social science discipline that looks at the production, distribution, and consumption of goods and services, primarily focusing on modern market processes. (10)

empirical verification The processes involved in checking our ideas through techniques of systematic observation, which can be repeated by others. (23)

errors Inaccurate information or data arising in the course of research. Errors may be due to gaps or bias in primary sources, cultural misunderstandings or biases, imperfect experimental designs, flawed questionnaires, sampling problems, etc. (36)

ethics Ethical or moral issues, which are inescapable in social research, include issues of ensuring the honesty of the researcher in acknowledging sources of ideas and in reporting findings, responsible and humane treatment of participants in experiments, and the confidentiality and anonymity of survey respondents and fieldwork informants and contacts. (40)

ethnocentrism Taking one's own values, beliefs, and social customs as the normal and universal basis for judging other cultures. (173)

ethnography The systematic observation and detailed description of the ways of life and physical culture of a human group. (160)

experimental group The group that is exposed to a particular condition or stimulus in an experiment in order to see whether it causes a change in the dependent variable. (137)

experimental study A study in which techniques are used to isolate, control, and manipulate the major variables in a hypothesis. (130)

explanatory study A study that attempts to explain why certain things happen or to account for patterns that have been observed to regularly occur. (35)

exploratory study A study that attempts to discover information on a topic about which there is very little prior research to guide or focus the researcher. (35)

external validity The degree to which research conclusions can be applied to other observable cases, events, or situations. (142)

face-to-face interview In survey research, applying a questionnaire through a trained interviewer speaking directly and in person with the interviewee. Although very expensive, this technique is considered to reduce rates of refusal. (121)

faith Unquestioned belief in ideas. (4)

field experiment A research technique conducted in real-life social settings where specially trained researchers "act out" specific, highly controlled events to see whether hypothesized reactions will occur among the members of the public who witness the events. (131)

field interview Improvised questioning of people during the course of fieldwork, often while observing other social processes. (170)

field study Research that involves studying social phenomena as they occur naturally in the real world, with a minimum of control, manipulation, or deception. (35)

folkways Deeply rooted human assumptions about social order. (4)

frequency distribution Quantitative summary of data in the form of a table presenting the types, values, or scores and their frequency of occurrence. (214)

generalizability The quality of research findings that can be applied to a broader range of groups and situations than the original research focus; an especially important issue in experimental research. (142)

geography A discipline that encompasses both physical and natural sciences; a study of the earth as the space in which humans live, and the interrelations between the physical, material environment, the effects and constraints of this environment on human life, and the effects of human activities on the environment. (10)

government documents Reports, papers, pamphlets, and books published by government agencies such as Statistics Canada, Health Canada, and Revenu Québec; often gathered together in a separate section of a library and increasingly available, at least in part, over the Internet. (202)

group-administered questionnaire Questionnaires that are distributed to respondents as a group; for example, in a meeting, in a classroom, at a workplace, or in a family setting. (117)

Hawthorne effect The situation in experimental research in which the control group reacts as if it were the experimental group by producing experimental effects in the absence of the experimental stimulus. Likewise, individuals may change their behaviour because they know they are being studied or the researcher is present. (149)

history The social science discipline engaged in the study of the human past through analysis of primary written documents and other human artifacts from the past. (10)

humanistic social research Various perspectives in the social sciences that emphasize ways in which the natural and the social sciences differ in their basic assumptions, approaches to research, and methods. (33)

hypothesis A statement that focuses on the possible relationships between variables, expressed in a way that promotes testing the statement through systematic observation. (26)

independent variable A variable identified in a hypothesis as one that, as it changes, leads to changes in another variable, the dependent variable. (27)

indirect or nonreactive research methods Research techniques that use the results of human activities, such as documents, artifacts, and physical signs or effects of action, as their data. (183)

individually administered questionnaire A type of self-administered questionnaire that the individual respondent completes in the presence of a member of the research team. (117)

inductive reasoning Thinking that begins with particular items of factual information and then tries to infer or develop broader, more general patterns or principles from them. (25)

inferential statistics Statistical techniques that help researchers decide between competing interpretations of data. (234)

informant (in fieldwork) A member of the group that is being studied who provides important information about the group. (174)

informed consent In experimental research and fieldwork, the opportunity for those who will become the focus of study to accept or reject becoming an object of research. (178)

internal validity The effective elimination of alternative causal factors apart from the experimental variable through an appropriately organized experimental setup. (142)

Internet directory A database of websites organized by subject matter categories, with a search menu allowing searches to range from the general to the specific. Directories may be very broad, general-interest ones or very specialized, focusing on a specific topic. (49)

interval level of measurement Measurement involving a defined scale or set of intervals between one score and another so that the size of differences between scores is quantified. (218)

intervening variable A variable that logically fits into a sequence between the independent and the dependent variable in complex relationships; the intervening variable acts as a dependent variable in relation to the independent variable, and as an independent variable in relation to the dependent variable. (30)

laboratory experiment An experiment conducted in a laboratory setting with a large range of controls in place. (131)

latent content In content analysis it refers to underlying or implicit meanings in a communication usually analyzed qualitatively, as distinct from *manifest content*. (186)

leading question In interviews or questionnaires, questions that are set up or that lead the respondent to answer in one way rather than another, producing biased results. (108)

library catalogue Lists of all or most of a library's book and other document holdings; originally in the form of card catalogues, but most libraries now have computerized catalogues available on the Internet. (49)

Library of Congress (LC) Classification System The main system used in North American libraries for classification of published materials. (50)

limitations Boundaries of research arising from the researcher's decisions about the focus and type of study, the resources available, the characteristics of the data involved, and any biases the researcher may have had. (245)

line graph A graph that shows changes in variables over a period of time; the horizontal axis shows units of time, and the vertical axis shows changes in the values of the variable or variables. (232)

longitudinal study A study that shows changes over time, usually either by tracking a particular group or by taking a series of "snapshots" of different groups at different time periods. (101)

mail survey A self-administered questionnaire sent by mail to respondents. (117)

manifest content In content analysis, the explicit terms or symbols used in a communication usually analyzed quantitatively, as distinguished from *latent content*. (186)

margin of error The difference between the true value of a population characteristic and the value provided by a sample from that population. (91)

matrix questions A group of questions set up with a grid of identical response categories. (113)

mean A measure of central tendency calculated by dividing the sum of a set of scores by the total number of scores. (219)

measurement A standardized system for describing changes in the value of a variable. (216)

measures of association Ways of measuring the strength and sometimes the direction (positive or negative) of a relationship between variables. (229)

median A measure of central tendency; the middle point in a frequency distribution arranged from lowest to highest values, including all values. (219)

MLA documentation style The system of formatting in-text citations, footnotes or endnotes, and references, as well as the general style and organization for research papers, articles, and books; developed by the Modern Language Association. (256)

mode A measure of central tendency; the most frequent group or case in a frequency distribution. (219)

moral authorities Sources of information whose correctness we tend to take for granted because of our respect for, or faith in, the source. (3)

mortality (in samples) Loss of members of a sample over the time of the research study. (92)

multistage cluster sampling A random sampling method in which sample members are selected in stages from broader sampling units. For example, in a national sample of wage earners, you might select various regions first, then various localities within these regions, then households within these localities. (83)

natural experiment A study that uses an uncontrolled real-life event, such as an accident or a change in law, and looks at human responses to the event as if it were an experimental variable. (131)

negative correlation A relationship pattern between variables in which high scores on one variable are related to low scores on another. For example, studies generally show that the higher the level of formal education, the lower the risk of lengthy unemployment. (228)

negatively skewed distribution A distribution curve with a long tail on the left side, indicating a concentration of cases toward the high end of the measured characteristic. (221)

nominal level of measurement Sorting data into cases or classes and counting the frequency in each class. (216)

nonparticipant observation Observation of group behaviour without entering into the activities of the group. (160)

nonrandom sample Any type of sample in which the selection of the individual units of the sample is not based on the principle of an equal, known chance of selection for each member of the population; also known as a nonprobability sample. (76)

normal distribution Frequency distributions that can graphically be represented as some type of symmetrical bell curve. (220)

objectivity An essential part of research requiring researchers to set aside their everyday beliefs and assumptions, follow the procedures of empirical investigation, and present their discoveries and methods honestly and openly. (23)

open coding The initial stage of qualitative data analysis; essentially, the researcher carefully reads all historical data or field notes, noting critical terms, key themes, or important events, thus developing a basic framework for initial analysis. (238)

open-ended question Questions in interviews and questionnaires that permit the respondents to answer in words of their own choosing. (105)

operational definition Definition of variables in terms of the kinds of measures or data collection, specifying the existence and variation of the variables in the research. (27, 103)

operationalizing Identifying the ways that changes in variables may be empirically specified. (27)

ordinal level of measurement Measurement involving ordering or ranking cases or scores without scaling or quantifying the degrees of difference between cases or scores. (218)

panel study A study that examines changes in the characteristics of the same group of people over a period of time. (101)

participant observation A method of field research whereby the researcher establishes a social position within the group being studied. (162)

peer-reviewed research Social science research that is evaluated for originality, adherence to accepted standards of good research, and contribution to knowledge; the review is usually undertaken by other specialists in the field and is set up to ensure that the reviewers do not know the identity of the researcher and vice versa. (200)

personal journal A record of the researcher's personal reactions to the experience of fieldwork; kept as a way to check for possible sources of bias. (171)

pie chart A circular chart representing a frequency distribution as percentages of the total. (231)

political science The social science discipline studying systems of government, political groups, actions, and processes. (11)

population A set of individuals, groups, documents, etc., that you are interested in studying and from which you need to take a sample. (74)

positive correlation/association A graphically presented or statistically measured relationship between variables in which, as the values of one variable increase, so do the values of the other variable. (228)

positively skewed distribution A distribution curve with a long tail on the right side, indicating a concentration of cases toward the low end of the measured characteristic. (221)

postcode Arranging a questionnaire so that survey data are coded after they have been collected. (211)

precode Arranging a questionnaire so that the data can be coded as they are being collected. (211)

prediction In science, a testable statement about future events, based on empirically verifiable causal understanding of the relationships between variables. (26)

pretest The process of checking whether a survey questionnaire or other data-gathering instrument will work properly prior to using it for research. (114)

primary data Data from original sources, such as eyewitness accounts, tape recordings of field interviews, and original letters or other documents. (196)

probe An interviewer's technique for encouraging interview respondents to elaborate on their answers. (122)

psychology The social science discipline that focuses on human behaviour and cognitive processes. (11)

purposive sampling Deliberately selecting sample units according to research needs and without using random sampling techniques; also known as judgmental sampling. (87)

qualitative content analysis Research techniques that describe the content of communications in an objective and systematic way that emphasizes the qualitative meanings of the communication and describes these meanings in a nonquantitative way. (186)

qualitative research A non-numerical approach to data gathering and analysis to discover underlying meanings and patterns; particularly prominent in field and historical research. (39)

qualitative variables Variables that are not measured numerically. (28)

quantitative content analysis Research techniques that describe the content of communication in an objective, systematic, and quantitative way. (186)

quantitative research A numerical approach to data gathering and analysis to describe and explain phenomena of interest; associated in particular with survey research and experimental research. (39)

quantitative variables Variables that are represented or measured numerically. (28)

questionnaire A written set of questions organized in a sequence appropriate to the purpose of a survey, interview, or psychological test. (100)

quota sampling A type of nonrandom sample in which subgroups are represented according to their proportions in the population. (87)

random assignment Placing individuals into experimental and control groups using probability selection techniques. (141)

random numbers Sets of numbers lacking any patterned sequence and generated by a computer program or a published list of random numbers. (79)

random sample A sample drawn in such a way that each and every member or unit of the population has an equal chance of being selected; also known as a probability sample. (76)

range The difference between the highest and lowest scores in a distribution; a measure of variability. (222)

ratio level of measurement A scale of measurement with a true zero point. (218)

raw data Research information that has been collected but not yet organized, summarized, or interpreted. (210)

references Documentation of sources used in the writing of research papers, articles, and books; the title of the list of references when the APA documentation style is used. (251)

reliability The degree to which a research instrument, such as a questionnaire or a coding system, produces consistent results with repeated use. (138)

replicate To repeat a research design in its entirety to see if the same results can be obtained; most often undertaken with experiments and surveys in the social sciences. (37)

research methods The techniques of systematic, empirical research that have become acceptable in the social sciences. (9)

research process The sequence or steps of systematic empirical research. (15)

research report The final written research document, summarizing the research methods, its findings, and its practical or scientific implications. (247)

response rate The percentage of completed questionnaires (or questions within a questionnaire) obtained in a survey. (117)

response set The tendency of respondents to unwittingly repeat the type of response from one question to another in a questionnaire. (113)

sample Groups of individuals or objects from a population that were actually studied and that become the basis of generalizations about the wider population. (36)

sampling frame A list of all or most of the individuals or objects in a population from which a sample is to be taken. (78)

scattergram A graph that shows the way scores on any two variables are distributed throughout the range of possible values; also known as a scatterplot. (228)

scientific method An approach to research that is based on logic, that uses procedures of systematic empirical investigation, and that is committed to open and honest communication of research results and ideas. (6)

secondary data analysis Techniques for analyzing, reorganizing, and reinterpreting data gathered by others. (196)

self-administered questionnaire A questionnaire that is actually filled in by the respondent either individually, as in a mail survey (individually administered questionnaire), or in a group, as in a class survey of students (group-administered questionnaire). (116)

semiparticipant observation Fieldwork in which there is limited involvement with the group or groups being researched. (161)

simple random sample A sample drawn by first numbering all the units in the sampling frame, and then using random numbers to determine which units will be selected for study. (78)

skepticism A basic value of science, leading researchers to doubt ideas or information that cannot be put to the test by systematic empirical research. (23)

snowball sample A nonrandom sample where the researcher first studies a few cases that fit the research needs and then asks those studied to refer the researcher to others who would fit these needs. (88)

social sciences The various disciplines that apply the principles of scientific reasoning to different aspects of human life, including anthropology, economics, geography, history, political science, psychology, and sociology. (2)

social survey A research technique that obtains information from a sample of individuals by administering a questionnaire to every member of the sample and then systematically analyzes the responses. (99)

sociology The social science discipline that studies human groups, social processes, and social change, focusing primarily on modern industrialized societies. (12)

special-purpose files In qualitative analysis, files dealing with such matters as reviews of other research rather than primary data or analyses and summaries of those data. (237)

spreadsheet A computer program developed for the purpose of inputting and analyzing numerical data. (213)

standard deviation A measure of variability calculated by taking the square root of the mean of the squared deviations from the mean, divided by the total number of cases. (223)

stratified random sample A random sample where the population is first subdivided into subgroups and then simple random samples are taken from the subgroups. (82)

subjective meaning The ways that people understand and experience the world and their lives in it; a major focus of qualitative research. (236)

synthesis A coherent picture or interpretation, which is the aim of qualitative research. (240)

systematic empirical research Research that follows the scientific method. (6)

systematic random sample A sample that begins with a random selection of the first unit and then selects at fixed intervals. (81)

telephone survey A survey questionnaire that is administered over the telephone. (121)

theory A logically coherent set of ideas that accounts for the empirical patterns discovered by empirical research. (25)

trend study A longitudinal study that tracks the same set of variables over time, for example, taking repeated samples of individuals with certain levels of education and checking the average level of unemployment at each level. (101)

triangulation Using several methods of data collection or different data sources to obtain information on the same phenomena, event, or process. (173)

true zero A characteristic of certain types of data, such as income, which can then be measured using the ratio level of measurement. (218)

univariate analysis The summary and organization of data on a single variable. (214)

unobtrusive measurement/research The use of physical evidence, nonparticipant observations, and archival or documentary evidence without the knowledge of participants. (183)

validity In an experiment or a survey, the ability of an instrument to measure what it is intended to measure. (138)

variable Any aspect or characteristic that varies from case to case or over different times of observation. (27)

variance The mean of the sum of squared deviations from a mean of a distribution; important for calculating standard deviation. (222)

verstehen The German sociologist Max Weber's term for the method of recreating and interpreting the subjective meaning of culture. (193)

works cited Documentation of sources mentioned in a research paper; the title of the list of references when the MLA documentation style is used. (251)

zero correlation Lack of relationship in a measured association between variables. (228)

Bibliography

Adler, E. S., & Clark, R. (2003). *How it's done: An invitation to social research* (2nd ed.). Toronto: Thomson/Wadsworth.

Agnew, N. M., & Pyke, S. W. (1994). *The science game: An introduction to research in the social sciences* (6th ed.). Englewood Cliffs, NJ: Prentice-Hall.

Alonso, W., & Starr, P. (1987). *The politics of numbers.* New York: Russell Sage Foundation.

Aries, P. (1962). *Centuries of childhood: A social history of family life* (R. Baldick, Trans.). New York: Vintage.

Babbie, E. R. (1986). *Observing ourselves: Essays in social research.* Belmont, CA: Wadsworth.

Babbie, E. R. (1990). *Survey research methods* (2nd ed.). Belmont, CA: Wadsworth.

Babbie, E. R. (1998). *The practice of social research* (8th ed.). Belmont, CA: Wadsworth.

Babbie, E. R., & Benaquisto, L. (2002). *Fundamentals of social research.* Scarborough, ON: Thomson Nelson.

Baker, T. L. (1999). *Doing social research* (3rd ed.). New York: McGraw-Hill.

Banks, M. (2001). *Visual methods in social research.* London: Sage.

Barnes, J. A. (1979). *Who should know what? Social science, privacy and ethics.* New York: Cambridge University Press.

Bart, P., & Frankel, L. (1986). *The student sociologist's handbook* (4th ed.). New York: Random House.

Baxter-Moore, N., Terrance, C., & Church, R. (1994). *Studying politics: An introduction to argument and analysis.* Toronto: Copp-Clark.

Becker, H. S. (1958). Problems of inference and proof in participation observation. *American Sociological Review, 23,* 652–660.

Becker, H. S. (1986). *Writing for social scientists: How to start and finish your thesis, book, or article.* Chicago: Chicago University Press.

Benjafield, J. G. (1994). *Thinking critically about research methods.* Boston: Allyn and Bacon.

Berg, B. L. (2004). *Qualitative research methods for the social sciences.* Boston: Allyn and Bacon.

Berger, J. (1973). *Ways of seeing.* New York: Viking.

Berry, J. W., & Ponce, J. A. (Eds.). (1994). *Ethnicity and culture in Canada: The research landscape.* Toronto: University of Toronto Press.

Bertaux, D. (Ed.). (1981). *Biography and society: The life history approach in the social sciences.* Newbury Park, CA: Sage.

Best, J. (2001). *Damned lies and statistics: Interpreting numbers from the media, politicians and activists.* Berkeley: University of California Press.

Best, J. (2008). *Stat-spotting: A field guide to identifying dubious data.* Berkeley: University of California Press.

Bibby, R. W. (2000). *Canada's teens: Today, yesterday, and tomorrow.* Toronto: Stoddart.

Blalock, H. M. (1979). *Social statistics* (2nd ed.). New York: McGraw-Hill.

Bloch, M. (1953). *The historian's craft* (P. Putnam, Trans.). New York: Vintage.

Bowen, R. W. (1992). *Graph it! How to make, read, and interpret graphs.* Englewood Cliffs, NJ: Prentice-Hall.

Bowers, J. W., & Courtright, J. A. (1984). *Communication research methods.* Glenview, IL: Scott, Freeman and Company.

Boyl, J. R., & Boekeloo, B. O. (2006). Perceived parental approval of drinking and its impact on problem drinking behaviors among first-year college students. *Journal of American College Health, 54,* 238–245.

Bryman, A. (2001). *Social science research methods.* Oxford and New York: Oxford University Press.

Bulmer, M. (Ed.). (1982). *Social research ethics.* London: Macmillan.

Burgess, R. G. (Ed.). (1982). *Field research.* Boston: George Allen and Unwin.

Burman, P. (1988). *Killing time, losing ground: Experiences of unemployment.* Toronto: Wall and Thompson.

Caldwell, S. (2004). *Statistics unplugged.* Toronto: Thomson/Wadsworth.

Campbell, D. T., & Stanley, J. C. (1963). *Experimental and quasi-experimental designs for research.* Chicago: Rand McNally.

Campbell, K., & Denov, M. (2004). The burden of innocence: Coping with a wrongful imprisonment. *Canadian Journal of Criminology and Criminal Justice, 46*(2), 139–164.

Cantril, H. (1940). *The invasion from Mars.* Princeton, NJ: Princeton University Press.

Christensen, L. B. (2004). *Experimental methodology.* Boston: Allyn and Bacon.

Collier, J., & Collier, M. (1986). *Visual anthropology: Photography as a research method* (Rev. ed.). Albuquerque: University of New Mexico.

Converse, J. M., & Schuman, H. (1974). *Conversations at random: Survey research as interviewers see it.* New York: Wiley.

Cook, T. D., & Campbell, D. T. (1979). *Quasi-experimentation: Design and analysis issues for field settings.* Chicago: Rand McNally.

Cooper, H. M. (1989). *Integrating research: A guide for literature reviews* (2nd ed.). Newbury Park, CA: Sage.

Cooper, M. (2008). *Designing effective web surveys.* Newbury Park, CA: Sage.

Cozby, P. C. (2004). *Methods in behavioral research* (8th ed.). Boston: McGraw-Hill.

Cuba, L. (1993). *A short guide to writing about social science* (2nd ed.). New York: HarperCollins.

Curry, G. N. (2003). Moving beyond postdevelopment: Facilitating indigenous alternatives for "development." *Economic Geography, 79*(4), 405–424.

Dale, A., Arber, S., & Procter, M. (1988). *Doing secondary analysis.* Boston: Unwin Hymen.

Davenport, P. (1988). Economics. In *The Canadian Encyclopedia* (2nd ed., pp. 648–650). Edmonton: Hurtig.

Davis, J. A. (1971). *Elementary survey analysis.* Englewood Cliffs, NJ: Prentice-Hall.

Denzin, N. K. (1989). *Interpretive biography.* Newbury Park, CA: Sage.

Denzin, N. K. (1989). *The research act: A theoretical introduction to sociological methods* (3rd ed.). Englewood Cliffs, NJ: Prentice-Hall.

Denzin, N. K. (1997). *Interpretive ethnography.* Thousand Oaks, CA: Sage.

Denzin, N. K., & Lincoln, Y. S. (Eds.). (2000). *Handbook of qualitative research.* Thousand Oaks, CA: Sage.

Denzin, N. K., & Lincoln, Y. S. (Eds.). (2003). *Collecting and interpreting qualitative materials.* Thousand Oaks, CA: Sage.

Dey, I. (1993). *Qualitative data analysis: A user-friendly guide for social scientists.* New York: Routledge.

Diamond, J., & Robinson, J. A. (Eds.). (2010). *The natural experiments of history.* Cambridge, MA: Bellknap Press.

Dillman, D. A. (2004). *Internet, mail, and mixed surveys: The tailored design method.* New York: John Wiley and Sons.

Dixon, B. R., Bouma, G. D., & Atkinson, G. B. J. (1987). *A handbook of social science research: A comprehensive and practical guide for students.* New York: Oxford University Press.

Douglas, J. (1985). *Creative interviewing.* Beverly Hills, CA: Sage.

Eastthorpe, G. (1974). *A history of research methods.* London: Longman.

Eichler, M. (1988). *Nonsexist research methods: A practical guide.* Boston: George Allen and Unwin.

Emerson, R. M. (Ed.). (1983). *Contemporary field research.* Boston: Little, Brown.

Eshel, N., Daelmans, B., Cabral de Mello, M., & Martines, J. (2006). Responsive parenting: Interventions and outcomes. *Bulletin of the World Health Organization, 84,* 991–999.

Falk, A., & Ichino, A. (2006). Clean evidence on peer effects. *Journal of Labor Economics, 24,* 39–58.

Farha, B. (2003). Stupid "Pet Psychic" tricks: Crossing over with Fifi and Fido on the Animal Planet network. *Skeptic Magazine, 10*(1), 20–23.

Feagin, J. R., Orum, A. M., & Sjoberg, C. (Eds.). (1991). *A case for the case study.* Chapel Hill, NC: University of North Carolina.

Festinger, L., Reicken, H. W., & Schachter, S. (1956). *When prophecy fails.* New York: Harper and Row.

Fetterman, D. M. (1989). *Ethnography step by step*. Newbury Park, CA: Sage.

Freedman, D., Pisani, R., Purves, R., & Adhikari, A. (1991). *Statistics* (2nd ed.). New York: W. W. Norton.

Fritzley, H. V., & Lee, K. (2003). Do young children always say yes to yes–no questions? A metadevelopmental study of the affirmation bias. *Child Development, 74*(5), 1297–1315.

Geertz, C. (1973). *The interpretation of cultures*. New York: Basic Books.

Genovese, J. E. C. (2002). Cognitive skills valued by educators: Historical content analysis of testing in Ohio. *Journal of Educational Research, 96*(2), 101–116.

Giere, R. N. (1991). *Understanding scientific reasoning* (3rd ed.). Fort Worth, TX: Holt, Reinhart and Winston.

Giffen, P. J. (1976). Official rates of crime and delinquency. In W. T. McGrath (Ed.), *Crime and its treatment in Canada* (2nd ed.). Toronto: Macmillan.

Gillham, B. (2000). *Case study research methods*. New York: Continuum.

Ginzberg, E., & Henry, F. (1984/1985, Winter). Confirming discrimination in the Toronto labour market. *Current Reading in Race Relations, 2*, 23–28.

Glicken, M. (2003). *Social research: A simple guide*. Boston: Allyn and Bacon.

Goffman, E. (1962). *Asylums: Essays on the social situation of mental patients and other inmates*. Chicago: Aldine.

Gold, G. (2003). Rediscovering place: Experiences of a quadriplegic anthropologist. *The Canadian Geographer, 47*(4), 467–480.

Golde, P. (Ed.). (1992). *Women in the field: Anthropological experiences* (2nd ed.). Berkeley: University of California Press.

Goldman, R. (1992). *Reading ads socially*. New York: Routledge.

Goode, E. (2002). Education, scientific knowledge, and belief in the paranormal. *Skeptical Inquirer, 26*(1), 24–28.

Gordon, R. L. (1980). *Interviewing: Strategy, techniques, and tactics* (3rd ed.). Homewood, IL: Dorsey.

Grabb, E. G. (1990). *Theories of social inequality: Classical and contemporary theorists* (2nd ed.). Toronto: Holt, Rinehart and Winston.

Graham, A. (1994). *Teach yourself statistics*. London: NTC Publishing.

Granger, L. (1988). Psychology. In *The Canadian Encyclopedia* (2nd ed., pp. 1778–1779). Edmonton: Hurtig.

Gray, G., & Guppy, N. (1994). *Successful surveys: Research methods and practice*. Toronto: Harcourt Brace.

Griffin, J. H. (1976). *Black like me*. New York: New American Library.

Grimes, M. D. (1991). *Class in twentieth-century American sociology: An analysis of theories and measurement strategies*. New York: Praeger.

Grinsell, M. M., Showalter, S., Gordo, K. A., & Norwood, V. F. (2006). Single kidney and sports participation: Perception versus reality. *Pediatrics, 118,* 1019–1028.

Gubrium, J. F. (Ed.). (1989). *The politics of field research: Beyond enlightenment.* Newbury Park, CA: Sage.

Guegue, N. (2003). Help on the web: The effect of the same first name between the sender and the receptor in a request made by e-mail. *The Psychological Record, 53*(3), 459–467.

Hall, E. T. (1959). *The silent language.* Garden City, NY: Doubleday.

Hall, L. M., Angus, J., Peter, E., O'Brien-Pallas, L., Wynn, F., & Donner, G. (2003). Media portrayal of nurses' perspectives and concerns in the SARS crisis in Toronto. *Journal of Nursing Scholarship, 35*(3), 211–217.

Hanawalt, B. (1993). *Growing up in medieval London: The experience of childhood in history.* New York: Oxford University Press.

Harvey, L., & MacDonald, M. (1993). *Doing sociology: A practical introduction.* London: Macmillan.

Heckelman, J. C., & Yates, A. J. (2003). And a hockey game broke out: Crime and punishment in the NHL. *Economic Inquiry, 41*(4), 705–713.

Heerweigh, D. (2005). Effects of personal salutations in e-mail invitations to participate in a Web survey. *Public Opinion Quarterly, 69,* 588–599.

Heller, F. (Ed.). (1986). *The use and abuse of social science.* Beverly Hills, CA: Sage.

Hollander, A. (1978). *Seeing through clothes.* New York: Viking.

Homans, R. (1980). The ethics of covert methods. *British Journal of Sociology, 31,* 46–57.

Hoover, K. R., & Donovan, T. (2004). *The elements of social scientific thinking* (8th ed.). Toronto: Thomson/Wadsworth.

Huff, D. (1993). *How to lie with statistics.* New York: W. W. Norton.

Hult, C. A. (2006). *Researching and writing across the curriculum* (3rd ed.). Toronto: Pearson-Longman.

Humphreys, L. (1975). *Tearoom trade: Impersonal sex in public places* (Enlarged ed.). Chicago: Aldine.

Hunt, L. (Ed.). (1989). *The new cultural history.* Berkeley: University of California Press.

Hunt, M. (1985). *Profiles of social research: The scientific study of human interactions.* New York: Russell Sage Foundation.

Hyman, H. H. (Ed.). (1991). *Taking society's measure: A personal history of survey research.* New York: Russell Sage Foundation.

Jackson, W. (1999). *Methods: Doing social research* (2nd ed.). Scarborough, ON: Prentice-Hall.

Jakowski, N. W. (Ed.). (2009). *E-research: Integration in scholarly practice.* New York: Routledge.

Joinson, A. N., & Reips, U. D. (2007). Personalized salutation, power of sender and response rates to Web-based surveys. *Computers in Human Behavior, 23*(3), 1372–1383.

Jorgenson, D. L. (1989). *Participant observation: A methodology for human studies.* Beverly Hills, CA: Sage.

Katz, J. (1972). *Experimentation with human beings.* New York: Russell Sage.

Katzer, J., Cook, K. H., & Crouch, W. W. (1982). *Evaluating information: A guide for users of social science research* (2nd ed.). Reading, MA: Addison-Wesley.

Keegan, J. (1995). *The battle for history: Re-fighting World War Two.* Toronto: Vintage Books.

Kilburn, H. Whitt. (2011). Survey research. In J. T. Ishiyama & M. Breuning (Eds.), *21st century political science: A reference handbook.* Thousand Oaks, CA: Sage. Retrieved from http://www.sage-ereference.com/21stcenturypolisci/Article_n61.html

Kirby, S., Sanders, L., Greeves, L., & Reed, C. (2006). *Experiencing research social change: Methods beyond the mainstream* (2nd ed.). Peterborough, ON: Broadview Press.

Krippendorf, K. (1980). *Content analysis: An introduction to its methodology.* Beverly Hills, CA: Sage.

Kuala, S. (1996). *InterViews: An introduction to qualitative research interviewing.* Thousand Oaks, CA: Sage.

Levin, J., & Fox, J. A. (2004). *Elementary statistics in social research: The essentials.* Toronto: Pearson.

Lewis, G. H. (Ed.). (1975). *Fist-fights in the kitchen: Manners and methods in social research.* Pacific Palisades, CA: Goodyear.

Lewis-Beck, M. S. (Ed.). (2004). *The Sage encyclopedia of social science research methods.* Thousand Oaks, CA: Sage.

Liebow, E. (1993). *Tell them who I am: The lives of homeless women.* New York: Free Press.

Light, R. J., & Pillemer, D. B. (1984). *Summing up: The science of reviewing research.* Cambridge, MA: Harvard University Press.

Lofland, J., & Lofland, L. (1984). *Analyzing social settings* (2nd ed.). Belmont, CA: Wadsworth.

Lowenthal, D. (1985). *The past is a foreign country.* New York: Cambridge University Press.

Macleod, J. (1997). *Ain't no makin' it: Aspirations and attainment in a low-income neighborhood* (2nd ed.). Boulder, CO: Westview.

MacMillan, Margaret. (2008). *The uses and abuses of history.* Toronto: Penguin.

Marius, R. A. (1995). *A short guide to writing about history.* New York: HarperCollins.

Marsh, C. (1982). *The survey method: The contribution of surveys to sociological explanation.* Boston: George Allen and Unwin.

Marsh, L. (1993). *Canadians in and out of work: A survey of economic classes and their relation to the labour market.* Toronto: Canadian Scholars' Press.

Martin, D. W. (1991). *Doing psychology experiments.* Monterey, CA: Brooks/Cole.

McCall, G. J. (1978). *Observing the law: Field methods in the study of crime and the criminal justice system.* New York: Free Press.

McDiarmid, G. (1971). *Teaching prejudice: A content analysis of social studies textbooks authorized for use in Ontario.* Toronto: Ontario Institute for Studies in Education.

McFarland, K., & Rhoades, D. (2003). Tips for professional writing. *Annals of the American Psychotherapy Association, 6*(3), 31–34.

McGuigan, F. J. (1990). *Experimental psychology: Methods of research* (5th ed.). Englewood Cliffs, NJ: Prentice-Hall.

McKie, C., & Thompson, K. (Eds.). (1990). *Canadian social trends.* Toronto: Thompson Educational Publishing.

McKillop, A. B. (1988). Historiography in English. In *The Canadian Encyclopedia* (2nd ed., pp. 993–994). Edmonton: Hurtig.

Merton, R. K., & Kendall, P. L. (1990). *The focused interview: A manual of problems and procedures* (2nd ed.). New York: The Free Press.

Miles, M. B., & Huberman, A. M. (1984). *Qualitative data analysis.* Thousand Oaks, CA: Sage.

Milgram, S. (1974). *Obedience to authority.* New York: Harper and Row.

Millar, W. (1994). A trend to a healthier lifestyle. In *Canadian social trends: A Canadian studies reader* (Vol. 2, pp. 91–93). Toronto: Thompson Educational Publishing.

Miller, D. (1998). *Material cultures: Why some things matter.* London: University College Press.

Miller, D. C. (1991). *Handbook of research design and social measurement.* Newbury Park, CA: Sage.

Miller, J. E. (2004). *The Chicago guide to writing about numbers.* Chicago: University of Chicago Press.

Mills, C. W. (1959). *The sociological imagination.* New York: Oxford University Press.

Moloney, M. F., Dietrich, A. S., Strickland, O., & Myerburg, S. (2002). Using Internet discussion boards as virtual focus groups. *Advances in Nursing Science, 26*(4), 274–287.

Monette, D. R., Sullivan, T. J., & DeJong, C. R. (1994). *Applied social research: Tool for the human services* (3rd ed.). Fort Worth, TX: Harcourt Brace.

Moore, D. S. (2006). *Statistics: Concepts and controversies* (6th ed.). New York: W. H. Freeman and Company.

Morgan, D. L. (2001). *Focus groups in social research.* Thousand Oaks, CA: Sage.

Morrison, D. E. (1998). *The search for a method: Focus groups and the development of mass communications research.* Luton, UK: University of Luton.

Neuman, W. L. (2011). *Basics of social research: Qualitative and quantitative approaches.* Boston: Allyn and Bacon.

Newman, K. S. (2004). *Rampage: The social roots of school shootings.* New York: Basic Books.

Newman, O. (1972). *Defensible space: People and design in the violent city.* London: Architectural Press.

Northey, M. (2002). *Making sense: A student's guide to research and writing: Psychology and the life sciences.* Don Mills: Oxford University Press.

Outhwaite, W., & Bottomore, T. (Eds.). (1993). *The Blackwell dictionary of twentieth-century social thought.* Oxford: Blackwell.

Palys, T. (1997). *Research decisions: Quantitative and qualitative perspectives* (3d ed.). Toronto: Harcourt Brace and Company.

Parliament, J. B. (1994). Labour force trends: Two decades in review. In *Canadian social trends: A Canadian studies reader* (Vol. 2, pp. 255–258). Toronto: Thompson Educational Publishing.

Paternoster, R., Bushway, S., Brame, R., & Apel, R. (2003). The effect of teenage employment on delinquency and problem behaviors. *Social Forces, 82*(1), 297–336.

Paulos, J. A. (1988). *Innumeracy: Mathematical illiteracy and its consequences.* New York: Hill and Wang.

Pawlchuck, D., Shaffir, W., & Mial, C. (Eds.). (2005). *Doing ethnography: Studying everyday life.* Toronto: Canadian Scholars' Press.

Pink, S. (2001). *Doing visual ethnography.* London: Sage.

Platek, R., Pierre, F. K., & Stevens, P. (1985). *Development and design of survey questionnaires* (Catalogue No. 12-519). Ottawa: Statistics Canada.

Porter, T. M. (1986). *The rise of statistical thinking: 1820–1900.* Princeton, NJ: Princeton University Press.

Preston, R. J., & Tremblay, M. A. (1988). Anthropology. In *The Canadian Encyclopedia* (2nd ed., pp. 82–83). Edmonton: Hurtig.

Rabinow, P., & Sullivan, W. (Eds.). (1979). *Interpretative social science.* Berkeley: University of California Press.

Randi, J. (2002). How to talk to the dead ('Twas brillig…). *Skeptic Magazine, 9*(3), 9–11.

Rathje, W. L. (1989). The three faces of garbage: Measurements, perceptions, behaviours. *Journal of Resource Management and Technology, 17,* 14–71.

Rathje, W. L. (1992). *Rubbish! The archaeology of garbage.* New York: HarperCollins.

Reeves, C. C. (1992). *Quantitative research for the behavioural sciences.* New York: Wiley.

Rehner, J. (1994). *Strategies for critical thinking.* Boston: Houghton Mifflin.

Reinharz, S. (1992). *Feminist methods in social research.* New York: Oxford University Press.

Reips, U. D. (2002). Standards for Internet-based experimenting. *Experimental Psychology, 49,* 243–256.

Ritchie, D. A. (Ed.). (2011). *The Oxford handbook of oral history.* Oxford: Oxford University Press.

Robbins, N. (2005). *Creating more effective graphs.* New York: Wiley.

Rosenberg, K. M., & Daly, H. B. (1993). *Foundations of behavioral research: A basic question approach.* Fort Worth, TX: Harcourt Brace.

Rosenthal, R. (1976). *Experimenter effects in behavioral research.* New York: Irvington.

Rowntree, D. (1981). *Statistics without tears: A primer for non-mathematicians.* New York: Charles Scribner's Sons.

Roy, F. (1988). Historiography in French. In *The Canadian Encyclopedia* (2nd ed., pp. 992–993). Edmonton: Hurtig.

Rutman, L. (Ed.). (1984). *Evaluation research methods: A basic guide* (2nd ed.). Beverly Hills, CA: Sage.

Sacco, V., & Johnson, H. (1994). Household property crime. In *Canadian social trends: A Canadian studies reader* (Vol. 2, pp. 393–395). Toronto: Thompson Educational Publishing.

Sanders, W. B. (1983). *The conduct of social research.* New York: CBS College Publishing.

Satin, A., & Shastry, W. (1993). *Survey sampling: A non-mathematical guide* (2nd ed., Catalogue No. 12-602). Ottawa: Statistics Canada.

Saxe, L., & Fine, M. (1981). *Social experiments: Methods for design and evaluation.* Beverly Hills, CA: Sage.

Schachter, S. (1959). *The psychology of affiliation: Experimental studies of the sources of gregariousness.* Stanford, CA: Stanford University Press.

Sechrest, L. (Ed.). (1979). *Unobtrusive measurement today.* San Francisco: Jossey-Bass.

Seyler, D. U. (1993). *Doing research: The complete research paper guide.* New York: McGraw-Hill.

Shaffir, W. B., & Stebbins, R. A. (Eds.). (1991). *Experiencing fieldwork: An inside view of qualitative research.* Newbury Park, CA: Sage.

Shaffir, W. B., Stebbins, R., & Turowetz, A. (Eds.). (1980). *Fieldwork experience.* New York: St. Martin's Press.

Shandasewi, L. (1990). *Focus groups: Theory and practice.* Newbury Park, CA: Sage.

Shermer, M. (2002). Skepticism as a virtue: An inquiry into the original meaning of the word "skeptic." *Scientific American, 286*(4), 37–38.

Shermer, M. (2003). Psychic for a day: Or how I learned Tarot cards, palm reading, astrology and mediumship in 24 hours. *Skeptic Magazine, 10*(1), 48–57.

Shively, W. P. (1990). *The craft of political research.* Englewood Cliffs, NJ: Prentice-Hall.

Sicher, J. E. (1992). *Planning ethically responsible research: A guide for social science students.* Newbury Park, CA: Sage.

Silverman, D. (1993). *Interpreting qualitative data.* Newbury Park, CA: Sage.

Sitton, T., Mehaffy, G., & Davis, Jr., O. L. (1983). *Oral history.* Austin: University of Texas Press.

Skouteris, H., & Kelly, L. (2006). Repeated-viewing and co-viewing of an animated video: An examination of factors that impact on young children's comprehension of video content. *Australian Journal of Early Childhood, 31,* 22–31.

Slonim, M. J. (1966). *Sampling.* New York: Simon and Schuster.

Sommer, B., & Sommer, R. (2002). *A practical guide to behavioural research: Tools and techniques* (5th ed.). New York: Oxford University Press.

Stangor, C. (2004). *Research methods for the behavioural sciences.* Boston: Houghton Miffin.

Stern, P. C. (1979). *Evaluating social science research.* New York: Oxford University Press.

Stewart, D. W. (1984). *Secondary research: Information sources and methods.* Beverly Hills, CA: Sage.

Strauss, A. (1987). *Qualitative analysis for social scientists.* New York: Cambridge University Press.

Strauss, A., & Corbin, J. (1998). *Basics of qualitative research: Techniques and procedures for developing grounded theory.* Thousand Oaks, CA: Sage.

Strike, C. (1990). AIDS in Canada. In C. Makie & K. Thompson (Eds.), *Canadian social trends* (pp. 83–85). Toronto: Thompson Educational Publishing.

Sudman, S., & Bradburn, N. M. (1983). *Asking questions: A practical guide to questionnaire design.* San Francisco: Jossey-Bass.

Sullivan, T. (1991). *Applied sociology: Research and critical thinking.* New York: Macmillan.

Survey Research Center, Institute for Social Research. (1976). *Interviewer's manual* (Rev. ed.). Ann Arbor: University of Michigan.

Taylor, K. W. (1990). *Social science research: Theory and practice.* Scarborough, ON: Nelson Canada.

Thompson, E. P. (1963). *The making of the English working class.* New York: Vintage.

Thompson, P. (1978). *The voice of the past: Oral history.* New York: Oxford University Press.

Thomsett, M. C. (1990). *The little black book of business statistics.* New York: American Management Association.

Tufte, E. R. (1983). *The visual display of quantitative information.* Cheshire, CT: Graphics Press.

Tutty, L. M., Rothery, M. A., & Grinnell, Jr., R. M. (1996). *Qualitative research for social workers.* Boston: Allyn and Bacon.

U.S. Department of Health and Human Services. (1992). *Survey measurement of drug use.* Washington, DC: Government Printing Office.

Utts, J. M. (1999). *Seeing through statistics* (2nd ed.). New York: Duxbury Press.

Vallee, F. G. (1988). Social science. In *The Canadian Encyclopedia* (2nd ed., p. 2031). Edmonton: Hurtig.

Van Maanen, J. (1988). *Tales of the field: On writing ethnography*. Chicago: University of Chicago Press.

Van Maanen, J. (1998). *Qualitative studies of organizations*. Thousand Oaks, CA: Sage.

Vaughan, D. (1995). *The* Challenger *launch decision: Risky technology, culture, and deviance*. Chicago: University of Chicago Press.

Wagner, J. (Ed.). (1979). *Images of information: Still photography in the social sciences*. Beverly Hills, CA: Sage.

Wargon, S. (1986). Using census data for research on the family in Canada. *Journal of Comparative Family Studies, 3,* 150–158.

Webb, E. T., Campbell, D. T., Schwartz, R. D., Sechrest, L., & Grove, J. B. (1981). *Nonreactive measures in the social sciences*. Boston: Houghton Mifflin.

Weber, M. (1958). *The Protestant ethic and the spirit of capitalism* (T. Parsons, Trans.). New York: Charles Scribner's Sons. (Original work published 1904–1905).

Weber, R. P. (1985). *Basic content analysis*. Beverly Hills, CA: Sage.

Whitaker, R. (1988). Political science. In *The Canadian Encyclopedia* (2nd ed., pp. 1711–1712). Edmonton: Hurtig.

Whyte, D. R., & Vallee, F. G. (1988). Sociology. In *The Canadian Encyclopedia* (2nd ed., pp. 2035–2037). Edmonton: Hurtig.

Whyte, W. F., & Whyte, K. K. (1984). *Learning from the field: A guide from experience*. Beverly Hills, CA: Sage.

Williams, B. (1978). *A sampler on sampling*. New York: Wiley.

Williams, F. (1992). *Reasoning with statistics: How to read quantitative research* (4th ed.). Fort Worth, TX: Harcourt Brace Jovanovich.

Williams, G. (1989). Enticing viewers: Sex and violence in *TV Guide* program advertisements. *Journalism Quarterly, 66,* 970–973.

Wolf, D. (1991). *The rebels: A brotherhood of outlaw bikers*. Toronto: University of Toronto Press.

Yegidis, B. L., & Weinbach, R. W. (1996). *Research methods for social workers*. Boston: Allyn and Bacon.

Yin, R. K. (2002). *Case study research* (3rd ed.). Thousand Oaks, CA: Sage.

Zeisel, H. (1985). *Say it with figures* (6th ed.). New York: Harper and Row.

Zeller, R., & Carmines, E. G. (1980). *Measurement in the social sciences: The link between theory and data*. New York: Cambridge University Press.

Zimbardo, P. G. (1972). Pathology of imprisonment. *Society, 9,* 4–6.

Zuckerman, H. (1978). Interviewing an ultra-elite. *Public Opinion Quarterly, 36,* 159–175.

Index

Note: Page references followed by "b," "f," and "t" refer to boxes, figures, and tables.

Coding sheets, 191t
Coding system, 190–91, 211
Common sense, 7t
Complexity, appreciation of, 172, 187
Computer-assisted telephone interviewing
    (CATI), 122, 271
Computers
    library card catalogues, 49–50
    library research, use in, 50–51
Conclusion
    defined, 271
    note of, in literature review, 65
    in research report, 249, 251, 266
Conditioning, 132
Confidence interval, 91, 271
Confidence level, 91, 271
Confidentiality, 126, 175, 253, 272
Consistent association, 31, 134
Constraints of research process
    ethics, 40b–41b, 44
    human data, problems with, 42
    the personal equation, 42–43
    resources, limit on, 40
    social context, 42–43
Content analysis
    advantages, 194–95, 204t
    coding system, 190–91, 191t
    defined, 272
    described, 183
    disadvantages, 194–96, 204t
    example of, 189–90
    interpretation, issue of, 196b
    latent content, 186
    manifest content, 186
    measures, 191
    qualitative content analysis, 192–94, 195
    quantitative content analysis, 186–87
    questions to ask about, 196b
    sampling, 192
    SAT scores, decline in, 186–87
    steps in, 195b
    units of meaning, 190
    working technique, 195b
Contingency (skip to) questions, 112–13,
    272
Control group, 138–40, 272
Convenience sample, 86, 90b, 166, 272
Correlation
    vs. causation, 230
    defined, 272
    described, 31
    negative correlation, 228
    Pearson's product moment correlation
        coefficient, 229
    positive correlation, 228
    quantitative data analysis, 235, 236
    zero correlation, 228
Correlation coefficient, 229
Covert observation, 178–79, 272
Critical approach, development of, 48
Cross-sectional surveys, 100, 272

Cross-tabulation (cross-tab), 226–28,
    226t, 227t, 272
Cumulative frequency distribution, 214–16, 272
Customary information, 3

D

The Daily (Statistics Canada), 199
Data
    available data, 196–98
    existing data (*See* Existing data)
    human data, problems with, 42
    primary data, 196
    raw data, 210
    secondary data, 196–98
Data analysis
    defined, 272
    described, 37–39, 38t
    fieldwork, 166–67, 170–71
    qualitative data analysis (*See* Qualitative data
        analysis)
    quantitative data analysis (*See* Quantitative
        data analysis)
    in research report, 251b, 253–55
    secondary data analysis, 196–98
    websites, 242
Database, 51–52, 199–200, 272
Data collection
    defined, 273
    described, 36, 38t
    fieldwork, 171
    methods, in research report, 245
Data files, 237, 273
Data interpretation, 177. *See also* Data analysis
Data summary, 212–216
Debriefing, 150, 153, 273
Deception, 150
Deductive reasoning, 25, 26t, 273
Dehoaxing, 150
Delinquency, and teenage employment, 198–99
Dependent variables, 27, 273
Depth, 172, 173
Descriptive statistics, 234, 273
Descriptive study
    defined, 273
    described, 35
    vs. exploratory study, 35–36
Dial-in polls, 98
Dichotomous question, 103, 109, 273
Directories, 53
Direct research methods, 183, 273
Discussion
    article, 56
    research report, 249, 251
Discussion groups, 5, 175
Dispersion, measures of, 222–24
    range, 222
    standard deviation, 222–23
    usefulness of, 222–24
    variance, 222–23
Divorce rates, under 25, 196–97

Meaningful understanding of subjects, 172
Measurement
    central tendency, 219–21
    characteristics of different levels, 218t
    content analysis, 191
    defined, 216, 277
    dispersion, 222–24
    interval level of measurement, 218
    levels of, 216–219, 218t
    nominal level of measurement, 216
    operationalization, 29b
    ordinal level of measurement, 218
    ratio level of measurement, 218
    true zero, 218
Measures of association, 229, 277
Median, 219–22, 277
Microfilm, 202
Milgram, Stanley, 150
MLA style
    article in academic journal, 264
    article in edited book, 263
    article in popular magazine, 264
    defined, 261
    described, 58, 261
    edited book, 263
    endnotes, 262
    essay paper format, 261
    examples of, 264
    footnotes, 262
    Internet sources, 266–67
    organization as author, 263
    in-text citations, 261–62
    single author, 262
    two or more authors, 262
    works cited, 262
Mode, 219–22, 277
Modern Language Association, 58
Monograph, 54–58
Moral authorities, 3, 277
Mortality (in samples), 92, 277
Movies, repeat viewing by young children, 137
Multistage cluster sampling, 83, 88b, 277
Mutually exclusive category items, 106

N

National Hockey League, 145
Natural experiment, 131, 145–46, 151t, 277
Nazi Germany, 21–22
Negative correlation, 228, 229, 230, 277
Negative items, avoidance of, 110
Negatively skewed distribution, 221, 221f
Negative skew, 277
Newsgroups, 5
Newspapers, 4, 46, 51, 57b
Nominal level of measurement, 216, 218, 277
Nonparticipant observation
    advantages, 169t
    data, type of, 166
    defined, 277
    described, 161–62

disadvantages, 169t
effort required, 165
location, identification of, 164
note taking, 165
observational skills, development of, 164–65
preparation for, 164
problems with, 165–66
question, choice of, 163
reliability, 165–66
sampling, 166
steps in, 163b
thick description, 166
time required, 165
use of, 162–63
validity, 166
Nonprobability. *See* Nonrandom (nonprobability) sampling procedure
Nonrandom (nonprobability) sampling procedure
    accidental sampling, 85–86
    convenience sampling, 86–87, 90b
    defined, 277
    described, 85
    purposive sampling, 87
    quota sampling, 87–88, 90b
    snowball sampling, 88, 90b
    types of samples, 89b–90b
    validity, effects on, 87
Nonrandom sample, 38t
Nonreactive methods. *See* Indirect (nonreactive) methods
Normal distribution, 221f, 224, 225f, 278
Note taking, 165, 170–71

O

Objectivity, 23, 278
Observation skills, 164–65
Online periodicals, 55
Open coding, 238, 278
Open-ended questions, 105–6, 278
Operational definition, 28, 104b–105b, 278
Operationalization
    defined, 278
    experimental research, 137–38
Oral history, 158b
Ordering of questions, 114
Ordinal level of measurement, 218, 278
Organization of literature review, 65–66

P

Panel study, 101, 278
Park, Robert, 158b
Participant observation
    advantages, 169t
    completion of fieldwork phase, 172
    data analysis, 171–72
    data collection, 171–72
    defined, 278
    described, 161–62

teenage employment and delinquency, 198–99
use of existing data, 198–99
Selection of research topic, 15, 46–47
Self-administered questionnaires
defined, 281
described, 117
group-administered questionnaire, 117
individually administered questionnaire, 117
mail survey, 117–18
response rate, 118
types of, 116–18
Semiparticipant observation
advantages, 169
completion of fieldwork phase, 172
data analysis, 171–72
data collection, 171–72
defined, 281
described, 166–67
disadvantages, 169
doing fieldwork, 170–71
entering the field, 170
focus, definition of, 167
vs. participant observation, 166–67
preparation for, 168–69
steps in, 168b
Sifting, 237–39
Simple random sample, 78–79, 89b
Skepticism, 23–24, 24b–25b, 281
Skewed distributions, 221f
Skinner, B.F., 132b
Skinner boxes, 132b
Skip to questions, 112–13
Snowball sample, 88, 90b, 281
Social context, 42
Social information, 3–4
Social science research. *See also* Scientific research
alternative views, 33
anthropology, 9, 9t, 157–58
bias in, 12
controversy and disagreement, 22
vs. everyday social inquiry, 3
empirical research, 8t
experiment, 132b
family planning research examples, 9t
humanistic view, 42–43
pioneering empirical research, examples of, 8t
process (*See* Research process)
qualitative research, 38t, 39
quantitative research, 38t, 39
unique features of, 12–13
uses of, 3–5
websites, 18–19
Social sciences
anthropology, 9–10, 9t, 157b–58b
critical view, 34
defined, 281
different views of, 35–36
disciplines in, 2, 9–12, 9t
economics, 2, 9t, 10–11

emergence of, 8–12
geography, 2, 9t, 10
history, 9, 10–11
humanistic view, 42–43
interpretative view, 42
political science, 9t, 11
postmodernist view, 42
psychology, 9t, 11–12
as science, 33
sociology, 9t, 12
web sites 18–19
Social surveys. *See also* Questionnaires
administration, 123b
advantages of, 127
aim and purpose of, 100–1
casual inquiry, 98–99
conducting a survey, 101b–2b
confidentiality of, 126
cross-sectional surveys, 100
defined, 281
disadvantages of, 127b
e-mail surveys, 118–19
ethical considerations, 125–26
examples of results, 103b–5b
Internet-based surveys, 118–21
key aspect of process, 101
longitudinal study, 101
mail, 117–18
main ingredients of, 99
panel study, 101
questions to ask about, 124–25
sample selection, 124
scientific inquiry, 98–100
sponsors of, 124
techniques of survey administration, 123b
trend study, 101
websites, 128–29
web-based surveys, 119–21
Sociology, 9t, 12, 281
Sorting, 237–39
Sources of information
abstracts, 56, 58, 60, 62
bibliographies, 50
books, 49, 54
citation indexes, 50, 54
class readings, 47
encyclopedias, 48
evaluating, 56b
government document, 50
indexes, 50
Internet, 3–5, 52–54
journal articles, 55–56, 58
magazines, 46, 51
newspapers, 49, 51
other sources, 52–53
semi-annuals, 10
textbooks, 47–48
Special-purpose files, 237, 281
Spreadsheets, 213b, 281
Stack number, 51
Stacks, 51